AMERICA CALLING

AMERICA CALLING

A Social History
of the Telephone to 1940

Claude S. Fischer

UNIVERSITY OF CALIFORNIA PRESS

Berkeley   Los Angeles   Oxford

University of California Press
Berkeley and Los Angeles, California

University of California Press, Ltd.
Oxford, England

© 1992 by
The Regents of the University of California

Library of Congress Cataloging-in-Publication Data

Fischer, Claude S., 1948–
    America calling: a social history of the telephone to 1940/
Claude S. Fischer.
        p.   cm.
    Includes bibliographical references and index.
    ISBN 0-520-07933-7 (alk. paper)
    1. Telephone—Social aspects—United States—History.   I. Title.
HE8817.F56   1992
302.23'5'0973–dc20   91–38355
                                                              CIP

Printed in the United States of America
1 2 3 4 5 6 7 8 9

To . . .
My mother, Rosette,
my sister, Cathy,
my daughter, Leah,
and most certainly, my wife, Ann Swidler,
for teaching me the value of the telephone

Contents

Photograph section follows page 153.

List of Figures

List of Tables

Preface

This is a social history of the telephone in North America up to World War II. I did not come to this topic because of a personal affinity for the telephone; friends and relatives will report that I, like many men, tend to be telephone-averse. I came to this topic because, as a sociologist, I am fascinated by the development of what we call modern society, in particular by how the material culture of our modern lives—our tools, our toys, our things—influences our experience. The telephone is, both functionally and emblematically, a key element of modern material culture. And yet, although we know some about its technical and business history, we know surprisingly little about its social history. *America Calling* explores that social history and might, I hope, stimulate similar explorations of other modern technologies.

This book describes how Americans in the years up to World War II encountered and employed the telephone. It addresses the social role of the home telephone: How did people learn of it? Who subscribed and why? How did people use it in their personal relations? What reactions did they have to it? What difference did having it make socially and psychologically? To answer some of these and similar questions, I and my assistants studied how the telephone industry advertised its residential service, how the telephone became available across the country and across different social groups, how its use evolved in three specific communities, which families adopted it and when, and how people reacted to the experience. For comparison, we also looked at some of the same issues in regard to the automobile. *America Calling* summarizes the results of this research in what I trust are plain terms. I have written the text with the general reader in mind and put the technical details in the back. The story of the telephone, as told here, has at its center, not a technology or a machine, but ordinary Americans encountering, mastering, and making commonplace one of the radically new technologies of the past few generations.

ACKNOWLEDGMENTS

Over the years of research summarized in this volume I have depended on many people and many institutions. Major financial support came from the following sources: the National Endowment for the Humanities (Grant RO-20612); the National Science Foundation (Grant SES83-09301); the Russell Sage Foundation; and the Center for Advanced Study in the Behavioral Sciences, Stanford, with a grant from the Andrew W. Mellon Foundation. At the University of California, Berkeley, I was assisted by the Committee on Research, the Institute of Urban and Regional Development, and the Institute of Transportation Studies. Organizations and individuals in the telephone industry were helpful, but no financial aid for the research came from the industry.

The team of Berkeley research assistants who worked longest on this project included Melanie Archer, John Chan, Steve Derné, and Barbara Loomis. I relied on these young scholars not only for much of the legwork, but also for advice and criticism. Important conceptual and procedural decisions emerged from our project discussions. Many other students helped as well: William Barnett, Keith Dierkx, Michelle Dillon, Sylvia Flatt, Barry Goetz, Molly Haggard, Kinuthia Macharia, Lisa Rhode, Mary Waters, and Laura Weide. I, of course, am responsible for the research and its interpretation.

The generous help of staff at the telephone industry archives was essential. They guided me through their materials and often provided copies of documents. These people include Norm Hawker, Ken Rolin, and Don Thrall at the Archives and Historical Research Center of the Telephone Pioneer Communications Museum of San Francisco; Mildred Ettlinger, Robert Garnet, and Robert Lewis at the AT&T Historical Archives, then in New York; Peg Chronister at the Museum of Independent Telephony in Abilene, Kansas; Stephanie Sykes and Nina Bederian-Gardner at Bell Canada Historical in Montreal; and Rita Lapka at Illinois Bell Information Center in Chicago. I also wish to thank the staffs at the National Museum of American History, the Antioch Historical Society, the Marin County

Historical Society, the Palo Alto Historical Society, and the newspaper collection of the University of California, Berkeley, Library.

Professor Glenn Carroll of the University of California, Berkeley, Business School, and Professor Jon Gjerde of the History Department each collaborated on part of the research reported here. Many people read and commented on this manuscript or on earlier papers from the project, including Therese Baker, Victoria Bonnell, Paul Burstein, Lawrence Busch, Robert Garnet, Elihu Gerson, Roland Marchand, Robert Pike, Lana Rakow, Everett Rogers, Mark Rose, Michael Schudson, Anne Scott, Ilan Solomon, Ann Swidler, Joel Tarr, Barry Wellman, Langdon Winner, and several anonymous readers. I did not always take their advice, but I learned from it.

I would also like to thank Naomi Schneider, Valeurie Friedman, and Marilyn Schwartz at the University of California Press for their strong commitment to this project. I thank, at Publication Services, Carol Elder for clarifying the book's prose and Chad Colburn for efficiently managing its production.

Finally, I thank those friends and colleagues who, after hearing about the telephone project, stopped chuckling long enough to urge me on. Most of all and always, Ann Swidler gave me as much encouragement as a professional colleague, a loving wife, and the harried mother of Avi and Leah could manage (that is, between taking phone calls).

CHAPTER ONE

Technology and Modern Life

In 1926 the Knights of Columbus Adult Education Committee proposed that its group meetings discuss the topic "Do modern inventions
help or mar character and health?" Among the specific questions the
committee posed were

"Does the telephone make men more active or more lazy?"

"Does the telephone break up home life and the old practice of
visiting friends?"

"Who can afford an automobile and under what conditions?"

"How can a man be master of an auto instead of it being his master?"

The Knights also considered whether modern comforts "softened"
people, high-rise living ruined character, electric lighting kept people
at home, and radio's "low-grade music" undermined morality. The
preamble to the questions declared that these inventions "are all indif-
ferent, of course; the point is to show the men that unless they individ-
ually master these things, the things will weaken them. The Church
is not opposed to progress, but the best Catholic thinkers realize that
moral education is not keeping up with material inventions."[1]

Worry about the moral implications of modern devices was especially appropriate in 1926, for middle-aged Americans had by then witnessed radical material changes in their lives. Despite the awe that many express about today's technological developments, the material innovations in our everyday lives are incremental compared to those around the turn of the century. Major improvements in food distribution and sanitation lengthened life and probably lowered the birth rate. Streetcars brought average Americans easy and cheap local travel. Telephone and radio permitted ordinary people to talk and hear over vast distances. Electric lighting gave them the nighttime hours. Add other innovations, such as elevators, movies, and refrigerators, and it becomes apparent that today's technical whirl is by comparison merely a slow waltz.[2]

The questions the Knights pondered were widely addressed. Many, especially representatives of business, gave rousing answers: Modern inventions liberated, empowered, and ennobled the average American. The American Telephone and Telegraph Company (AT&T) issued a public relations announcement in 1916 entitled "The Kingdom of the Subscriber." It declared:

> In the development of the telephone system, the subscriber is the dominant factor. His ever-growing requirements inspire invention, lead to endless scientific research, and make necessary vast improvements and extensions. . . .
>
> The telephone cannot think or talk for you, but it carries your thought where you will. It's yours to use. . . .
>
> The telephone is essentially democratic; it carries the voice of the child and the grown-up with equal speed and directness. . . .
>
> It is not only the implement of the individual, but it fulfills the needs of all the people.[3]

Less self-interested parties made similar claims. In 1881 *Scientific American* lauded the telegraph for having promoted a "kinship of humanity." Forty years later a journalist extolled the radio for "achieving the task of making us feel together, think together, live together."[4] The author of *The Romance of the Automobile Industry* declared in 1916 that the "mission of the automobile is to increase personal efficiency; to make happier the lot of people who have led isolated lives in the country and congested lives in the city; to serve as an equalizer and a balance." Many urban planners and farm women, to take two disparate groups, shared similar images of the automobile as a liberator.[5]

But others, notably ministers and sociologists—in those days not always distinguishable—warned that these inventions sapped Americans' moral fiber. In 1896 the Presbyterian Assembly condemned bicycling on Sundays for enticing parishioners away from church—a forecast of complaints about the automobile. Booth Tarkington's fictional automobile manufacturer in *The Magnificent Ambersons* reflects:

> With all their speed forward they may be a step backward in civilization—in spiritual civilization. It may be that they will not add to the beauty of the world, nor to the life of men's souls. I am not sure. But automobiles have come, and they bring a greater change in our life than most of us suspect. They are here, and almost all outward things are going to be different because of what they bring. . . . I think men's minds are going to be changed in subtle ways because of automobiles; just how, though, I could hardly guess.

Robert and Helen Lynd, the former a cleric turned sociologist, claimed in their classic *Middletown* (1929) that the automobile and the enticements it brought within reach—roadhouses, movies, and the like—undermined the family and encouraged promiscuity. College administrators in the 1920s argued that automobiles distracted students from their studies and led many to drop out. Observers worried less often about the telephone, but some objected that it encouraged too much familiarity and incivility and that it undermined neighborhood solidarity.[6]

These comments, whether by industry representatives or viewers-with-alarm, reflected genuine and widespread concerns, at least by elites, about the social implications of modern inventions. The concerns are, in turn, rooted in a larger meditation in Western societies about modernity.

MODERN TIMES

Modernity is an omnibus concept and, like the omnibus of the nineteenth century, carries a variety of riders—an eclectic assortment of ideas about economic, social, and cultural changes over the past several generations.* Most sociologists and historians writing about

---

*Large bodies of literature in sociology, history, and the humanties address the concepts of modern, modernization, and modernity. They identify many sociocultural

modernization focus on industrial and commercial development: the rise of the factory, market, or corporation, and the increase in affluence. Others stress changes in social organization, such as the evolution of the nation-state and the small household. Still others emphasize alterations in culture and psyche, for example, the growth of individualism, sentimentality, or self-absorption. Modernization theorists also differ about when in the past three centuries the critical transformations happened. Most, however, implicitly agree that modernity comes as a coordinated set of changes. Whichever change is depicted as the conductor of this omnibus, the rest inevitably come along for the ride, for modernization is a global process.[7]

Contemporary writers follow the path trod by the founders of social science, theorists such as Emile Durkheim, Max Weber, Karl Marx, Ferdinand Tonniës, and Georg Simmel. Living from the mid-nineteenth century through the early years of the twentieth and surrounded by severe disjunctures in material culture, they believed that a new society was being born. The theorists largely concentrated on changes in economic organization, but much of their attention also turned to social life—to personal relations, family, and community. Modernity in these spheres followed in part from changes in how people made a living, but modernization also directly transformed private life. The growth of cities, wider communication, more material goods, mass media, and the specialization of land use and institutions—these kinds of changes, the early social theorists argued, altered personal ties, community life, and culture. More specifically, modernization fostered individualism and interpersonal alienation, abraded the bonds of social groups, and bred skepticism in place of faith. Some theorists described these developments as the liberation of individuals from the shackles of oppressive communities, others as

---

attributes as *the* trait that distinguishes the modern from the premodern (never mind the postmodern): rationality, individualism, secularism, organization (*Gemeinschaft,* usually defined as "society," as opposed to *Gesellschaft,* "community"). The conceptual statements usually beg an empirical question by assuming that this property is more common now than it was "then" (whenever and wherever "then" was). Since I am concerned precisely about the empirical assumptions, my usage is simple. By modernity I mean the style of social life and culture typical of twentieth-century America, as contrasted to earlier eras, especially the nineteenth century. Some, especially those who locate the great transition a few centuries ago, will find that a misuse of the term. Presumably, whatever the criterion is, it nevertheless ought to have become more evident over the past four generations.

the isolation of individuals from loving communities. Two sides of the same coin. Much, perhaps most, of modern sociology and increasingly of the field of social history involves variations on this motif.[8]

Modernization theory, by now implicit in the language used to discuss contemporary society, is open to several criticisms. Critics debate whether such transformations really happened. The assumption that economic, social, and psychological changes would occur together is debatable. Charles Tilly, for example, has challenged the theoretical assumption that " 'social change' is a coherent general phenomenon, explicable *en bloc*." Darrett Rutman has poked fun at the tendency of his colleagues in the field of history to locate the "lost community" ever backward in time: "Some have said we lost it when we disembarked John Winthrop from the *Arbella*" — all of which "has made us appear to be classic absent-minded professors regularly losing our valuables."[9]

Still, the concerns addressed by modernization theorists, and in simpler forms by nonacademics like the 1926 Knights of Columbus, are profound. The material culture of twentieth-century society differs strikingly from that of earlier eras. How has that difference altered the personal lives of ordinary people? In this book I am concerned with the manner in which turn-of-the-century technologies made a difference to North Americans' ways of life, in particular to community and personal relations. I use the telephone as a specific instance of that material change, bringing in the automobile for comparison.

The results of this inquiry suggest, in broad strokes, that while a material change as fundamental as the telephone alters the conditions of daily life, it does not determine the basic character of that life. Instead, people turn new devices to various purposes, even ones that the producers could hardly have foreseen or desired. As much as people adapt their lives to the changed circumstances created by a new technology, they also adapt that technology to their lives. The telephone did not radically alter American ways of life; rather, Americans used it to more vigorously pursue their characteristic ways of life.

The next section of this chapter pursues theoretical issues in the study of technology. Some readers may wish to turn to a later section of this chapter — "Why the Telephone?" — where explicit discussion of the telephone begins (p. 21).

DOES TECHNOLOGY DRIVE
SOCIAL CHANGE?

Technological change in the personal sphere is a central dynamic of all theories of modernity.[10] Today's instruments of daily life—food preservatives, artificial fabrics, cars, and so on—are at least necessary, if not sufficient, conditions for what we consider modern society. Interest in whether and how such technologies alter social life generated a field of study, "technology and society."

Once a sociology of technology focused on these matters. It flourished until the early 1950s under the leadership of the University of Chicago's William F. Ogburn, but "passed into oblivion in slightly more than two decades." Currently, scholarship on technology rests largely in the hands of historians and economists, although a band of more sociologically oriented scholars are active. Historians have superbly documented the technological developments that mark Western modernization. Yet they usually write on the social sources of technological change (for example, how national cultures shaped the development of trolley systems) rather than on the technological sources of social change. Economists tend to focus on immediate and straightforward applications of technical advances. Neither group, and few scholars generally, have looked closely at how the use of major technologies affects personal and social life.[11] There are important exceptions. Most noteworthy are several historians who have studied housework technologies. They have striven to understand how vacuum cleaners, stoves, and the like altered the lifestyles and well-being of American women.[12] In general, however, scholars have neglected the social role of technology and left "theorizing"—that is, accounting for the influence of technology on social life—to the older Ogburn approaches or to common sense.

Others, quite different, have eagerly addressed the social implications of technology. These, loosely termed "culture critics," contend that technology has created a modern *mentalité*. They have posed some challenging ideas. Where Ogburn and others saw the nuts and bolts of a technology, they see its symbolism and sensibility.

But both perspectives on technology are problematic. Our way of thinking about the *causal* link between technology and social action impedes our understanding of technology's role. Even the language we employ can be a problem, as in the common use of the word *impact* to describe the consequences of technological change.

Defining Technology

The dictionary defines technology as applied science. Some have construed it more broadly, as "practical arts," the knowledge for making artifacts, or even the entire set of ways that people organize themselves to attain their wants. Put that broadly, the concept comes to subsume almost all human culture, including magic. As the label stretches — as it becomes, for example, a synonym for rationality — "technology" becomes less a subject of study and more a rhetorical term.[13]

Let us restrict the idea to the more tangible, physical aspects of technology, to devices and their systems of use. And since this study concerns the everyday domestic sphere, technology here is similar to the idea of material culture. For some people, items of material culture, such as refrigerators, bicycles, telephones, phonograph records, and air conditioners, may seem too mundane for serious study. Yet Siegfried Giedion offers another viewpoint in the opening pages of *Mechanization Takes Command*:

> We shall deal here with humble things, things not usually granted earnest consideration, or at least not valued for their historical import. But no more in history than in painting is it the impressiveness of the subject that matters. The sun is mirrored even in a coffee spoon.
>
> In their aggregate, the humble objects of which we shall speak have shaken our mode of living to its very roots. Modest things of daily life, they accumulate into forces acting upon whoever moves within the orbit of our civilization.[14]

The prosaic objects of our culture form the instruments *with* which and the conditions *within* which we enact some of the most profound conduct of our lives: dealing with family, friends, and ourselves.

For most culture critics these objects are the focus of concern. The key question usually is: What has the automobile, or the television, or the skyscraper, or whatever *thing*, done to us? Of course, a material object itself, lying bare on the ground, is of no interest. As historian Thomas Hughes has emphasized, there is a "system" around a functioning technology — a commercial broadcasting system around the television; appliance, electrical, and food-packaging systems around the refrigerator. References to the material object, as in "the diffusion of the automobile," are shorthand for the larger system.[15] The point is not merely a matter of lexicon. Separable parts of a technological system may have separable consequences. Television, for example, can be analyzed by its specific content — such as the sexual titillation, violence, and commercials it broadcasts — or by its technical features — such as the flickering of images, dissociation of place, and passivity of watching.

Intellectual approaches to technology and society can be divided into two broad classes: those that treat a technology as an external, exogenous, or autonomous "force" that "impacts" social life and alters history, and those that treat a technology as the embodiment or symptom of a deeper cultural "logic," representing or transmitting the cultural ethos that determines history.[16] Each approach is problematic.

Impact Analysis

The older, Ogburn analysis is a "billiard-ball" model, in which a technological development rolls in from outside and "impacts" elements of society, which in turn "impact" one another. Effects cascade, each weaker than the last, until the force dissipates. So, for example, the automobile reduced the demand for horses, which reduced the demand for feed grain, which increased the land available for planting edible grains, which reduced the price of food, and so on. A classic illustration is Lyn White's argument that the invention of the stirrup led, by a series of intermediate steps, to feudalism.[17]

Economic rationality is an implicit assumption in the billiard-ball metaphor. A technology is considered imperative to the extent that it is rational to adopt it. Adopting it in turn alters related calculations, leading to further changes in action. The model allows for unintended consequences, particularly during Ogburn's famous "cultural lag" (a period of dislocation when changes in social practice have not yet accommodated the new material culture), but change largely follows the logic of comparative advantage among devices. More contemporary versions of this impact model appear in the literature on technology assessment.[18] Such thinking about technology is deterministic: Rationality requires that devices be used in the most efficient fashion.

Critics have challenged the assumption that technological change comes from outside society as part of an autonomous scientific development and that application of a device follows straightforwardly from its instrumental logic. Instead, these critics contend that particular social groups develop technologies for particular purposes — such as entrepreneurs for profits and the military for warfare. The developers or other groups, operating under distinctive social and cultural constraints, then influence whether and how consumers use the new tools.[19] Some scholars have argued, for example, that the automobile, tire, and oil industries, through various financial stratagems, killed the electric streetcar in the United States to promote automobile and bus transportation.[20] In this view technological change is better understood as a force called up and manipulated by actors in society. Historian George Daniels puts the challenge broadly:

> No single invention—and no group of them taken together in isolation from nontechnological elements—ever changed the direction in which a society was going. . . . [Moreover,] the direction in which the society is going determines the nature of its technological innovations. . . .
> Habits seem to grow out of other habits far more directly than they do out of gadgets.[21]

Against the metaphor of ricocheting billiard balls, we have perhaps the metaphor of a great river of history drawing into it technological flotsam and jetsam, which may in turn occasionally jam up and alter the water's flow, but only slightly.

Others reject technological determinism less completely, granting that material items have consequences, but claiming that those consequences are socially conditioned. Societies experience technological developments differently according to their structure and culture. For example, John P. McKay has shown how the trolley system developed more slowly but more securely in Europe than in the United States. Others have argued that France's autocratic centralism retarded the diffusion of the telephone.[22] More generally, historians of technology often explain that a technological development may have unfolded otherwise were it not for social, political, or cultural circumstances. For instance, some historians of housework contend that American households might have developed communal cooking and laundering facilities with their neighbors, but instead most individual American families own small industrial plants of ovens and washers, expensive machines that are idle 90 percent of the time. This is not economically efficient, critics contend; rather, it is the outcome of American institutions and culture. (More on this "social constructivism" perspective later.) The blunt conclusion from the last generation of scholarship is that the whig analysis of technology cannot hold. The ideas that technologies develop from the logical unfolding of scientific rationality, that they find places in society according to principles of economic optimization, that their use must be comparatively advantageous to all, and that the only deviation from this rationality is the brief period of social disruption labeled "cultural lag"—this model has long been rejected as conceptually and empirically insufficient.

But another form of determinism has arisen: the "impact-imprint" model. According to this school of thought, new technologies alter history, not by their economic logic, but by the cultural and psychological transfer of their essential qualities to their users. A technology "imprints" itself on personal and collective psyches.

Stephen Kern's *The Culture of Time and Space, 1880–1918*, which illustrates this approach, is a well-received and thoughtful analysis of space-transcending technologies developed before World War I: the telegraph, telephone, bicycle, and automobile. Together, Kern contends, these new technologies "eradicated" space and shrank time, thus creating "the vast extended present of simultaneity." Without barriers of space and time, we moderns can reach and be reached from all places instantly, an experience leading to heightened alertness and tension.

The crux of Kern's argument is that the essences of the technologies—the speed of the bicycle and automobile, the instancy of the telegraph and telephone—transfer to their users. For example, Kern cites a 1910 book on the telephone (subsidized, it turns out, by AT&T) claiming that with its use "has come a new habit of mind. The slow and sluggish mood has been sloughed off . . . [and] life has become more tense, alert, vivid." Similarly, he quotes a French author on how driving an automobile builds skills of attention and fast reaction. The technologies passed on their instancy and speed to the users and, through them and through artists, to the wider culture.[23]

But how can a technology pass on its properties? Ultimately, the argument rests on metaphor become reality. At points, Kern lays out a plausible causal explanation. For example, he contends that unexpected telephone calls at home promote anxiety and feelings of helplessness.[24] He does not, however, pursue this kind of speculation consistently. Had he done so, he might have found that it did not always lead in the same direction. The telephone might also promote calm because its calls reassure us that our appointments are set and our loved ones are safe. Kern might also have more consistently compared the psychological consequences of these technologies with those of their precursors. While he compares the suddenness and demand of the telephone call to the leisureliness of the letter, he does not compare it to the surprise and awkwardness of an unexpected visitor at the door. The power of Kern's general argument rests ultimately on the impact-imprint metaphor: The jarring ring of the telephone manifests itself in a jarred and nervous psyche.

Kern's analysis also raises issues of evidence. Most of his material comes from literary and artistic works, suggestive and significant to be sure, but not to be taken at face value.* Even more, he and his sources typically reason from the properties of the technologies to the uses of them and then to the consequences. For example, the essence of the automobile is speed; it is used in a speedy way; thus its users' lives are speedier. Instead of reasoning from the properties of the tools, however, one might look at what people do with the tools. In the case of the automobile, one could reason that the replacement of the horse and train by the automobile would have sped up users' experiences. This may sometimes be so, but not always or perhaps even mostly. Touring by car rather than train probably led, according to a historian of touring, to a more leisurely pace. People could pull over and enjoy the countryside, "smell the roses." Similarly, farmers who replaced their horses with motor vehicles could travel faster to market, but many apparently used the saved time to sleep in longer on market day.[25] Kern's *Space and Time* exemplifies a mode of thinking about technology that, while more sophisticated than the earlier simple technological determinism, is still deterministic.

Joshua Meyerowitz's *No Sense of Place* presents a similar logic. In this award-winning volume Meyerowitz combines McLuhanesque insights with some sociology to create an argument both similar to and different from Kern's. Electronic media "lead to a nearly total dissociation of physical place and social 'place.' When we communicate through telephone, radio, television, or computer, where we are physically no longer determines where and who we are socially."[26] All places become like all others; cultural distinctions among places are erased, privacy is reduced, and areas of life previously sheltered from public view—the "backstage"—are revealed. Like Kern, Meyerowitz reasons from the properties of the technologies to their consequences: Electronic media are "place-less," so people lose their sense of place.

The problems of this approach are similar to Kern's. Meyerowitz, for example, argues that, unlike letter writers, telephone callers can

---

*By which I mean: Artists do not simply mirror their society. Instead of merely describing reality, they often "play" with reality by, for example, depicting escapes from it, ironic twists on it, fears about it, or romanticizations of it.

pierce other people's facades by hearing sounds in the background of the other party. Thus the telephone breaks down privacy. But why not instead compare the telephone call to the personal visit or to the front-stoop conversation? If telephone calls have replaced more face-to-face talks than letters, then the telephone has increased privacy. On empirical issues Meyerowitz relies on "common sense" or news stories for evidence and produces very few historical accounts. To take a minor illustration, Meyerowitz argues that "electronic messages . . . steal into places like thieves in the night. . . . Indeed, were we not so accustomed to television and radio and telephone messages invading our homes, they might be the recurring subjects of nightmares and horror films." Perhaps. But while accounts of early telephony (pronounced teh-LEH-feh-nee) suggest a wide range of reactions, including wonder and distaste, they do not indicate that early users had nightmares about invading messages.

The two forms of technological determinism reviewed here differ. The older one was "hard," simple, and mechanistic; the newer is "soft," complex, and psychocultural. But both are deterministic. A technology enters a society from outside and "impacts" social life. Both describe a form of cultural lag, during which sets of adaptive problems arise because we, by nature or by historical experience, are unable to use a new technology to meet our needs and instead are used by it. Ironically, because the newer form of determinism is more cultural and thus more holistic (and thus also in some ways like the "symptomatic" approach discussed in the following section), it typically describes a convergence of similar effects—for example, in Meyerowitz's electronic media and placelessness. Different specific technologies change us in the same ways. This logic can be even more deterministic than that of Ogburn, since his analysis contains the possibility that specific cause-and-effect trajectories may diverge. In either case, such impact analyses ought to be abandoned. The first is too rationalized, mechanical, and lacking in social context. The latter is too reliant on imagery rather than evidence. It suffers from what historian David Hackett Fischer labels "the fallacy of identity."[27] Indeed, we should abandon the word *impact*. The metaphor misleads.

Symptomatic Approaches

"Symptomatic" analysts, to use literary critic Raymond Williams's term, describe technologies not as intrusions into a culture but as expressions of it. Langdon Winner uses the term "technological politics" for a theory that "insists that the *entire structure* of the technological order be the subject of critical inquiry. It is only minimally interested in the questions of 'use' and 'misuse,' finding in such notions an attempt to obfuscate technology's systematic (rather than incidental) effects on the world at large." Typically, the underlying *Geist*, or spirit, is an increasing rationalization of life, carrying with it mechanization, inauthenticity, and similar sweeping changes. Specific material goods are in essence manifestations of this fundamental *Geist*.[28]

Much of Lewis Mumford's later writings are in this vein, for example:

> During the last two centuries, a power-centered technics has taken command of one activity after another. By now a large part of the population of this planet feels uneasy, indeed deprived and neglected unless it is securely tied to the megamachine: to an assembly line, a conveyor belt, a motor car, a radio or a television station, a computer, or a space capsule. . . . Every autonomous activity, one located mainly in the human organism or in the social group, has either been bulldozed out of existence or reshaped . . . to conform to the requirements of the machine.[29]

More popular writings, such as those of Ellul on *technique* and Schumacher in *Small Is Beautiful*, also describe a deep force that spawns a homogeneous set of technologies.

A specific technology matters little. It may be the actual instrument of a deeper process or just a sign of it, a synecdoche for all technology. Leo Marx has shown how nineteenth-century American Romantics used the railroad as an emblem for social change. More recently, writers have held that other technologies, such as the engine, assembly line, and automobile, epitomize deeper conditions such as cultural modernity.[30]

The symptomatic approach raises its own problems. The causal logic is usually opaque: How does a *Geist* shape psyche and culture? Do people learn, say, rationalization, by using specific devices? Or, is using a device the expression of rationalization learned in other ways, say, through mass media? The approach carries a major assumption about technology that seems both logically and empirically unwarranted: that modern technologies form a coherent, consistent whole—a contention that follows almost necessarily from the idea of an underlying process. Jennifer Stack has pointed out that "by assuming, and therefore searching for, only correspondences [of technologies with the *Geist*] writers deny the possibility that a technology might embody elements that truly contradict the essence of the totality or simply express something other than the essence."[31] This holism appears in several forms.

One form is the implicit claim that these technologies operate in parallel with homogeneous effects. Mumford makes that claim in his list of devices that people cling to, and others make it in arguments that modern technologies generally lead to routinization or that they necessarily alienate users from nature. But do all these modern tools operate in parallel? Perhaps not. Take, as another example, philosopher Albert Borgmann's inquiry on *Technology and the Character of Contemporary Life*. He defines modern technology as "the typical way in which one in the modern era takes up with reality," a truly global definition. Borgmann then distinguishes modern (1700 to now) *devices* from largely premodern *focal things*. Things are objects whose operations we understand and that can "center and illuminate our lives" — like fireplaces, violins, and national parks. They are good. Devices are objects whose internal workings are mysteries and that merely deliver some end to us — like central heating, stereos, and motor homes. They are bad. (The evaluations are explicit in Borgmann's book.)[32] One immediate problem, among others, is that Borgmann equates so many diverse objects — toasters to telemetry — and asserts that they all deeply affect relations and psyches in the same way.*

There is little theoretical and less empirical reason to lump these diverse objects into a single category *a priori* and to assume parallelism. Such an action forecloses rather than broadens scholarly inquiry. (It assumes a "myth of cultural integration."[33]) The various uses of different technologies may clash with one another. Perhaps, for example, movies helped bring people into public spaces more, but television reversed that. Or take the idea of routinization. Some have suggested that the railroads enforced a rigidity about time through their fixed schedules. If so, the automobile must have contradicted this trend by allowing people to come and go as they pleased. Or take housework. Ruth Cowan has persuasively argued that some household appliances brought functions into the home and others extruded functions

---

*Other problems include the difficulty any other observer would have in distinguishing a focal thing from a device, the evident subjectivity of the distinction. As in many other cultural critiques, we have a catalog of class prejudices. Violins, Borgmann claims, are focal, because he presumably can play and enjoy them; the operations of stereos are alienating mysteries. Of course, for others, the reverse is true. Similarly, computers are mere devices to Borgmann, although to many they are engrossing and fulfilling, constituting a focus of community.

from it. Or, finally, take the set of technologies Malcolm Willey and Stuart Rice call "agencies of communication," some of which they claim increased cultural standardization (radio, movies) and some of which they claim reduced it (telephone, automobile).[34] If even within such narrow sets of technologies there could be such varieties of possible consequences, how can we assume homogeneous consequences across the hodgepodge of modern tools?*

Another corollary is the assumption that the several effects of any device operate in parallel and are the same for all people. A technology could, instead, have contradictory consequences or different ones for different groups. For example, farmers' use of the automobile may have simultaneously solidified rural communities by increasing local interaction and weakened them by allowing farm families to tour distant locales. And use of the automobile may have increased the social mobility of blacks in the South more than that of whites. The workplace computer may both degrade the skills of middle managers and upgrade those of secretaries.[35]

Another dubious corollary is that technology has cumulative effects: The more of the cause, the more of the consequence; for example, the more powerful computers are, the more "placelessness" there is, to use Meyerowitz's term. Sometimes this may be so, but often it probably is not. When televisions were scarce, for instance, family members and even neighbors came together to watch, but as televisions became common, it seems that people increasingly watched them alone. Similarly, early washing machines may have encouraged collective housework, drawing homemakers to laundromats, but the later, cheaper machines probably encouraged privatization of housework by allowing homemakers to do the wash at home.[36]

Since those writing in the symptomatic mode assume that history has a grand direction, they often tend to extrapolate developments almost *ad infinitum*. Video games provide a cautionary tale. In the early 1980s many commentators projected the PacMan-ization of American youth. Yet the video craze collapsed almost as fast as it grew (and then it rebounded with Nintendo games, but perhaps only for a while).

---

*Sigmund Freud made a similar point in *Civilizations and Its Discontents*: "Is there, then, no positive gain in pleasure, no unequivocal increase in my feeling of happiness, if I can, as often as I please, hear the voice of a child of mine who is living hundreds of miles away . . . ? [But] if there had been no railway to conquer distances, my child never would have left his native town and I should need no telephone to hear his voice. . . . " (translated by James Strachey, Norton edition, 1962, p. 35).

Claims about the computerization of the American home appear to be similarly mistaken.[37]

The symptomatic approach widens our view of technology from simply mechanical and instrumental attributes to the cultural and symbolic contexts within which devices are developed and employed. It reinforces the need to incorporate social context into our explanations. In some ways, however, this approach is more problematic than simple technological determinism. Because its proponents locate the source of change in a global *Geist* and therefore disdain serious attention to any particular technology, this approach cannot explain how people come to use a technology and thereby change their lives. Its holism may conceal and confuse matters more than the piecemeal nature of technological determinism.

Social Constructivism

Several historians and sociologists, particularly European scholars, have in recent years formalized an approach that stresses the indeterminacy of technological change. Mechanical properties do not predestine the development and employment of an innovation. Instead, struggles and negotiations among interested parties shape that history. Inventors, investors, competitors, organized customers, agencies of government, the media, and others conflict over how an innovation will develop. The outcome is a particular definition and a structure for the new technology, perhaps even a "reinvention" of the device. The story could always have been otherwise if the struggles had proceeded differently. That is why the same devices may have different histories and uses in different nations. I have already mentioned the example of streetcar systems. Similarly, radio frequencies became privately owned franchises broadcasting commercially sponsored entertainment in the United States because of social conditions and political arguments specific to this country. (Critics of a more deterministic bent might rejoin, however, that such national differences in radio operations pale in comparison to their similarities.)[38]

This perspective brings us closer to incorporating end users into the analysis. Carolyn Marvin, for example, describes debates among electrical experts of the late nineteenth century about the social implications of lights and telephones and what ought to be done to manage those implications. Users are represented in "negotiations" that reshape innovations and channel their use by interest groups and ultimately by the purchase decisions of individual customers and the actual use to which those individuals put the technology. By this process, the technology is transformed into something different. In the case of the telephone, we will see how AT&T leaders, pressed in part by consumers, eventually tried to redefine their product from a totally practical service into a "comfort," a luxury, of the modern lifestyle.[39]

Most social constructivism has concentrated on the producers, marketers, or experts of a technological system. I intend to go further, to emphasize the mass users of technology, to go to what Ruth Schwartz Cowan has labeled the "consumption junction"—the point at which the final consumers choose, employ, and experience a technology. What we ultimately need, as Cowan argues and illustrates with the history of stoves, is a focus on the consumer if we are really to understand the social implications of technology.[40]

A User Heuristic: From the Consumer's Viewpoint

Once we have understood the genesis of a technology, its development and promotion, we can begin looking at consequences. Here we should ask: Who adopted the device? With what intention? How did they use it? What role did it play in their lives? How did using it alter their lives? This angle, an extension of social constructivism, emphasizes human agency and intentionality among end users. People are neither "impacted" by an external force, nor are they the unconscious pawns of a cultural *Geist*. Instead of being manipulated, they manipulate. We assume that users have purposes they mean the technology to serve, and—this is a point of method—that users can understand and tell us about those ends and means.*

This rational, individualistic model is, by itself, inadequate. Social and cultural conditions largely determine people's ends, be those ends the desire to be entertained, or to see family, or to appear *au courant*. Moreover, social and cultural conditions limit people's choices. People choose within obvious constraints, such as the income they have and the costs they face. They also choose within the constraints of their information, their skills, formal and informal rules, and the like. So, for example, teenagers who do not understand pregnancy cannot

---

*This discussion is akin to von Hippel's on "users as innovators." In the arena of producer goods he documents how often users develop innovations for a technology, modifications that are later commercialized by the original manufacturers (von Hippel, *Sources of Innovation*).

reasonably choose a birth control device, older people unfamiliar with electronics will shy away from computers, and men exposed to cultural images that depict cooking as feminine may be unable to master oven controls. People also choose within the constraints imposed by the distribution system of the technology. If telephone services are not provided in their community, people cannot use them. Alternatively, distributors can force people to use a new technology by eliminating other options, as, for example, when banks make it hard to use human tellers and thus constrain customers to use automatic tellers.* The sensibility of users can thus operate only within narrow social and cultural limits.

From this perspective the consequences of a technology are, initially and most simply, the ends that users seek. People, however, have multiple, often contradictory, purposes, so that use of a technology may have nonobvious consequences. In particular, some technologies can alter the trade-offs among people's goals and yield paradoxical results or even no evident effects at all. For example, the nature of the urban housing market means that many Americans must trade proximity to their jobs for spacious homes farther away. Some urban scholars suggest that most Americans have used automobiles not to shorten their work trips but to move farther away from their jobs and thereby purchase larger but cheaper housing. Thus the automobile may have led, not to shorter commutes, but to more spacious housing. Similarly, some historians suggest that the mechanization of housework saved American homemakers considerable time, but most women used the time savings not to gain respite but to attain even greater cleanliness, and thus they ended up devoting the same amount of time to housework as they had before. As a final example, most Americans may have decided that the time they saved using modern transportation to keep in touch with their kin should be spent, not for more frequent contact with those relatives, but for the same frequency of contact at greater distances. More generally, people can put technologies to various ends—which may include keeping some activities just as they were. In these ways, some major technologies may have few direct and overt consequences.

So far, I have addressed intended consequences, but new technologies may also have second- and third-order consequences that are unintended. Individuals directly experience the unintended con-

---

*This point was suggested by Ilan Solomon.

sequences of their own choices. For example, spending money on a new device means limiting other expenditures. Touring by automobile exposes travelers to new cultural influences.

More interesting and less controllable, individuals indirectly experience the unintended collective consequences of *others'* use. Over the years, shopping by automobile probably encouraged the dispersal of stores and so perhaps increased everyone's need to have an automobile. As more people use telephones to get services, service providers reorganize to deal with calls and perhaps thereby pressure nonsubscribers to get telephones. These examples illustrate one kind of collective by-product of adopting a new technology: An optional device becomes necessary. Other collective consequences include what economists call "externalities," such as the increased demand for oil because of the automobile or the decline in slide–rule skills because of the calculator. These reverberations can be paradoxical. For example, congestion on streetcars may have encouraged Americans to switch to automobiles for commuting, which eventually led to yet another form of traffic congestion.[41]

These externalities illustrate that a technology can be both a *tool* for an individual user and, aggregated, become a *structure* that constrains the individual. Individuals may not choose to watch television, but they must still contend with television in popular culture, children's fantasy lives, politics, public schedules (at least one presidential inauguration has been worked around the Superbowl), and so on. At either level of analysis, individual or structural, the center of the process is the purposeful user employing, rejecting, or modifying technologies to his or her ends, but doing so within circumstances that may in some instances be so constraining as to leave little choice at all.

This "heuristic," or instructive tool for thinking about technology, may be closer to the instrumental model I described earlier than to the symptomatic model, but it emphasizes the users rather than the imperative properties of the technology,* stresses social ends and social contexts, and denies the determinism of the billiard–ball metaphor.

One implication of this perspective is that empirical, historical research is of critical importance. If we can neither deduce a technology's social role from its manifest properties nor easily extrapolate it from a cultural *Geist*, if it matters more what individual users choose

---

*It is therefore possible for people to "misuse" a technology, at least from the point of view of its providers, as we shall see in the case of the telephone.

to do with a device and how these choices aggregate, then we must look closely at the histories of specific technologies. (Oliver Wendell Holmes once wrote that on some points "a page of history is worth a volume of logic.") Of course, we always seek to simplify, to group together specific instances, or find a few underlying dimensions (for Kern, the key category is space-transcending implements; for Meyerowitz, electronics). And so we should. But until we have reason—or better, evidence—to the contrary, we should assume that each technology may be used differently and play a different social role, and that different people may use the same technology to different effect.

This, too, is a serious problem in the field: the shortage of reliable evidence, compared to the plenty of impression, anecdote, and abstracted inference. Borgmann again illustrates. He purposely eschews any empirical literature and concentrates instead on philosophical discourse (in part because he distrusts social science as a technological, alienating "device"). Instead of research, he says, we should rely on our "common intuitions." This is a mistake, since few intuitions are so common as to be indisputable and even common intuitions are often false (for example, the world is flat, "blood tells," and so on). Borgmann's essay, like most supposedly theoretical discourses, rests on many empirical assumptions, some plausible, some dubious, most unexamined.[42] But even less polemical writers rely often on impression in place of hard evidence. For example, many a scholar has repeated the claim that the railroad companies developed standard time zones to rationalize their work. Recent research shows, however, that scientists were the ones who pushed the standardization; the railroads were not terribly interested.[43]

We need to study how specific devices were introduced and adopted, what people used them for, how that use changed as the technology evolved, how those uses altered other actions, how patterns of use changed the context for other actors, and so on. (Again, social constructivists have explored some of these concerns in concrete case studies.) To address questions about twentieth-century modernity, such studies ought to examine the key technologies of the transition, such as the automobile, assembly line, radio, and refrigerator. Historians have documented the development of many of those technologies, but have rarely described their social roles (the research on housework being an exception). Once we understand how the technologies emerged, we need to ask a few key questions: First, why and how did individuals use the technology? Second, how did using it alter other, less immediate aspects of their lives? Third, how did the collective use of a technology and the collective responses to it alter social structure and culture?

In this book I try to follow this "use heuristic" as far as the evidence will allow. Although the next two chapters take the common social constructivist path, examining how producers of the telephone developed and marketed their service, later chapters turn more toward Cowan's "consumption junction." They focus on users as individuals and as communities. Since the users were, as years passed, increasingly a mass population, the way of studying them becomes more sociological and statistical rather than historical and biographical. (More on method in a later section.) But the intent is always to discern how the average user reacted to and employed the technology.

WHY THE TELEPHONE?

Concerns about modernity, technology, and community motivated this study. I wanted to understand an aspect of the coming of modern society by examining the technological changes that were integral to it. According to the argument I have just laid out, this requires diligent empirical study of how people adopted and used a specific technology. Many technologies could and ought to be studied. Producers used some new technologies to alter goods production and delivery—new machines, materials, communication systems, control processes, and the like—with profound consequences for work and the economy.[44] I chose to focus, however, on a technology that people used daily in private life, a technology that may have affected social relations, community, and culture.

That still leaves a wide range of technologies. Figure 1 shows how a few key consumer products spread in the twentieth century. There are many other possibilities as well. Recent women's history scholars, for example, have studied technologies used for food preparation and cleaning. I further narrowed my choice to point-to-point, space-transcending technologies, such as the railroad, automobile, telephone, and streetcar.

The ability to travel and speak across space changed fundamentally between 1850 and 1950: from horsepower for the few to railroads, streetcars, bicycles, and automobiles for the masses; from military semaphore to business telegraphs and then telephones for the masses.

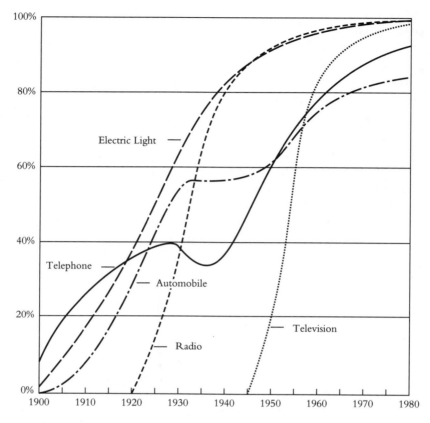

Note: Smoothed lines.

FIGURE I. U.S. HOUSEHOLDS WITH SELECTED CONSUMER GOODS, 1900–1980. This figure shows how several domestic technologies spread among Americans in the twentieth century. Slowed in part by the Depression, the telephone and automobile did not diffuse as rapidly as the three electronic devices. (Source: U.S. Bureau of the Census, *Historical Statistics* and *Statistical Abstract 1990*.)

These new technologies undergirded other material changes, such as increasing production and the rise of national markets. To the classical sociologists—most explicitly, to Emile Durkheim[45]—the multiplication and extension of interpersonal contacts were crucial to the development of modern society. More interaction generated economic and social specialization, brought cultures together and accentuated their discords, and shifted the bases of social solidarity from bloodline and place to occupation and taste. If we understand this change in social interaction, classical theory suggests, we understand much of modern society. Finally, to understand changes in private life and personal relations, it is appropriate to examine the means by which people conducted those relations.

Of the several space-transcending technologies, I selected the telephone, for two major reasons. First, the telephone captures most cleanly the magnification of social contact, without the complications of freight hauling or commuting involved in, say, the automobile or railroad. In 1875 Americans who wanted to send a message had to travel or use an intermediary who traveled; the messages were brief and one-way; the range and volume of communication were severely limited. (Use of the telegraph was highly restricted to business and rare emergencies.) In 1925 most Americans could speak to one another across town or across country quickly, back and forth, and fully. The possibilities of personal communication expanded vastly. How did people adopt and adapt to such a drastically new condition of social life?

The second reason is that among the space-transcending technologies of this era the telephone has been studied least. (Since I began this research in the early 1980s, some serious work has appeared. See the bibliographic essay in Appendix A.) In truth, none of these technologies has been studied *sociologically* in any depth. Compared to the shelves of research on, for example, television and its consequences, even the automobile is a mystery. Moreover, except for a few business historians, scholars have all but ignored the telephone. Why? Perhaps the moment of the telephone's notoriety preceded the era of social research. Or perhaps few social problems seem tied to the telephone. Or as one literary analyst has suggested, perhaps the telephone belongs to the class of "anonymous objects . . . so imbedded in daily routine as to have become undifferentiated from the rest of our immediate landscape."[46]

I want, then, to understand the introduction of the telephone, the uses to which people put it, and its evolving social role in daily life. To understand these developments, one must do more than reason forward from the properties of the telephone; one must study the historical process itself. One must do more than catalog the commentaries of contemporary observers; one must look at the conduct of daily life itself. One must do more than study telephone use today; one must examine change over time. The telephone began as a novelty, became business's substitute for the telegraph, and then evolved into a mass product, an everyday device for handling chores and having conversations. The role of the telephone unfolded over time. To what effect?

Such assessments can best be done by establishing some benchmarks for comparison. I chose to compare the social history of the telephone to that of the automobile. Although as an object and a system the automobile differs greatly from the telephone—gas and electric services are more like the telephone in form—from the user's point of view they are comparable. The automobile provides some of the same space-transcending functions of the telephone, albeit more slowly. Where possible, therefore, I contrast the diffusion and social uses of the telephone to those of the automobile.

THE TELEPHONE'S SOCIAL ROLE:
SOME SPECULATIONS

Despite the paucity of research, there have been some speculations about the social implications of the telephone. Ithiel de Sola Pool, one of the few researchers in this field, compiled a long list of forecasts made before 1940 about the telephone's role. Commentators predicted a range of consequences, from the disappearance of regional dialects to the elimination of written records for historians.[47]

Two topics illustrate the range of the weightier claims. One: Some have argued that use of the telephone altered the physical layout of American cities. Because telephone conversations erase the "friction of space"—the time and cost of crossing distances—they also reduce the importance of central location. Businesses and people can therefore more easily move to the urban periphery.[48] Two: Some serious commentators, as well as many industry representatives, have described the telephone as a force for democracy, because it permits citizens to communicate, to collaborate, and even to conspire uncontrolled by a central authority.[49] As intriguing as these and many other speculations are, we have very little, if any, solid evidence on their plausibility, much less their factuality.

This study looks more closely at a few other sets of speculations. One is the broad concern over whether the telephone has expanded or diminished personal relations. The industry itself said that telephone calls enriched social ties, offering "gaiety, solace, and security," even making of America "a nation of neighbors." Less interested parties, as well, described the telephone as a device that worked on behalf of social attachments.[50] The most common claims were that the telephone allowed rural people to overcome isolation, perhaps even saving many farm wives from insanity (see Chapter 4). Others, however, charge that the telephone provides but an echo of true human communication. "It brought people into close contact but obliged them to 'live at wider distances' and created a palpable emptiness across which voices seemed uniquely disembodied and remote," writes Stephen Kern. It is, in such views, an impersonal instrument whose use spreads impersonality.[51]

A second and widespread conviction is that telephone use weakens local ties in favor of extralocal contacts and national interests. Some make this claim approvingly, stating that the telephone is "an antidote to provincialism." Increased communication promises to advance contact among cultures, to help bring "the brotherhood of man." But for others the telephone is yet another of modernity's blows against local *Gemeinschaft*, the close community. We get larger "electronic neighborhoods... but shallower kinds of community." Ron Westrum has argued that devices such as the telephone "allow the destruction of community because they encourage far-flung operations and far-flung relationships." At an even deeper level the telephone contributes to placelessness, and without rootedness both community and identity are at risk.[52]

Few have argued against the delocalization claim, but Malcolm Willey and Stuart Rice did so in the most comprehensive study of the new communications' effects, a monograph published in 1933 for President Hoover's Commission on Social Trends. They argued that people use the telephone, like other point-to-point media, to augment local ties much more than extralocal ones and that calling strengthens localities against homogenizing cultural forces, such as movies and radio. "The telephone replaced the back fence and so was local in its influence," as another author put it.[53]

A third general concern has been for the subjective implications of telephone use. Many have ruminated on subtle psychological effects, for example, the possible creation of an alert, tense, "speedy" frame of mind. People are on edge, conscious that a call may occur at any instant, always impatient because the telephone has trained them to expect immediate results. Yet others describe the telephone as providing a calming sense of security.[54] Similarly, commentators have worried about privacy and "privatism." Carolyn Marvin wrote: "The telephone was the first electric medium to enter the home and unsettle customary ways of dividing the private person and family from the more public setting of the community." One common complaint in the nineteenth century was that the telephone permitted intrusion into the domestic circle by solicitors, purveyors of inferior music, eavesdropping operators, and even wire-transmitted germs. Among some communication theorists the telephone's intrinsic social psychological character wears away privacy: Messages come unbidden; background sounds reveal intimacies of the home to the caller; speakers cannot prepare for or reflect upon the discussion as they can in letters; callers' voices are disembodied from context; and so on.[55]

Others, however, blame telephone use, as well as television watching, suburban backyards, and the like, for creating "a general withdrawal into self-pursuit and privatism." One concern in the earliest days was that the telephone allowed people to conceal from community scrutiny inappropriate activities, such as illicit romances or liquor purchases. With the telephone and other devices people need public spaces less often and thus disengage from public life, burrowing into familial cocoons.[56]

These speculations revolve around what might be called the first-order consequences of telephone use: what its use means for the users. There are also second-order consequences: what widespread use of the telephone means for others and for the community. For example, at some point people with telephones began to assume that others would be instantly reachable. As Willey and Rice put it in 1933, "to be without a telephone or a telephone listing is to suffer a curious isolation in the telephonic age."[57]

There is little confirmation of the validity of these speculations, either in reports by contemporary observers or, much less, in systematic comparative evidence. The claims depend on an analysis of the inherent "logic" of the telephone, on impressions (not always unbiased), on anecdotes and second-hand tales. The dominance of opinion over evidence in this area is illustrated by a trivial example that came to intrigue me. Repeatedly, writers claimed that the telephone made construction of skyscrapers possible. The first instance of this claim seems to have been in 1902, and the latest I found was in 1989. Its greatest publicist was AT&T's chief engineer in the early 1900s, John J. Carty. A telephone was useful in managing construction high above the ground, he argued, but was even more important in solving the messenger problem:

Take . . . any of the giant office buildings. How many messages do you suppose go in and out of those buildings every day[?] Suppose there was [*sic*] no telephone and every message had to be carried by a personal messenger. How much room do you think the necessary elevators would leave for offices? Such structures would be an economic impossibility.

This contention lacks both evidence and plausibility. The historical timing is off, and other means of sending messages — pneumatic tubes, for example — were available. Yet this claim has been repeated for over 80 years without serious examination.[58] If we know so little about such a simple, material issue, consider how little we really know about the role of the telephone in personal relations, families, and community life.

Claims about the automobile's role are numerous, and some of them — especially those dealing with the changing physical layout of North American cities — have been well researched. But many claims about the automobile's role in the social lives of its users are as contradictory and as undocumented as those about the telephone.

Many blame or credit the automobile for the decline of local attachments in favor of placeless ties, whether for better — "the unshackling of the age-old bonds of locality," according to Robert Heilbroner — or for worse. A few, conversely, claim that the automobile instead abetted a retreat from urban cosmopolitanism into suburban provincialism.[59] Many commentators, particularly in the 1920s, lamented that the automobile undermined the family by permitting its members to pursue their pleasures at movies, roadhouses, campsites, and lovers' lanes. More recently, others say that the automobile encouraged extreme familism, an encapsulated privatism.[60] For some observers the automobile has been a tool for women's liberation (and another antidote to farmwife insanity), but for others it helped shackle women to their domestic chores.[61] Unfortunately, for many of these speculations, and especially for the seamier ones, there is but one major source of historical evidence, itself sometimes debatable, a chapter in the Lynds' *Middletown*.[62]

This quick review of speculations about the telephone and automobile suggests at least two points: that these technologies may have affected basic features of American life and that we have few facts about these phenomena.

We will look closely at the telephone in the development of modern American life, making brief comparisons to the automobile. We will dwell most on personal relations, local community, and subjective reactions. We do *not* ask what the "impacts" or "effects" of the telephone were. That is the wrong language, a mechanical language that implies that human actions are impelled by external forces when they are really the outcomes of actors making purposeful choices under constraints. Instead, we ask who adopted the telephone, when, where, how, and why; for what ends; and to what uses. By these uses—and by the second-order constraints generated by common use of telephones— we can understand what role the telephone played in modernization.

We may discover negative answers to these questions. We may find that the role of the telephone or automobile in these spheres was negligible, that relations, local ties, families would have been little different without the devices. Historian Daniel Boorstin asserts that "the telephone was only a convenience, permitting Americans to do more casually and with less effort what they had already been doing before."[63] That would be a fascinating conclusion because it would imply that people can assimilate drastic alterations in material conditions—here, the capacity to talk instantly with almost anyone—and continue the same social patterns they had before. It would show a powerful tendency toward homeostasis. Indeed, most of the evidence we will review suggests that Americans assimilated the telephone easily, even becoming nonchalant about it by the 1920s. It also suggests that Americans used this device to pursue their ends, not "more casually," but more aggressively and fully.

The next section discusses the methods used to pursue these questions. In the section after that, the reader will find an outline of the book. Chapter 2 begins the study with a summary history of the telephone and automobile in North America.

A NOTE ON METHOD

This study spans history and sociology, two disciplines that have grown closer in the past generation. Many historians have realized that they do far more than simply narrate, that their stories convey causal explanations, even if only implicitly. Many sociologists have abandoned the naive model of a physical science, realizing instead that their discipline, like the other life sciences, describes and explains historical events. Thus the work of historians and sociologists has converged in the study of certain issues—for example, mobilization in great revolutions, the adaptation of immigrant groups—in ways that sometimes make it difficult to divine the authors' pedigrees.

Yet a gap remains. Sociologists and historians differ in intent, historians usually seeking to provide a fully realized account for an event and sociologists usually seeking to extract general principles. Rhetorical styles vary. Sociologists usually persuade by weaving subtle and complex correlations into a simple, plausible, theoretical fabric. Historians more often rely on narrative structure, story lines featuring flesh-and-blood actors rather than bodiless attributes. "Historians want readers to remark that things became *really different* and for a coherent set of reasons—and to remember this in something like a story form" writes the social historian John Model. "Historians simplify reality for literary reasons, and then aim to overcome that simplification with concreteness (hence, quotations; hence, examples) and evocation."[64] Preferred causal explanations often differ, with historians more commonly stressing human agency, sociologists more likely attributing action to structural circumstances. Standards of evidence diverge. A first-hand account that historians might consider concrete and contextualized sociologists might dismiss as "anecdotal," that is, idiosyncratic and biased. A statistical pattern of covariation that sociologists might hail as revealing historians might dismiss as an abstracted conflation of diverse cases, without context, and lacking in any persuasive cause-and-effect narrative.

I am interested in a historical "moment" for its intrinsic significance and for its ability to reveal, in a general way, how people deal with changing material conditions. My sociological heritage, however, will be obvious. The reader will find more attention to the accurate generalization than to the telling anecdote, more effort to organize an argument than to establish a chronology, more persuasion by weight of data than by the logic of narrative. Nevertheless, I use a combination of typical historians' and sociologists' methods and hope that the outcome will inform both schools.

My general strategy was to combine several levels and modes of investigation to understand how Americans adopted and used the telephone—and the automobile—in the years up to World War II. These were the years during which the two technologies became staples of middle-class American life. In this period we could observe people coming to know, adopt, use, and adapt to the innovations.

The research includes a study of how the telephone industry marketed its product. How did the vendors, whose livelihoods were at stake, comprehend the public demand for the technology? They were not, as we shall see, always accurate in their perceptions. Nevertheless, their knowledge of the market, the advertising they designed, and the consumer responses they surveyed all provide indirect evidence of popular reaction. The next step, still closer to the user orientation, is an analysis of the patterns of diffusion: Who adopted the telephone, where, and when? By examining adoption patterns, we may, admittedly with some error, infer motivations and uses. Yet another strategy is to trace the integration of the technology into daily life: Where does the telephone appear in regular activities? How do people use it? What can we infer then about its social role?

These general approaches are translated into several concrete studies. The major ones are

1.   A history of how the telephone industry marketed its product to North American households, with special attention to rural and working-class customers. This study draws largely on industry and government archives: publications, reports, internal correspondence, and the like.

2.   Statistical analyses of state-level data on telephone and automobile adoption, assessing the factors that apparently encouraged or discouraged diffusion.

3.   The largest and most complex segment, a triad of community studies, reported first in Chapter 5, on three towns in the San Francisco Bay Area—Antioch, Palo Alto, and San Rafael. The research included a few different components: (a) a social history of each town from 1890 to 1940, focusing on community social life; (b) an account of how the two technologies entered each town; (c) statistical analyses of telephone and automobile diffusion; and (d) statistical analyses of social change.

4.   A statistical analysis of who adopted the telephone when. We drew samples of households from each of our three towns for five years, selected from the period of 1900 to 1936, and by linking telephone directory entries to census or city directory lists were able to find out what sorts of households were most or least likely to adopt the telephone in which year. We also used a national survey conducted during World War I and a census of Iowa farmers in 1924.

5.   Oral histories with 35 elderly people living in the three towns (described more fully in Chapter 8). It would be valuable to have first-hand accounts written by typical Americans about their encounters with telephones in the early twentieth century. But, besides the problem long noted by social historians that few ordinary people leave memoirs and diaries, getting or using a telephone was not, as we shall

see, a remarkable event. Even our elderly interviewees had to be encouraged to think about it.

Because there are few first-hand accounts of everyday life generations ago, and because the claims of interested parties must be viewed with caution, we rely in many places on sociological data. These data, such as censuses and surveys, speak only indirectly about individual action, hide personalities, and require interpretation, yet they are more representative and systematic than the—yes—anecdotal evidence one must otherwise rely upon.

These are the major components of the research, augmented by other bits here and there. The research includes both conventional archival research and conventional econometric analyses. The specific methodologies are described in the appropriate chapters, in appendixes, or in related articles.*

A GUIDE TO THE BOOK

America Calling generally moves from the telephone industry to the user to the social role of the telephone, from the national to the local to the personal level.

Chapter 2 presents a brief, nontechnical history of the telephone in North America. Chapter 3 explores the various ways that the telephone industry, especially AT&T, marketed its service to households, exploring the manner in which the industry understood or misunderstood subscribers' use of the telephone. Chapter 4 tracks the diffusion of the telephone across the United States, assessing the factors that encouraged or retarded its spread. It also contrasts the telephone's

---

*A major concern for some historians may be the collaborative nature of the local histories. (I personally gathered almost all the material from industry archives.) The tradition in history is that the lone researcher fingers each scrap of parchment to judge its authenticity and to place it in context. The time and effort required plainly narrow the scope of any single historian's research. Although an important standard, this value must be traded off against other research values, such as the desirability of *comparing* cases, without which it is difficult to draw any general conclusions. (On the value of multiple, comparative studies, see, for example, Dykstra and Silag, "Doing Local History.") In this study I compromised by focusing on three communities and by assigning each of my research assistants to do the primary research on a single town. Such collaborative research seems atypical in history, but when it is supervised by a single scholar and parallel guidelines are followed, as in our research, this approach can be fruitful. Where the community stories coincide, we draw confidence in making generalizations; where they diverge, we are stimulated to seek out the sources of difference.

diffusion in rural as opposed to urban areas and in the working as opposed to the middle class. Chapter 5 further examines diffusion, but at the level of the local community and the household. It recounts the response to the telephone in Antioch, Palo Alto, and San Rafael and then uses census data to determine which households in those towns adopted the telephone in which years. In most of these studies we use the automobile as a comparative benchmark.

Chapter 6 employs a variety of evidence, from etiquette manuals to counts of advertisements, to chart how the telephone became an accepted part of everyday life. Chapter 7 looks at social change in our three towns, focusing on localism: Did residents become less involved in and less attached to their towns as the half-century passed? Chapter 8 looks more closely at individuals, asking how they reacted to the telephone and how they used it in their personal lives. In that context the chapter also analyzes the differences between men and women in regard to the telephone. Chapter 9 outlines telephone history from 1940 and summarizes the findings and implications of this study.

CHAPTER TWO

The Telephone in America

Alexander Graham Bell's fabled first words over the telephone, "Mr. Watson, come here, I want you," may not have been as dramatic as those dit-dotted by Samuel Morse during the first major exhibition of the telegraph, "What hath God wrought," but telephony's early years contained great drama nevertheless. Tinkerers and scientists raced to improve the primitive device; entrepreneurs struggled to rescue a failing company that would grow into a great industrial empire; its leaders battled attackers to secure their monopoly; gritty linemen risked their lives in blizzards to keep the wires humming; and telephone operators bravely stayed at their switchboards during fires and floods to make calls that barely averted tragedy. Such drama is the stuff of most telephone histories. Even skeptics must acknowledge the accomplishments of North America's telephone pioneers. They built an outstanding industry and public service.

Our purpose here, however, is to understand how the telephone system developed in America from 1876 to 1940. Consistent with the theoretical charge of the previous chapter, we take the perspective of residential consumers rather than engineers concerned with the machinery, corporate executives concerned with financial issues, or the thousands of telephone workers on the line. We will, for example, track the changing cost of telephone service to the average user. The technical, economic, and even political developments in the telephone story remain critical, of course, since they shaped the kinds of options available to consumers. By World War II consumers in the United States and most of Canada had to choose between a high-quality but expensive service from a private monopoly or no home telephone at all. How that situation came about is our immediate subject.

Much of telephone history has been recounted by others (see Appendix A for a bibliographical essay). This chapter will be only a sketch of that history.* It is of more than bibliographical importance that many sources we might draw upon were creations of the industry's public relations representatives. The effectiveness of those men is part of the telephone story, but it also makes separating fact from fluff difficult.[1]

After a brief review of communications in the years before the telephone, this chapter divides telephone history into distinctive commercial and regulatory periods. The first period spans the earliest years, when Alexander Graham Bell's** associates established the telephone system and secured the business. During the second period, up to 1893, Bell had a monopoly on telephones in the United States. During the third, from 1894 until about World War I, Bell struggled against many competing telephone companies; at one point it owned fewer than half the telephones in the United States. And during the fourth, from World War I to 1940, the industry consolidated: Competition almost ceased as the companies divided the telephone business among themselves under the hegemony of AT&T and the supervision of federal and state regulators. The telephone system remained essentially the same as it was on the eve of World War II until the court-ordered dismemberment of AT&T forty-four years later. This company, which was founded in 1875 and eventually grew into AT&T, underwent several reorganizations and name alterations. For simplicity's sake I refer to it as Bell and AT&T. The chapter concludes with a brief review of comparative cases: the telephone in Canada and Europe and the automobile.

---

*Well-known matters are not separately footnoted here, only specific issues or quotations.

**I use "Bell" to refer to AT&T, its system of companies, and its precursors. I use "Alexander Graham Bell" to refer to the man.

BEFORE THE TELEPHONE

Before the 1870s there was but one way for a North American to convey a message: If a yell couldn't reach the intended recipient, the message would have to be delivered by a mail carrier, a messenger, or most likely the sender. For intercity messages one might use the telegraph — although these were rarely employed for nonbusiness purposes — but that would still require a trip to the telegraph office. Shortly before Alexander Graham Bell called Mr. Watson, a few companies experimented with "home telegraphs." These systems usually consisted of a signaling device with several switch settings. Depending on the setting or the turns of a crank, subscribers could buzz the central telegraph office for a doctor, the police, a horse and buggy, or a messenger. A Bridgeport, Connecticut, company even set up a home-to-home telegraphing system. These nascent systems quickly succumbed to the telephone. In Dundas, Ontario, for example, the entrepreneur who started a "district telegraph" in January 1878 replaced it with a telephone system nine months later. Although such services continued until the turn of the century, their only lasting legacies are the stock ticker and the switchboard.[2]

The telephone, therefore, meant a radical change in personal communication. For most business activity, however, the import was less obvious. Businessmen relied on letters and telegrams, often with complex codes, to produce written records of their transactions. For them, voice transmission, scratchy and often indistinct, could be an adjunct at best. Or so it seemed to many people, such as inventor Elisha Gray, who set aside his ideas about voice telegraphy to work on improving the signaling telegraph and thereby played Wallace to Alexander Graham Bell's Darwin.[3]

FOUNDING THE TELEPHONE INDUSTRY

Like Gray, Alexander Graham Bell had been trying to improve the telegraph when he constructed the first telephone in March of 1876. That month he filed his patent claim, later to be a matter of legal dispute, and in May he showed the primitive device at the Centennial Exposition in Philadelphia. Alexander Graham Bell and his associates spent much of the next year or so giving demonstrations around the country of this "wonder," sometimes borrowing telegraph wires for long-distance calls (and sometimes failing). Watson would, for example, sing over the telephone to an audience gathered elsewhere in town. In 1877 a New York poster announced "An Entertainment of the Sunday School of Old John St. M. E. Church," including recitations, singing, and an exhibition of "Prof. Bell's Speaking and Singing Telephone." Admission was 25 cents.[4] These stunts garnered considerable publicity and awe as journalists relayed the news around the world.

Making a business of what was a novelty was more difficult. The backers of Alexander Graham Bell's telegraph work were his father-in-law, Gardiner Hubbard, and the father of one of his speech students, Thomas Sanders. In July 1877 the three men reorganized as the Bell Telephone Company, with Hubbard as trustee, and began seriously marketing the device. Initially, they leased pairs of telephones for simple two-point communications, commonly between two buildings of a business or between a businessman's home and office. The opening of the first telephone exchange, or switchboard, in New Haven in January 1878 was a profound step. Any subscriber could now be connected to any other.

The key financial decision, one of great long-term import, was Hubbard's determination that the company, as the exclusive builder of telephones, would lease the instruments and license local providers of telephone service. Bell thus controlled both the service and the consumers' equipment. (It is as if gas companies exclusively leased stoves and furnaces or electric utilities were the sole lessors of lamps.[5]) In this way Hubbard attracted franchisees around the country who used their own capital to rent telephones, string wires, build switchboards, and sell interconnections. Bell provided the instruments and technical advice and, in turn, collected rental fees. Over the years the company used its leverage on license renewals to set rates and to dictate technical and other features of the service. This close supervision allowed the company to convert a confederation of local franchisees into a "system" of local "Bell Operating Companies" acting in concert. Eventually, AT&T replaced the rents it charged with stock ownership in the local companies and, using this leverage, set common nationwide policies. But in the earliest years perhaps dozens of entrepreneurs in towns across America—some rounded up by Watson himself on marketing trips—made individual licensing agreements with Hubbard.[6]

By mid-1878 the telephone business was in ferment. About 10,000 Bell instruments were in use throughout the nation, but Bell now had serious competition. Western Union, already located in telegraph offices almost everywhere, adopted telephones designed by Thomas Edison and Elisha Gray to offer a competing service. Bell sued Western Union for patent infringement and hurriedly founded exchanges around the country to preempt markets. At the end of 1879 the contestants settled: Western Union conceded Bell all patent rights and instruments. In return, Bell agreed to renounce telegraph service, to pay Western Union 20 percent of gross receipts for a time, and to grant the telegraph company partial interest in a few local Bell companies. The resolution left Bell in early 1880 with about 60,000 subscribers in exchanges scattered about the country and a monopoly on the telephone business. (About 30 years later, Bell briefly absorbed Western Union until pressured by the federal government to sell it off.)

THE ERA OF MONOPOLY: 1880–1893

The typical telephone system of the 1880s was a cumbersome affair. (See Photo 1.) The instrument itself was a set of three boxes. The top box held a magneto generator, a crank, and a bell. The middle box had a speaker tube protruding forward and a receiver tube hanging from the side. The third box contained a wet-cell battery that needed to be refilled periodically and occasionally leaked. A caller turned the crank to signal the switchboard operator; the signal mechanically released a shutter on the switchboard in the central office, showing the origin of the call. The operator plugged her headset into the designated socket and asked the caller whom he or she was seeking. Then the operator rang the desired party and connected the two by wires and plugs in the switchboard. The two parties talked, usually loudly and with accompanying static, and then hung up. In some systems the caller cranked again to signal the end of the conversation. In others the operator listened in periodically to find out when the conversation was over so that she could disconnect the plugs.

The race to build exchanges, rapid adoption by businessmen, and other changes raised some technical problems in the 1880s. Edward J. Hall, considered "the most far-seeing, all around competent and efficient telephone man of his day," complained from his franchise in Buffalo as early as February 1880 of too much business and too many calls to provide subscribers adequate service.[7] One consequence of growth was increasing congestion at the switchboards. Spaghetti-like masses of wires crisscrossed the boards, which in turn grew in number, size, and complexity beyond the capacities of the operators struggling to reach around one another. Temporary solutions did not solve the problem, especially in the large urban centers, until the late 1890s.[8] In some places new electric and streetcar power lines created intolerable interference on the adjacent telephone lines. Some observers believe that this problem stunted telephone development in the late 1880s. (This nuisance recurred in rural America with the construction of power lines by the Rural Electrification Administration in the 1930s.)

Bell responded to the challenges by rebuilding its hardware. It eventually replaced single iron or steel wires (a system in which the electrical circuit was completed through the ground) with pairs of copper wires that returned the current. Bell also replaced wet batteries with a common-system battery; the power for all telephones on a line now came from the central exchange. In addition, Bell eventually developed new switchboards and procedures to alleviate switchboard congestion. These and other technical developments completely revamped much of Bell's telephone system by the early 1900s. Company leaders sought to develop high-quality service—clear sound, instant access, and the like—for the urban business customers they courted. To this end they rounded up as many telephone patents as possible, sponsored further research, and pooled the practical experience of their franchisees. Theodore N. Vail, as general manager and then president until 1887, used Bell's temporary patent monopoly to secure a technical and organizational edge over all future competitors, especially by developing long-distance service.

Although not favored, like Alexander Graham Bell, by a Hollywood biography, Theodore N. Vail is a figure of mythic stature in the telephone industry and in American corporate history. Beginning as a lowly telegrapher, Vail deployed his organizational skills and modern methods to rise to superintendent of the federal Railway Mail Service. Hubbard lured Vail, then 33, away to manage the fledgling Bell company in 1878. For several years Vail pressed aggressive expansion, patent protection, and business reorganization. In 1887, by then president, Vail resigned after conflicts with a more cautious board of financial officers. He succeeded in several business ventures around the world, but kept abreast of the telephone industry. Vail would come back.[9]

Vail's policy of establishing high-quality service meant that costs were high, especially in the larger cities where the complexities of switching were most difficult. The minimum flat rate in central Los

Angeles in 1888, for example, was $4 per month plus two cents a connection after the fortieth call. This rate equaled about 10 percent of the average nonfarm employee's wages. That same year Boston subscribers paid a minimum flat rate of $6 a month.[10] In addition, Bell's affiliates took every advantage of their monopoly to levy what the market could bear. For example, when the competing telephone exchange closed in San Francisco in 1880, the Bell local raised its charges from $40 to $60 a year. The local manager justified the move: "The increase was made because the public always expects to be 'cinched' when opposing corporations consolidate and it was too good an opportunity to lose. (Moreover, it would have been wrong to disappoint the confiding public.)"[11] Conflicts with irate customers arose, the most famous of which was an 18-month boycott of telephones organized in Rochester, New York, in 1886.[12] Bell's rates began to drop as 1894 approached, probably because of the competition it anticipated when its patents expired, although Bell claimed that improved technology explained the drop in charges. By 1895 the Los Angeles rate was down by 38 percent to $2.50 a month—7 percent of wages—plus two cents a call. Even then, telephone service remained expensive.[13]

The common practice during this era and beyond was to charge customers a flat-rate for the telephone service, allowing unlimited calls. During the 1880s local Bell companies repeatedly debated and experimented with a message-rate formula, charging by the call. One argument in favor of this approach was that it would permit the basic rental fee to be lowered and thereby encourage small users, such as families, to subscribe. Edward J. Hall was a leading proponent, labeled by some the "father of the message-rate system." Another reason for a pricing change, more favored by Vail and others, was that ending flat-rate service would discourage use, and thus "cut off all the superfluous business that tends to make the operation of the business so unremunerative." Existing customers, however, resisted the change by complaining, by petitioning the town officials who issued permits for telephone poles, or, as in Rochester, by boycotting the telephone service. Not until after the era of monopoly did message-rate service become common, although still not universal, in Bell's largest exchanges.[14]

Vail's agenda went beyond securing a technical monopoly. Through various devices he centralized control of the Bell System and its affiliates. Doing so was complex, since local situations varied widely. Each regional operating company had to deal with many governments to secure permits, to fend off complaints about the unsightliness of the wires, and sometimes to negotiate rates. Still, standardized policies, as well as a superior technology, helped brace the Bell system against challenge. Vail's successors after 1887 were, in retrospect, more interested in extracting monopoly profits from the system than in securing its future. That shift in priorities would become evident when the patents expired.

Strategic disagreements about pricing policy arose inside Bell, in part from different visions of the telephone's potential. It was not at all obvious whom the telephone would serve and how. As Sidney Aronson has noted, "[T]he inventor and his backers . . . faced the formidable task of inventing uses for the telephone and impressing them on others."[15] During the first few decades of telephony, industry marketers devised a variety of applications, including transmitting sermons, broadcasting news, providing wake-up calls, and many other experiments. As late as the 1910s, the trade journal *Telephony* had an index entry under "Telephone, novel uses of."* The industry spent considerable time, especially in the nineteenth century, simply introducing the public to the instrument and dispelling suspicions about it. (See Chapter 3 on marketing.)

Industry leaders approached telephony from their experiences with telegraphy. Alexander Graham Bell and his backers were initially trying to improve the telegraph. Theodore Vail came from a family involved in telegraphy and had been a telegrapher. Many local telephone entrepreneurs had started out selling telegraph service. An important exception was Edward J. Hall, the message-rate enthusiast, who started in his family's brick business after earning a degree in metallurgy. Hall established the first Buffalo telephone exchange, left the telephone business a few years later, and was lured back by Vail to manage long-distance development in 1885 and then Southern Bell for over 10 years.[16] Because telegraphy defined the background of most executives,[17] and because Americans in the nineteenth century used the telegraph almost exclusively as a business tool, it was logical that Bell used the telegraphy model to define the telephone as a device for business as well.

Who were the first telephone subscribers? Physicians were notable among the early users. The telephone allowed them to hear of emer-

---

*Included under that entry in volume 71, for example, are "degree conferred by telephone, dispatching tugs in harbor service, gauging water by telephone, telephoning in an aeroplane."

gencies quickly and to check in at their offices when they were away. Druggists typically had telephones, as well.[18] But businessmen formed the primary market.

Bell found some businessmen hesitant to replace the telegraph with the telephone because they valued a written record. Nevertheless, some manufacturers, lawyers, bankers, and the like—and later small shopkeepers—adopted the technology. In 1891 the New York and New Jersey Telephone Company served 937 physicians and hospitals, 401 drug stores, 363 liquor stores, 315 livery stables, 162 metalworking plants, 146 lawyers, 126 contractors, 100 printing shops—7322 commercial customers all told—but only 1442 residences. Residences with telephones were typically those of doctors or of business owners or managers.[19]

One issue for Bell was whether it could fruitfully expand into the general residential market (that is, beyond the households of the business elite). In late 1883, noting that "the Telephone business has passed its experimental stage," Vail surveyed affiliates around the country, asking, among other questions: "Is it desirable and what would be the most practical way, to provide a service which would be in the reach of families, etc.?" His aide summarized the responses:

> There would seem to be but one opinion on this query and that is, that it is *most desirable*. The difficulty which presents itself is the manner in which the desired end should be reached. It is admitted that a great increase in the business would occur by the introduction of a rate and system, whereby the Telephone would be made universal so to speak, amongst families, and several modes [are] suggested . . . [including more pay-telephone stations, party lines, and lower residential rates]. It would appear from many of the answers to this query, "that a reduction in royalty" would be a necessity. . . .[20]

There was the rub: Locals would have to reduce their rates, and to ease that reduction Bell would have to lower its charges on the locals. Except for a handful of populists in this era—notably Edward Hall of Buffalo; John I. Sabin, later president of Pacific Telephone; and Angus Hibbard of Chicago—the consensus was that any increased business would not make up for the profits lost by reducing rates, even in a measured-rate system. At the time many also believed that operating costs per subscriber increased as the number of customers increased because of the technical complications of interconnection.[21] Only later did industry analysts appreciate that, as a network, telephones became more attractive as more people subscribed and that there might be economies of scale. George Ladd, president of Pacific Telephone in 1883, expressed the conservative position. He wrote to Vail that he opposed the reduction of residential rates because it could not pay and customers would not be grateful: "I am opposed to low rates unless made necessary by competition. . . . Cheaper service will simply multiply the nuisance of wires and poles and excite [political pressure to put wires underground], without materially improving profits or permanently improving relations with the public."[22] Residential service was therefore a stepchild in the system.

This attitude, later described even by Bell's friends as arrogant, predominated in the company. In 1906, for example, New England Bell commissioned an attorney to study telephone service in the Midwest. In its earlier history, he reported, "the public interest received scant attention" from Bell companies. They "were almost, if not quite, inexcusably slow in coming to an intelligent apprehension of the public need and desire for increased and improved telephone service."[23]

Bell managers were also skeptical about providing service in smaller communities. Businessmen in several small California towns, for example, appealed to Pacific Telephone for service but were turned away. In a few cases local entrepreneurs built bootleg systems, risking lawsuits.[24] AT&T focused on providing big-city businesses with high-quality service, including long-distance calling, at high prices. Its representatives later explained that the pressures of escalating demand and technical renovations prevented the company from pursuing wider markets until the mid-1890s.[25] Still, most Bell managers saw few possibilities for expansion, and nearly none for greater profit, in the general residential market or even the business market outside the major centers.

Between 1880 and 1893 the number of telephones in the United States grew from about 60,000—roughly one per thousand people— to about 260,000—or one per 250 people. The vast majority, more than two-thirds, were located in businesses.[26] This expansion, while dramatic in the early years, slowed after 1883, perhaps because of the technical problems or, just as likely, because of predatory monopoly pricing.

THE ERA OF COMPETITION: 1894 TO WORLD WAR I

Bell's key American patents expired in 1893 and 1894. Within a decade literally thousands of new telephone ventures emerged across the

United States, many in competition with Bell, others tilling territories Bell had previously ignored.* Rates plummeted and telephone subscriptions soared. (Figure 2 displays the growth in the number of telephones and automobiles in the United States from 1894 to 1940.) In the convulsions of rapid expansion and fierce competition, AT&T reorganized and government regulation developed. By the eve of World War I the modern telephone system had formed, one with local-service monopolies and a long-distance Bell monopoly, operating under the effective domination of AT&T and the regulation of state commissions.

The "independents" were a varied lot. They included commercial ventures, a few of which survive today, such as General Telephone and United Communications. In 1902 roughly 3000 such companies operated in the United States. Most of the new concerns, however, were not profit-oriented but instead were mutual companies established by shareholder-subscribers, usually under the initiative of a small-town doctor or merchant. Many were even simpler "farmer lines" running among a few dozen or a handful of homesteads, some using barbed-wire fences for connections. In 1902 roughly 6000 mutuals were counted by the U.S. Census. The Liberty (Tennessee) Home Telephone Company illustrates the breed. Organized in 1910 by a circle of family and friends who complained about the high rates of the existing telephone provider, Liberty charged $25, plus a telephone, a pole, and some labor to join, and then assessed $7 a year as a flat annual charge.[27] Many of these telephone services were marginal operations—cheaply built, poorly maintained, short of capital, lacking managerial skills. Some were little more than financial scams. Fewer, starting off with the latest technology, such as automatic dialing, met or surpassed local Bell quality.

Despite their marginality, the independents severely challenged Bell. They entered territories and markets that Bell had ignored and satisfied many customers who would accept lower quality to get cheaper service. By serving rural hinterlands, the independents could even contest Bell for the larger towns, claiming that only they could provide connections to outlying subscribers. In addition, independents represented a hometown challenge to a national company that many viewed as a "foreign" exploiter. By 1902, 52 percent of towns with

---

*In Canada, Bell's monopoly ended in 1885 when the Supreme Court ruled that key patents were irrelevant to Canada. Bell's initial response there was aggressively to preempt all markets against potential competitors and redouble efforts to build long-distance lines—a trial run of its U.S. action several years later.

a population over 4000 had two or more telephone companies. The larger independents banded together in protective associations against the "Bell Octopus" and tried to establish a competing long-distance service. It ultimately failed for, among other reasons, an inability to crack Bell's monopolies in the largest cities and difficulties in raising capital, problems abetted by AT&T sabotage.[28]

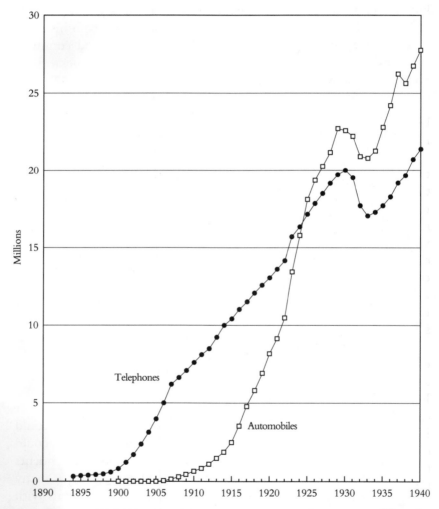

FIGURE 2. TELEPHONES AND AUTOMOBILES, 1894–1940. The telephone spread rapidly when competition began in 1893, but its growth slowed as AT&T reconsolidated control following the return of Theodore Vail in 1907. By the mid-twenties, the upstart automobile had spread to more American households. (Source: United States Bureau of the Census, 1975.)

Bell fought back. From 1894 to the 1900s it competed everywhere. Bell founded exchanges in as many areas as possible to preempt competitors, even at the cost of draining its capital. It slashed prices where forced. The independents often priced their service not only below Bell's but below true cost. Bell could match that and suffer the losses longer because of its "deep pockets." For example, John I. Sabin of Pacific Telephone reported in 1895 that his innovation of a cheap, 10-telephone party line had undercut competitors:

> Our ten-party lines in small places . . . are giving satisfactory services, and tie up to us for one year at least, a number of small users of the telephone, and these small users who have never had telephones are the ones that form the nucleus of the ordinary opposition exchange.

In 1899 the St. Louis manager similarly reported that a 10-party line had blocked competition.[29] Bell companies selectively and temporarily slashed prices for other levels of service as well.

Bell also leveraged its monopoly on long-distance service. Refusing to connect its lines to those of any independent, Bell could advertise that it alone provided access outside the local community. (In some regions independents combined to provide medium-distance toll calling, but they never came close to matching Bell's reach. One reason was that Bell interests used financial and political leverage to block independents' access to New York and Chicago.) And, over time, long-distance increasingly provided a subsidy to cover local losses. Bell had other tools, too. It continued patent suits, ran a publicity war against the independents, and had its subsidiary, Western Electric, refuse to sell equipment to them.[30]

In places, the battle was fierce, with spying, sabotage, secret purchases of competitors, bribery of city officials, financial subversion, and other "dirty tricks." In one community, people charged that Bell subsidized a grocery store to compete with one subscribing to an independent line. The local butchers then refused to supply the Bell grocery with meat. In South Omaha, Bell workers cut down competitors' lines—a crude ploy that was probably used elsewhere, too. It is likely that the independents were equally guilty of such tactics, but Bell was probably more often successful. The most celebrated case was Bell's secret control of the Kellogg Company, a key supplier of telephone equipment to independents. Not only did AT&T earn profits by this ownership, but it could also have had Kellogg purposely lose key patent lawsuits to Bell. The invalidation of Kellogg's equipment would then have exposed its independent clients to lawsuits as well. The courts voided the sale of Kellogg to Bell in 1909.[31]

Despite the ferocity of Bell's response in the early 1900s, it had lost half the market to independents by 1907. More precisely, in 1893 all 266,000 legal telephones in the nation were Bell's; 14 years later, Bell had made over 3 million new installations, but the independents and mutuals had done about the same. Moreover, the competitive pressure had driven down Bell's returns on investment sharply[32] and forced it deeply into debt to finance expansion. Financial groups in New York and Boston contested control over this increasingly precarious operation and in 1907 J. P. Morgan's ring won. They brought Theodore N. Vail back to head the Bell System, and he took immediate and dramatic charge.

Vail abruptly changed strategies. He curtailed expenses by slowing down expansion, easing price competition, and conceding markets to the independents. The largest case of strategic withdrawal occurred in Canada. In 1907 Bell Canada left the prairie provinces to government control, thereby reducing federal political pressure on Bell to serve rural Canadians. In more favorable markets Vail tried to buy as many competitors as possible and make interconnection agreements with small operations, such as farmer lines, thus absorbing them de facto. These agreements also kept the small systems from allying with the larger independents. Within five years Bell owned 58 percent of U.S. telephones and was clearly in ascendence.

Vail also reversed Bell policy and accepted government regulation as a way to organize a coherent, "universal" communications system—under Bell's leadership. Although state and federal regulation compelled AT&T to serve some markets and constrained its prices, this government involvement largely worked in Bell's competitive interests. By supporting high-quality technology, consolidation, and uniform pricing, the regulators, wittingly or not, pressed Bell's advantages.[33] Bell's leadership was further sustained by its investments in long-distance service and technical research. Vail also launched aggressive public relations efforts designed to blunt calls for government ownership, improve public attitudes toward AT&T, and promote his notion of a regulated but private monopoly (see Chapter 3).

AT&T's purchase of Western Union in 1910, complaints by some competitors about its purchases of other independents, and the general Progressive movement against corporate power led authorities in Washington to focus on AT&T as an antitrust target. AT&T forestalled court action (by about a half-century), by issuing in late 1913 the Kingsbury Commitment, a statement to the attorney general agreeing to divest itself of Western Union, to forgo the purchase of other telephone companies without government approval, and to permit non–Bell companies interconnection with its lines. This compromise initiated a process over the next decade or so, worked out with occasional federal assistance, of accommodation between Bell and the independents. This strategy granted Bell dominance but protected the financial investments of the large independents. As a result, a kind of senior–junior partnership developed; for example, Bell provided technical, legal, and political support for many independent companies' rate increases over the next decades. After World War I, Bell also continued to buy independents (discussed later). This Bell-dominated system of private telephone operations, along with the three government telephone companies in Canada's Prairie provinces, defined the North American telephone system until 1984.

Telephone consumers in the years before World War I received technically superior service, with more options, at vastly lower prices than they had before 1894. The instrument itself, on a common-battery system in large urban areas by the 1910s, was slimmer and easier to use. (See Photo 2.) For example, picking up the headset could now signal the switchboard operator. Improvements in underground cables made conversations more distinct, while other technical developments after the turn of the century increased the effective distance and efficiency of the service. In particular, Bell extended the range, capacity, and speed of long-distance calling, culminating in the first coast-to-coast call in January 1915, from New York to the World's Fair in San Francisco. (See Photo 6.)

The most important new option offered customers during the competitive era was the party line. With two, four, or sometimes more customers on the same wire, it was possible to charge considerably less than for an individual line. Customers with moderate incomes apparently valued the service, despite the inconvenience of busy lines, bells ringing for another party, and occasional eavesdropping. The mutual companies and farmer lines eagerly provided party-line service. Bell operating companies varied greatly, however. In 1901 only 4 percent of Detroit customers were on party lines, compared to 66 percent in Minneapolis. For most AT&T leaders, party lines, with their low rates, technical complications, and disagreeable clientele—Vail once called them "the worst class of people"—were a necessary evil to enable Bell to underprice the competition and preempt their markets.* For a few others, such as John Sabin and Angus Hibbard, party lines were vehicles for a mass and ultimately more profitable telephone system. Such devices—Sabin experimented with 10-party incoming-only lines and 30-party lines—would bring even the working class into the system and create a taste for telephone use. As Hibbard saw it, "Thousands of people got the telephone habit and if you get it once, you never get over it."[34]

One option Bell did not provide was automatic dialing. Many independents did, and it served them as a useful competitive device, despite its often balky performance. Bell resisted automatic dialing because, it claimed, the technology was not sufficiently developed and it depended too much on users' skills, or because, critics claimed, Bell had too much money sunk into the existing equipment. Although AT&T publicists extolled the humanity, service, and occasional bravery of operators, many customers preferred the privacy of their own dialing. Operators *did* occasionally listen in.[35]

The major change consumers encountered in this era was a radical drop in cost. In New York City, for example, Bell's annual rates plummeted from $150 for 1000 calls to $99 in 1899, $78 in 1905, and $51 in 1915—a reduction of more than two-thirds, adjusting for inflation. AT&T reported that its average rates, nationally, for residential service dropped from $56 in 1894 to $24 in 1909. (Internally, AT&T sharply cut its rental charges to local companies. It later abolished rental fees in return for a percentage of profits from each local.) Price wars in some areas helped consumers. Increasing use of measured- rather than flat-rate service, although resented by many customers, helped drive prices down, as did the various party lines. In 1900, for example, John I. Sabin, then heading Chicago Bell, installed 10-party, nickel-a-call telephones, which reportedly doubled subscription rates.[36]

---

*Chief Engineer Joseph P. Davis had another rationale for party lines in 1899:

There seems to be two distinct fields for two-party lines: 1st) To provide a service which shall be so limited in amount and of such inferior character, that—while (at a low price) it will attract new subscribers—will not satisfy them permanently, but will act as an educator and feeder for the more profitable grades of service. To accomplish its purpose, an extremely low rate must be affected, entailing to this company considerable loss, which loss may be considered a charge against procuring of new business. The object of this service will not be accomplished unless the service is unsatisfactory, when once taken. It therefore requires that enough subscribers be placed upon a line to make them dissatisfied and desirous of a less limited and better service. (Letter to Pres. Hudson, 6 February 1899; in Box 1284, AT&THA.)

Rates varied greatly from place to place and time to time, especially depending on competition, but Table 1 suggests the range of prices for simple monthly residential service in cities. Rates tended to be highest in the larger cities and lowest in areas independents served or where local regulators held rates down, as in Los Angeles in 1909. In this era a half-gallon of milk cost about 15 cents and a pound of bacon about 20 cents. The average wage-earner in manufacturing made only about $40 a month, and the average federal employee made about $90. The monthly charge for basic telephone service in, say, Omaha, represented 5 and 2 percent, respectively, of their wages. Telephones were cheaper than before—half or less in cost relative to income than before 1894—but still expensive for many, at least in the cities.

With the drop in rates, telephone adoption accelerated rapidly. The increase in the number of American telephones had slowed between 1885 and 1894 to an annual compound rate of roughly 6 percent. Between 1894 and 1907 it raced away at roughly 30 percent per year. Between 1907 and 1917 the rate slowed again to about 6 percent per year. The correlation of expansion with vigorous competition and price-cutting is plain. Much of the expansion was in the small-town and rural areas (except in the South) that Bell had never served. The

TABLE I

COST OF BASIC TELEPHONE SERVICE IN SELECTED CITIES, 1900–1912.[*]

Year	Place	Type of Service	Basic Rate/Month	In 1990 $$
1900	Chicago	4-party	$3.33	$52.50
1903	Los Angeles	10-party	1.50	21.50
1907	Manhattan	2-party	3.50	48.50
1909	Los Angeles	4-party	1.50	21.50
1912	Boston	4-party	1.25	16.50
1912	Savannah, GA	4-party	2.50	33.00
1912	Omaha, NE	4-party	2.00	27.50
1912	Denver	4-party	2.50	33.00
1912	Litchfield, MN	4-party	1.00	13.25

[*]The estimates are drawn from: Chicago: Mahon, *The Telephone in Chicago,* 27; Manhattan: New York State, *Report of the Committee;* Los Angeles: "Telephone on the Pacific Coast," Box 1045, AT&THA, 10–28; 1912 data: The United States Bureau of the Census, *Telephones and Telegraphs . . . 1912,* 52ff.

independents and cooperatives provided country customers with cheap, if cantankerous, service, and subscriptions boomed (see Chapter 4).

Public phones also spurred telephone use. Many were free to the users. Stores, particularly pharmacies that adopted telephones early to accommodate physicians, offered customers free use of their (flat-rate) telephones as a public service and a commercial inducement. Eventually, many druggists tired of the crowding around their telephones, and the telephone companies tired of nonsubscribers clogging their lines. By the turn of the century nickel-in-the-slot public telephones were common throughout major urban areas.[37]

In the roughly twenty years between the expiration of Bell's monopoly and World War I telephone service was transformed from a business tool and a luxury good to a common utility. Most North Americans still could not afford or would not invest in a home telephone, but middle-class urban households and millions of farm families did have service. The rest, largely the urban working class, were familiar with and occasionally used the telephone in a neighbor's home or at a local store.

WORLD WAR I TO WORLD WAR II: CONSOLIDATION AND DEPRESSION

Between the world wars the telephone industry steadily consolidated what it persuaded regulators was a "natural monopoly" and standardized its service. During the Depression, however, it faced its first period of retrenchment. The gains in residential subscribers during the booming twenties were almost all wiped out between 1930 and 1933. Even AT&T had to plead for customers. Not until 1939 did the industry fully recover. By then the technological environment had changed with the proliferation of cars and radios. The New Deal had changed the regulatory atmosphere in the United States, too.

For one year during World War I the federal government took over the telephone and telegraph industries. This worked to the benefit of the industry because the government instituted price increases that the companies had been unable to gain on their own. This, in turn, both raised baseline rates and helped discredit nationalization. Also the wartime experience of coordination between AT&T and the independents accelerated the unification of the industry. After the war a Senate committee determined that telephony was a "natural monopoly," and a House committee declared that telephone competition "was an endless annoyance." The Willis-Graham Act of 1921 amended the Kingsbury Commitment, empowering the Interstate Communications Commission to waive antitrust limitations on purchases of telephone companies. In the next thirteen years Bell was the buyer in 223 of 234 such purchases. AT&T also swapped local franchises with independents, so that the number of communities with competition dwindled rapidly. By the late 1920s, even independent telephone men declared that competition had been an unfortunate error. By 1930, 80 percent of telephones in the United States were Bell telephones, and 98 percent of the remainder connected to Bell lines.[38]

The federal blessing on consolidation coincided with increasing state regulation of telephone service. While many state commissions focused on holding down rate increases, their efforts to tie outlying subscribers into Bell lines meant that small mutuals had to accept Bell's technical standards and Bell's rates. Many small companies simply joined Bell. Bell and the larger independents cooperated in confronting state regulatory commissions. Also, many of those commissions had to rely on the information presented by the industry because they lacked the resources Bell and its allies had to investigate the conditions of telephone service. Regulation had become a comfortable environment for the telephone companies.

The New Deal brought to Washington much more aggressive watchdogs. The 1934 Federal Communications Act tried to address some concerns about oligopoly, but largely ratified the emerging system. The act authorized the Federal Communications Commission (FCC) to oversee rates, to approve construction and consolidation, and to investigate the industry. It also declared a national goal:

> [to] make available, so far as possible, to all the people of the United States, a rapid, efficient, nation-wide, and world-wide wire and radio communications service with adequate facilities at reasonable charges.[39]

A hostile investigation of AT&T by the FCC in 1938 documented the increasing "collusion" or "cooperation" in the industry. It also documented the extent to which the Bell system had evolved from a federation of local companies to a centralized national operation.[40]

Personnel and material shortages during World War I had caused the government-run telephone service to suffer and rates to rise. (For example, the government established an installation charge, long desired by the industry, that stayed when the system was reprivatized.) After the war, shortages due to unleashed civilian consumption, inflation, and labor unrest also hampered telephone service. The companies struggled to keep up. For example, Bell Canada had to use party lines in Montreal for the first time just to meet demand. Within a few years, however, telephone service, long-distance in particular, was again expanding and improving. Sound became clearer, carrying capacity expanded, costs dropped, and the time needed to place a toll call shrank from an average of seven minutes in 1920 to about one minute in 1928.[41]

Operators had served Bell publicists as a romantic selling point for many years, but with labor shortages, union demands, and inflation, AT&T began automating its major exchanges. This move shortened connecting time by a meaningful 5.3 seconds. By 1929 conversion to dial telephones was about one-fourth complete.

In 1929 most residential customers had party lines. Nationally, 36 percent of Bell subscribers shared two-party and 27 percent shared four-party lines. This system was especially common in the largest cities and in rural areas. A relatively few subscribers had extensions, which composed about 8 to 10 percent of telephones in the major cities. In the 1920s city folk took their calls on small desk sets with separate speakers and earphones, but near the end of the decade Bell started selling the "French" phone, with one horn-shaped instrument housing a speaker and a receiver on either end. By the end of the decade Bell serviced nearly 300,000 pay telephones across the country.[42] (See Photo 3.)

The war and the inflation that followed increased the costs of basic telephone service, although precise assessment is difficult given the great variation in local rates. The Bell System alone had hundreds of different rate schedules. In 1923 the average monthly charge for flat-rate four-party service ranged from a median of $1.67 in the smaller towns to $3.00 in the largest.* A private line cost 50 or 75 cents more. Non-Bell charges were typically lower. Yet a half-gallon of milk by then cost 28 cents and a pound of bacon 40 cents; workers in manufacturing averaged $105 and federal employees $138 a month in earnings. In "real" dollars, therefore, telephone rates had modestly declined over the previous 15 years.[43] Nominal telephone rates stayed roughly the same over the next 15 years, but other living costs dropped (milk

---

*About $13.25 and $22.65 in 1990 dollars.

dropped to under 21 cents and bacon to under 23 cents by 1933). So, the "real" costs of subscribing rose again during the Depression.[44]

Before the Depression residential subscription increased steadily, from 35 percent of United States households in 1920 to 42 percent in 1929. Since farm families were disconnecting their telephones in the same decade (see Chapter 4), the rate at which nonfarm families were subscribing was even greater—great enough that an AT&T expert worried in 1924 about the "surprising" gains in residential customers. The diffusion was geographically uneven, with subscription rates ranging from 22 percent to 65 percent of families across the largest 100 urban areas.[45]

In the late 1920s AT&T leaders worried, however, that the rate of adoption was not great enough, because more American families were buying cars, electricity, and radios than were buying telephone service.[46] Their subsequent efforts to widen the telephone market unexpectedly derailed when the Depression arrived. Between 1930, when the Depression hit home, and 1933 over two-and-a-half million households disconnected their residential telephones, a 20 percent loss.* The subscription rate dropped from 41 to 31 percent of U. S. households. The industry shifted its sales efforts from marketing long-distance service and extensions to persuading old customers not to disconnect. Companies mobilized all employees—that is, the 77 percent who still had jobs—in this effort. Many worked on their own time as companies cut back hours.[47] Companies also delayed plant improvements. (But, in a controversial move, AT&T kept its dividends to stockholders constant.) Local managers used various interim strategies, including leaving delinquent subscribers' telephones connected in hopes that they would eventually start paying again.

In the mid-1930s telephone subscription rose once more. By 1939 it had recovered to 1930 levels. Wartime demand for telephone service escalated rapidly, but supplies were limited. By the end of World War II Bell had large backlogs of requests. That backlog, along with greater affluence and federal aid to rural telephone systems, led to a 62 percent rate of residential subscription in the United States by 1950, 80 percent by 1962, and 90 percent by 1970 (see Figure 1 in Chapter 1). A home telephone became common not just for the middle class but for almost all Americans.

---

*Business telephones declined by 11 percent, automobile registrations by 10 percent, and electric service by only 1 percent.

COMPARISONS

To place this history in context, it is useful to contrast the telephone in the United States to, first, the telephone in comparable countries, and second, to other technologies.

The Telephone Elsewhere

The Canadian history of the telephone is most similar to that of the United States, so much so that I generally incorporate it in this study. Still, a few differences are worth noting. Bell was quick to establish itself at least in the urban regions of Canada, but the courts overruled its patent monopoly years before its expiration in the United States and so Bell faced Canadian competition earlier. Bell in Canada also encountered considerable pressure to serve rural areas. The federal government launched a major investigation of Bell's service in 1905, which, although it never yielded legislation, discomfited AT&T. In 1907, when Vail pared back the company's expansion, he withdrew Bell from the Western provinces, except for British Columbia. In Bell's stead, Manitoba, Alberta, and Saskatchewan established government telephone systems. These provided a subsidized, cheap service for the largely rural population of the prairie provinces, the most success-ful being Saskatchewan's combination of small cooperatives handling local calls and the province handling the toll lines. The socialized sys-tems suffered badly during the agricultural depression that began in the 1920s.

Canadian rates tended to be lower than those in the United States. The proportion of all Canadian households with service was, at least before the Depression, about the same as in the United States, although higher in cities and lower in the countryside. In 1930 Canadians had 10.1 residential telephones per 100 persons (from 5.4 on Prince Edward Island to 14.7 in British Columbia) compared to 10.6 in the United States.[48]

In Europe telephone systems generally began as scattered private enterprises, but government postal and telegraph administrations soon absorbed them. In Scandinavia government-owned trunk and urban systems linked rural telephone cooperatives. Typically, European offi-cials viewed the telephone as a public resource that could, for example, be used to connect dispersed communities, replacing telegraph oper-ators in villages. The administrations did not market telephone service as a consumer good. Their focus was largely on meeting publicly defined needs, such as responding to emergencies, and on keeping rates low. They struggled to raise capital for construction from general state revenues. Some European administrations restricted telephone expansion to protect their original missions of postal and telegraph service.

Over the years, AT&T claimed that government monopolies in Europe had not matched the success of private monopoly in the United States. A 1916 ad titled "Best and Cheapest Service in the World," for example, contrasted American and European telephony on access, promptness, and price.[49] Certainly, the diffusion of the telephone in the United States far exceeded that of Europe in the early years. In 1910 the United States had about eight telephones per 100 people, Canada three-and-a-half, the Scandinavian countries roughly three, and Germany and the United Kingdom under two. The reason for these differences, however, is less clear. Of course, the relative wealth of the nations clearly favored America. Also, the comparative advantage of the United States was greatest in rural areas, and it emerged most rapidly in the early 1900s — both suggesting that competitive free-for-all, not Bell's private, largely urban, monopoly, created the American advantage. In the United States isolated farm families linked themselves up by telephone, but in most European countries telephone administrations placed a single telephone or a handful of stations in village centers. In 1927 the United States had 15 telephones per 100 residents, Canada 13, Scandinavia about 8, Switzerland 5, Germany 4, and the United Kingdom 3, but again the advantage of the United States was considerably less in the large cities than in the small ones.[50]

Indeed, by 1927, 20 years after the heyday of competition and Vail's decision to substitute consolidation for confrontation, the number of telephones in the United States was no greater than one would have expected given its wealth. While America's 15 telephones per 100 people far exceeded that of European countries, so did its gross national product. The United States' 93 telephones per $1000 of GNP was not exceptional. The Scandinavian nations, for example, exceeded 180 telephones per $1000 GNP, Germany had 141, and Switzerland had 118. Moreover, Americans' household telephones were disproportionately likely to be on party rather than individual lines.[51]

North American and European telephony also differed greatly during the Depression. The United States and Canada lost ground in telephones per capita, while the European nations gained, sometimes substantially, in both the large cities and the countryside. One explanation is that in North America before the Depression even economically marginal households had adopted the telephone, and it was they who dropped off quickly during the crisis. But Scandinavian and Swiss big-city subscription rates were similar to those of the United States, and yet they expanded substantially during the Depression, equaling or exceeding rates in North American large cities in 1937. Other explanations for European success during the "Slump" include the greater diffusion in North America of competing technologies, especially the automobile, and that European telephone systems persisted because they were government programs. Governments may even have expanded telephone construction to provide jobs.

In the 1980s the number of telephones per capita in the United States was again just about what one would expect given its affluence. The United States in 1987 had 46 telephones per million dollars of gross domestic product; Canada had 43. Denmark and France were high at 64 and 53; Ireland and Italy were low at about 40.[52] Thus, both history and contemporary data cast doubt on claims of American—or AT&T—exceptionality in the diffusion of the telephone. Governmental monopoly seemingly did as well as private monopoly in producing telephones (although these data do not speak to the quality of the service). What would have happened had the unbridled competition of the early 1900s in the United States continued we will never know.

Electrification and the Automobile

The delivery of electricity and the delivery of telephone service have some similar structural properties. Both involve networks, both rely on attaining a critical mass of customers, and both tend to be delivered by local monopolies. The history of electrification, like that of the telephone, was shaped by the political context. And their American histories have some parallels. In the United States electric utilities at first ignored the home market in favor of industrial customers, so much so that electrification of homes spread more slowly here than in northern Europe. Similarly, the electric utilities welcomed regulation, even before Theodore Vail did and for similar reasons. Although there was no electrical giant comparable to AT&T, the political environment for the American electric companies was like that of the telephone industry.[53] If one might speak of "technological style," as does historian Thomas Hughes,[54] to characterize variations on technological development, both electricity and the telephone might well share an especially American style.

Although Europeans developed and popularized the first true automobiles, the United States soon became the preeminent automobile-using nation.[55] The automobile descended from the bicycle: Bicycle makers predominated among early automobile manufacturers and drew upon bicycle technology to build cars; automobile marketing followed that of the bicycle, with dealerships and annual models; and at first the automobile appealed to the same clientele—well-to-do city folk interested in recreation. A few drivers also had pragmatic purposes, particularly physicians. Later in the 1900s, when prices dropped, popular demand developed, most notably among farmers. Pleasure driving remained a key motive for taking the car out, but Americans increasingly put it to practical use—farmers for hauling, city-dwellers for commuting (by 1923, 52 percent of car owners used their cars to commute[56]), and both for shopping.

The early automobile industry included hundreds of companies. The technologies that made a horseless carriage move were so varied that no single invention or patent could be exclusive. In any event the leading companies developed a general patent-sharing arrangement. Continuous technological development and rabid competition reduced the number of important domestic manufacturers to a handful by the 1920s. The key step was Henry Ford's introduction in 1908 of the massively successful Model-T, a cheap car that was also a good car. The "Tin Lizzie" dropped in price from $825 in 1908 to $260 in 1925 (about $11,930 and $2,000 in 1990 dollars), pulling much of the industry with it. While only about 1 percent of American families owned automobiles in 1910, 26 percent did in 1920.[57]

Local governments quickly involved themselves. After briefly considering banning automobiles, they became regulators and facilitators, introducing speed limits, licensing rules, traffic controls, and the like. They literally undergirded the automobile by building and maintaining roads. Encouraged by the "Good Roads" movement that bicyclists had started, local and state governments surfaced, widened, and extended roads. The federal government became significantly involved around World War I. In 1913, 8 percent of taxes collected in the United States paved and maintained rural highways; by 1930, 15 percent did. At state and federal levels user fees (mostly the gasoline tax) paid for less than half of what was spent on roads; at lower levels user fees paid for almost none of the street work and traffic control. General taxes supported the automobile.[58]

After World War I the automobile spread feverishly (see Figure 2). By 1930 about 60 percent of American families had automobiles (compared to 41 percent who had telephones). The stiff, ungainly, balky, puttering, and in winter chilling automobile of the early 1910s had given way to the closed sedan, complete with heating, suspension, cushy tires, electric starters, and extra power. So popular was the automobile in the 1920s that the industry became its own major competitor as used car sales threatened to eclipse new car sales. To continue new car sales, the industry adopted more aggressive advertising, credit plans, and annual model changes. Automobile ownership sagged somewhat during the Depression, down to 55 percent of families in 1935, but not nearly as much as did telephone subscriptions.

The histories of the telephone and automobile have noteworthy similarities and differences. Both emerged from "parent" technologies, the telegraph and bicycle respectively, and that inheritance shaped their early histories. Both started out as expensive devices favored by the well-off and by physicians and then quickly grew popular with farmers. On the other hand, the telephone was initially a business tool leased from a single provider, while the automobile was a toy bought from a variety of producers. Eventually, other uses for both developed, but this evolution occurred more rapidly with the car. Competition, including price competition, spurred automobile diffusion, whereas monopoly and monopoly pricing apparently hindered telephone diffusion. (It took about 10 years, from 1905 to 1915, for the number of automobiles in the United States to go from 1 per thousand Americans to 20 per thousand, but twenty years, from 1880 to 1900, for telephones to spread as much—and this despite the much higher costs of the automobile.) But another, probably key, difference was the role of government.

The government's role in U.S. telephony was small. In some ways government modestly encouraged telephone diffusion by holding rates down, pressuring companies to serve outlying areas, and requiring interconnection. In other ways government discouraged telephone diffusion by sustaining Bell's technical standards. But there was no direct subsidy to telephone service; one way or another, users paid the full freight of service and then some. The government's role in automobile diffusion was substantial, easing traffic with laws, lights, police, and parking, and directly subsidizing automobile use by building improved roads, from streets to intercity highways.[59] This difference did not escape people in the telephone industry. In 1928 Vice President Page of AT&T noted that the automobile industry financed the campaigns for good roads and that "the money spent on good roads was the greatest subsidy an industry ever had."[60]

Why this contrast developed is a large question that goes beyond the current study. Perhaps the concentration of telephone service in the hands of one company made it logical to keep its production costs, such as wire-stringing, internal, while the multiplicity of automobile producers made its costs, such as road maintenance, seem to be public goods. Perhaps the automobile posed so many social problems so early—accidents, fights over rights of way, and so forth—that government intervention appeared necessary. Perhaps the coalition of interests around the automobile, from manufacturers to suburban home builders to farmers, was more politically powerful than any similar telephone coalition. Perhaps the costs of automobile infrastructure were so high that only government could handle them. In any case, the automobile was early and heavily subsidized by the general treasury. This situation contrasts, also, with the fate of the electric streetcar. Until the streetcar lines failed and were taken over by municipal governments, they and their passengers had to absorb almost all the costs of their infrastructure, even as they competed with the automobile.[61]

As for the telephone companies, they originally avoided government as long as they could. They would develop the telephone system on their own. How they sought to do so, how they tried to sell telephony, is the subject of the next chapter.

CHAPTER THREE

Educating the Public

The first step in understanding which Americans used the telephone and how is to observe how the sellers of the new technology tried to persuade Americans to use it. This chapter focuses on the evolution of the marketing of telephony. The marketing men (almost all sellers were men) were closest to the technology and its customers. Their livelihoods depended on developing its role in society. Their advertising influenced people's perceptions of the telephone. Finally, because salesmen were sensitive to consumer opinion, their discussions provide an indirect view of the public's reaction to the telephone.

This view, however, has distortions. Business decisions do not perfectly reflect market conditions. None of the great disasters of marketing history, such as the Edsel and new formula Coca–Cola, would have happened if that were so. Nor would important innovations languish on shelves for years, as did, for example, the typewriter.[1] Telephone men operated with their own habits, vanities, misinformation, and stereotypes. Yet where we find such distortions, we gain, through the back door, another glimpse of the public's encounter with the telephone. Such misperception forms an important part of this chapter's story.

The evolution of telephone salesmanship, especially Bell's, is evident in the records of the industry. Marketing developed in response to internal changes such as new strategic thinking, to the evolution of the advertising profession, and to customer practices. Initially, Alexander Graham Bell and his associates devoted their efforts largely to devising every possible means of bringing the telephone to the public's attention.

They often unveiled the device in flamboyant demonstrations, usually involving the broadcast of music and speeches from one place to an audience in another. It was appropriate to the late nineteenth century's "theatre of science," as Christopher Armstrong and H. V. Nelles have called it. Some telephone stunts, though lacking the deadly thrills of automobile races and airplane derring-do, were dramatic. When an audience of bishops and priests in Quebec City in 1877 heard a voice singing, "Thou are so near and yet so far" they stood up and sang back into the box. In Hamilton, Ontario, that year exhibitors set up 15 telephones in the local WCTU hall for the public's inspection, but a "disorderly mob of 400 burst into the Hall, tripped over the wires and broke the connections. One woman seized a phone, clapped it to her ear, heard nothing, and stalked out shouting, 'Humbug.'" For decades, telephone publicists contrived demonstrations to promote various services, such as long-distance. In 1907 Pacific Telephone and Telegraph (PT&T) arranged to have the California-Stanford "Big Game" play-by-play reports called into Palo Alto; in 1916 it helped Stanford University's President Lyman address from California an alumni gathering in New York City; and that same year PT&T made it possible for Mrs. Roy Pike of San Francisco to stage a smashing dinner party: Her husband spoke from New York cross-country to her guests, each of whom had a telephone beside his or her dinner plate. In the same spirit, AT&T welcomed William Randolph Hearst's offer in 1915 to show himself in a newsreel using the telephone to call New York from San Simeon.[2]

In the earliest years telephone men had to introduce the telephone and demonstrate its utility face-to-face. Chicago executive Angus Hibbard recalled, "In the first place, we had to go into a city or town and tell its inhabitants what the telephone was." This included convincing non-English-speakers that the instrument "spoke" their languages. The wonder of it lasted for years, especially in remote communities. When PT&T opened an exchange in 1897 in Lodi, California, people came to the town drugstore from miles away to try it. One wrote in a letter:

> I can now state in all truthfulness that I have talked over a talkaphone. It was an interesting experience and made me realize more fully than ever that we are living in a wonderful age. Although I had seen a talkaphone before I never tried one before this morning. I could hear the voice very clearly although the speaker was a long distance from me. It was like a voice from another world. Here I was speaking to a person far away from me whom I could hear as though he was at my side and yet I could not see him.[3]

Once they had shown that the wonder really did work, the telephone men faced a still harder task: persuading people to pay for it.

In the early twentieth century, AT&T both adopted more evocative sales messages that associated the telephone with practical ends such as saving time and initiated major public relations campaigns. By the late 1920s, Bell moved to softer themes, most notably linking the telephone to sociability. Later in the century these themes would appear as the "Reach Out and Touch Someone" campaigns.

The discussion of marketing begins with an overview of the techniques the industry used to publicize and sell telephone service. Then we examine several aspects of telephone marketing, in roughly the historical order they emerged: suggesting purposes for which people might find the telephone useful, instructing people on how to use the device, and nurturing goodwill toward the industry. In these ways, the industry men tried to shape the public's understanding and use of the telephone. Next, we look in more detail at the industry's attempts to manage customer use of the telephone for social conversation. We close with a comparative note on automobile marketing.

PUBLICITY AND SALES TECHNIQUES

> [The Bell System] had to invent the business uses of the telephone and convince people *that they were uses*. It had no help along this line.
>
> As the uses were created it had to invent multiplied means of satisfying them.
>
> It built up the *telephone habit* in cities like New York and Chicago and then it had to *cope satisfactorily* with the business conditions it had created.
>
> It has from the start *created the need* of the *telephone* and then *supplied it* (Bell System Advertisement, 1909; italics in original).[4]

Many in the telephone industry believed, as this Bell advertisement declares, that they had invented the telephone's uses. In 1909, the same year as that ad, Theodore Vail told a New York legislative committee that telephone development had been slow at first "because the public had to be educated; but as soon as the public was educated to the necessity and advantage of the telephone, then the growth became very fast indeed. . . . Today no man thinks of going into business or having a residence or anything else unless he has a telephone connection; but it took a long time to educate them [sic] up to that point." Yet, two decades later, a Bell publicist expressed doubts about the effectiveness of this education. He attributed an upsurge in disconnections to "the fact that we have never thoroughly educated the public to the possibilities of the use of the telephone . . . ." He recommended telling "the new subscriber what to do with his telephone . . . and to make him ashamed to consider such a thing as ever again doing without it."[5]

Both telephone publicists and advertising men emphasized the phrase "educating the public." This generally entailed creating desires for a product but also included other persuasive activities, such as instructing consumers on how to use the product and fostering goodwill toward the producer.* The sales techniques of the telephone industry, AT&T in particular, and the advertising industry evolved simultaneously.

The earliest education tactics focused on *free publicity,* the creation of news about the telephone, through such events as the musical demonstrations staged by Alexander Graham Bell and Thomas Watson. As the industry grew, local managers handled their own publicity, often through personal relations with newspaper men. In 1903, AT&T's President Fish hired the Publicity Bureau of Boston, perhaps the country's first public relations firm. Over the next few years, the Bureau reported having successfully placed scores of unattributed stories in newspapers: telephones on the farm, telephones in church life, a new switchboard at the Hotel Astor, financial problems of independent companies, the high cost of providing adequate service, the heroic work of repair crews, and the like. Vail, however, preferred direct advertising and dispensed with the agency when he took over. By then, one the Bureau's officers, J. D. Ellsworth, had joined AT&T and was continuing the publicity efforts in-house.[6] Independent telephone

---

*In an example from another industry, an official of the Colgate toothpaste company credited its early advertising with "teaching the public the habit of caring for their teeth" (Strasser, *Satisfaction Guaranteed,* 95).

companies also had gained much attention through comfortable ties with small-town editors in their service areas.[7]

The second major tactic used by the telephone industry to educate consumers was explicit advertising. Before Vail's return in 1907, local companies inserted notices in the press that were often little more than announcements of service or listings of business subscribers. They also posted billboards, distributed flyers, and gave away promotional items such as blotters to advertise service or to comfort restless customers. Straightforward ads soliciting subscribers also appeared—such as a drawing of a Victorian woman in full bustle and regalia who was saving time by phoning for her household's needs—but they were uncommon. (See Photos 7–9.) When Vail took over, he hired the leading advertising firm of N. W. Ayer & Son for Bell's campaigns. Advertising of all kinds increased, but AT&T focused on using national advertising, such as ads in popular magazines, to curry public support in political fights with the independents and with state and federal governments. The local Bell companies retained some discretion over their advertising, and budgets for newspaper ads varied considerably from town to town. Nevertheless, by the 1900s, AT&T headquarters was advising the locals and providing them with sophisticated ad copy. In 1927, A. W. Page moved from editing the magazine *World's Work* to heading publicity at AT&T, where he increased national advertising both for public relations and sales.[8]

Finally, telephone companies at times sold subscriptions door-to-door. In the nineteenth century, local entrepreneurs sometimes needed to twist arms. The first manager of the Eureka, California, exchange recalled that "my hardest work was getting my first twelve subscribers [the minimum for an exchange]. I called several times on Thomas F. Ricks who had a livery stable at the corner of 4th and F Street. He finally said he would sign to get rid of me." Whenever demand for service exceeded the supply of lines and telephones, local companies laid off solicitors. In the late 1920s, however, AT&T started a push for new subscribers and mobilized all employees, down to installers and operators, to sell service, even if only to friends and family. Independents had similar programs. The Depression accelerated these solicitation campaigns. Illinois Bell, for example, added over 100 new salesmen in 1931 and 1932 and encouraged its line workers to sell subscriptions both on their own time and while performing their jobs.[9]

Bell and others in the industry applied these three major tools of "education"—publicity, advertising, and soliciting—to somewhat different ends: to sell basic service some of the time, to sell auxiliary services like extension telephones and long-distance most of the time, and, especially after the return of Theodore Vail in 1907, to cultivate popular goodwill almost all the time. In particular, it accentuated the search for new subscribers in some eras and discouraged it in others. During the latter times, as in the early 1920s, salesmen pressed existing customers to upgrade service, and publicists focused on shaping political opinion. In 1924, for example, Bell headquarters cautioned locals that new business should be sought only where service already existed, in order to avoid having to construct new exchanges. However, shortly thereafter, partly under the impetus of A. W. Page and partly impelled by the realization that automobiles had overtaken telephones in American households, AT&T began again to search for new customers, in part by initiating the all-employee sales campaigns. Solicitation of basic service became even more important during the Depression, not only to get new customers but especially to keep the existing ones.[10]

These were the methods of the telephone industry's education. Its ends were to sell basic service some of the time, to sell auxiliary services like extension telephones and long distance most of the time, and, especially after the return of Theodore Vail in 1907, to cultivate popular goodwill almost all the time.

FINDING USES

As sociologist Sidney Aronson has pointed out, Alexander Graham Bell and his backers needed to convince people that their "toy" was a useful tool. Before the first telephone exchange in 1878, they suggested that subscribers could use it to talk between two fixed points, such as a house and a stable or a front office and a workshop. Later, they promoted the telephone as a replacement for the telegraph, allowing business messages or signals to be sent more easily and without an operator.[11] For decades, most telephone men—particularly those in marketing—believed that to sell their product they had to find, or to create, uses for it. In 1906 *Printers' Ink,* the advertising industry journal, noted: "Telephones are universal nowadays . . . [in] business. But the public is blind to many of their uses and needs to be reminded by special telephone arguments." In 1911 *Printers' Ink* applauded New York Bell for devising numerous new uses for the telephone. In 1916 the independents' journal, *Telephony,* ran a series urging its readers to follow Bell's lead in advertising the telephone's varied uses. In 1926 a Bell marketing man attributed recent sales success to the development of "new and additional uses," particularly for business subscribers. Yet, as late as 1929, another Bell "commercial engineer" (i.e., sales manager) attributed weakened demand to the industry's failure to teach the customer "what to do with his telephone."[12]

The search for uses often led to practices that would be novelties by today's standards. Telephone entrepreneurs in the early years broadcast news, concerts, church services, weather reports, and stores' sales announcements over their lines. Although these services disappeared quickly in most places, they lingered longer on rural party lines, especially weather reports and emergency news. Telephone companies also offered sports results, train arrival times, wake-up calls, and night watchman call-ins. Industry journals publicized inventive uses of the telephone, such as sales by telephone, get-out-the-vote campaigns, lullabies to put babies to sleep, and long-distance Christian Science healing.[13] (This process is similar to the computer industry's scramble to find uses for the home computer—storing recipes, balancing checkbooks, and so on.)

To sell more mundane applications, telephone vendors targeted businessmen first and most often. The telephone, they claimed, would increase efficiency, save time, and impress customers. The titles of some 1910 advertisements indicate the range of suggested uses: "Sixth Sense"—long-distance calls are effective for business:

"Multiplication of Power"

"Don't Write"—personal conversation is more powerful

"Man for the Moment"—business contacts can be reached in emergencies

"Business Competitors" will get your customers

"Call Ahead for Business"—save time by making appointments

"Business May Need More Than One Telephone"

"Save Time by Telephone"

"Telephone to the Country" for orders and reservations

"In Touch with His World"—the businessman can be reached on vacation

"Telephone Against Time"—a telephone is instantaneous

"At a Psychological Moment"—a telephone can deliver a business message quickly

"The Voice of Success"[14]

Sales pitches targeted different messages at different businessmen— the possibilities of new customers for store owners, the advantages of calling ahead for traveling salesmen, the ability to receive sudden orders for wholesalers, and so on. Businessmen were also reminded that they could call their wives for example, when they were going to be late or when they were bringing a colleague home for dinner unexpectedly.

Although the focus of this book is on the history of the residential telephone, it is important to understand that in its earliest years the industry paid only secondary attention to marketing residential service. Moreover, those residential sales efforts that were made before the 1920s emphasized the "business" of the household, the ways in which the telephone could help the affluent household manager accomplish her tasks. An 1878 circular in New Haven stated that "your wife may order your dinner, a hack, your family physician, etc., all by Telephone without leaving the house or trusting servants or messengers to do it." (It got almost no response.) A 1904 handbook for solicitors in Illinois suggested putting these arguments to potential customers:

> While residential telephone service may not directly save money for the household, yet, in an indirect way, it accomplishes the same thing by saving time, labor, and drudgery, and in making the whole household run more smoothly. It is always on duty, shops in all weather, corrects mistakes, and hastens deliveries. It saves letter writing, orders the dinner, invites the guests, reserves the tickets, and calls the carriage. It makes appointments, changes the time, cancels them altogether and renews them. It calls the expressman, calls the cab, and instructs the office. It invites one's friends, asks them to stay away, asks them to hurry and enables them to invite in return. . . .

A description of how the telephone could help in emergencies then followed. A 1910 series of advertisements suggested that, besides "Marketing by Telephone," the homemaker could telephone seam- stresses, florists, theatres, inns, rental agents, coal dealers, schools, and so on.* Other ads from this period suggested using the telephone

---

*While the telephone company urged women to shop by telephone, at least one popular exponent of "household engineering," Christine Fredrick, warned against it: "The telephone habit, *as generally practiced,* makes for extravagance, encourages

to convey messages of moderate urgency (calling for a plumber or sending an announcement to the press), to convey invitations (to an impromptu party, for a fourth at bridge), and to keep "In Touch with Friends and Relatives." A few ads alluded to the modernity of the telephone ("It's up to the Times!"), but the basic theme clearly was household management. The idea of using the telephone for social purposes such as chatting with kin and friends appeared only as a minor concern in the marketing literature. Indeed, a 1910 ad entitled "The Telephone at Christmas" recommended the telephone as an aid in holiday preparations, not as a means for giving season's greetings. This balance between practical and social uses of the home telephone would not shift until after World War I (see the section on the discovery of sociability later in this chapter).[15]

As early as the first decade of the twentieth century, Bell's marketing strategy included efforts to induce current business and residential subscribers to upgrade their service, by moving from party to individual lines, leasing extensions, and calling long-distance. These upgrades were profitable items, could usually be provided without new construction, and meant dealing with existing customers rather than reaching out to new and often financially suspect clients. The degree to which sales managers emphasized these items depended on available capacity, plant costs, and political concerns. In 1928 and 1929, for example, when capacity was strained, *Pacemaker,* PT&T's in-house sales magazine, encouraged employees to push the extra services, especially extensions. With the loss of subscribers in the Depression, however, selling basic service or simply keeping existing customers became more important than selling extras.[16]

For promoting extensions, Bell turned to advertising as well as direct solicitation. A 1905 ad read: "A woman slaves enough without having to tramp up and down the stairs several times a day to answer the telephone. . . . " An ad from the 1920s read: " my heart stood still. . . . I heard stealthy voices . . . someone tinkering with a lock . . . a muffled footstep . . . saw a shadow flit by my window. . . . I reached over to the stand by the bedside and seized—no, not a revolver—a telephone." Bell does not seem to have been very successful in these

---

hand-to-mouth buying . . . and increases the cost of doing business to the retailer" (*Household Engineering,* 329; italics in original). Similarly, in the early part of the century, the New York City Board of Education recommended, in its home economics literature, against shopping for food by telephone (Strasser, *Satisfaction Guaranteed,* 265).

campaigns to sell extras. One door-to-door salesman who worked in the 1930s told us that even though he was instructed to stress extensions, he leased few because customers considered them to be luxuries. The statistics bear him out; fewer than 10 percent of residential customers had extensions before the Depression. Neither did the companies have much success in selling another upgrade, individual-line service, in these pre-War years.[17]

Meanwhile, long-distance service provided Bell with an unanswerable advantage over the independents during the competitive era, and AT&T stressed it often, particularly in public relations. Sales advertising encouraged both business and, especially after World War I, household use of long-distance calling. A 1915 ad addressing businessmen read, "You fishermen who feel these warm days of Spring luring you to your favorite stream.... You can adjust affairs before leaving, ascertain the condition of streams, secure accommodations, and always be in touch with business and home." Ads invited residential subscribers to check in with family and to send greetings on holidays. The 80 percent of subscribers who rarely called long-distance, even by the late 1920s, represented a large and untapped resource. Whenever trunk lines got congested, however, the company stressed selling other special services, such as extensions and PBXs (private branch exchanges—i.e., customers' own switchboards).[18]

Telephone salesmen, in Bell and perhaps more so out of Bell, were inventive and energetic in trying to find or create uses for the telephone (of course, the claim that they had created the need for telephony was self-serving, even if only unconsciously). They presented two basic uses of the telephone to urban Americans through World War I.* The primary use was defined as the conduct of business or related activities by businessmen. Management of the household by the home executive, the housewife, was regarded as a secondary use. To these early salesmen, the telephone was a practical device for attaining practical ends.

MANAGING USE

In the first few decades of telephony, companies faced not only the task of finding uses but also that of managing customers' use of the telephone, showing them how to work it and how to use proper etiquette.

---

*Telephone marketing presentations aimed at farmers were more diverse; see Chapter 4.

Through circulars, ads, and notices on telephone directories, companies instructed subscribers on the operation of the machine. In the nineteenth century companies advised customers first to turn the crank three times and then to lift the receiver, and in the twentieth century to lift the receiver, put their fingers in dial holes, turn clockwise, and so forth.[19] A Canadian notice in 1896 instructed:

> To Listen: Place the telephone fairly against the ear, with an upward motion, so that the lower extremity or lobe of the ear is gathered in, into the cavity of the telephone; in this position it will be found to fit snugly and comfortably—the lobe of the ear acting as a cushion and at the same time closing out all ulterior sounds, thus enabling the voice to be heard with clearness and precision.

A California instruction warned, "Speak directly into the mouthpiece keeping mustache out of the opening." Other notices cautioned users to answer promptly, speak directly into the transmitter, avoid banging the receiver when hanging up, and so forth. As late as the mid-1920s, the independent Southern Indiana Telephone Company placed ads and stories in the press asking their customers to turn the cranks more rapidly and to answer promptly.[20] (See Photo 12.)

Telephone companies undertook not only mechanical but also social education. Many industry people complained of profanity, yelling, and abuse on the telephone. Through notices, direct chastisement of customers by employees, and occasional legal action, the companies sought to improve telephone courtesy. A 1910 Bell ad entitled "Dr. Jekyll and Mr. Hyde at the Telephone" concluded, in bold italics, that "the marvelous growth of the Bell System has made use of the telephone universal and misuse a matter of public concern." In the same period, AT&T distributed cards labeled "The Telephone Pledge," to be attached to instruments and reading, "I believe in the Golden Rule and will try to be as Courteous and Considerate over the Telephone as if Face to Face." Companies cut off service to abusers and obtained legislation that fined or even jailed profane customers.[21]

The industry at times even tried to script appropriate conversation. Among its many efforts to instill propriety, Bell subsidized a 1910 contest in *Telephone Engineer* for the best hortatory essay. The winning entry, later distributed by AT&T to its local companies for insertion in directories, read in part:

> Would you rush into an office or up to the door of a residence and blurt out "Hello! Hello! Who am I talking to?" No, one should open conversations with phrases such as "Mr. Wood, of Curtis and Sons, wishes to talk with Mr. White . . . " without any unnecessary and undignified "Hello's."

AT&T tried at first to suppress "hello" as a vulgarity. It failed decisively, so much so that it later endorsed the nickname, *hello-girls,* for its operators.[22]

In 1926, a telephone magazine raised the subtler issue of who should end a conversation, the caller or receiver. It determined that the caller should signal the end, unless a man and a woman were speaking, in which case termination was to be the prerogative of the "second sex."[23]

A common concern of Bell companies, independents, and rural mutual lines alike was teaching party-line etiquette. They repeatedly cautioned subscribers not to eavesdrop, both for reasons of privacy and to reduce the drain on the electrical current caused by so many open connections. If the testimonies of oral histories and the repetitiveness of the pleas are to be believed, the rule of privacy was perhaps as often breached as honored. One Indiana woman reminisced that "when the ring occurred, everybody in the community would take down [the receiver] and would listen."*[24] The companies also tried to teach customers to avoid occupying the line with long conversations. They printed notices, had operators intervene, and sent warning letters to particularly talkative customers. In some places, companies imposed time limits that seemed to help, although it is unclear how strictly operators enforced such limits. Informal understandings may have done the most to minimize the problems of party-line use. Another Indiana woman recalled that telephone conversations ceased if someone broke in to ask for the line and then recommenced after the caller had finished. No doubt, the standard for the appropriate length of calls varied from place to place and from household to household. As an elderly woman recalled to communications historian Lana Rakow, "There was definitely a telephone etiquette that some parents tried to teach in their homes, when we had those large party lines. The children did not play with the phone and try to make calls. And you didn't keep the phone for long periods of time because others might need it."[25] The elderly people we interviewed varied in the level of strictness they recalled their parents maintaining in holding the party line.

The struggle of the telephone companies to civilize their customers yields various interpretations. It may imply that telephone subscribers were particularly difficult people, or it may suggest that the com-

---

*Eavesdropping remains an issue in a few places. The manager of the Hot Springs, Montana, telephone office told the *Wall Street Journal* in 1986 that "in a small community like this, if you don't want someone to know it, don't say it on the phone" (30 October 1986, 1).

panies' executives held especially high standards of decorum for use of their product. In either case, if the chronic battles, including court cases, are any indication, the industry met with only moderate success in its attempt to manage its customers in the early years of the telephone.[26]

SUSTAINING GOODWILL

AT&T led the development of corporate public relations in America. It spent much of its advertising budget and other resources to "teach the public in a big broad way what the fundamental requirements for a good telephone system are and to show that the underlying policies of the Bell System are economically sound, and are for the best interests of the public," or in the plainer words of Vice President Kingsbury in 1915, to "educate the people . . . that . . . we seek to be absolutely fair and honest, that we are above graft, pettiness, discrimination. . . . "[27]

In the nineteenth century, the American public generally regarded Bell with suspicion and hostility, as it did many large corporate trusts. AT&T's efforts to maintain its patents through massive legal suits reinforced its image as an arrogant outside monopoly. Bell paid a price for this image during the early stages of its battle with home-grown independents.[28]

As noted earlier, President Fish inaugurated a policy of systematic public relations in the early 1900s by hiring the Publicity Bureau. The Bureau and President Fish believed that obtaining free publicity was the best method for increasing public approval.[29] Although more blatant story planting eventually declined, AT&T publicists continued to suggest article topics to journalists until at least 1940, leading to features in major national magazines and books on telephony.[30]

Free publicity, as well as supportive editorials, often resulted from alliances between telephone companies and newspapers. In some smaller communities, editors had financial stakes in the telephone independents. Elsewhere, local telephone managers routinely provided free telephones and other services to the press. In 1908, for instance, PT&T applauded itself for saving $50,000 per year by trimming such free service to California newspapers. In 1914 the advertising manager of Southwestern Bell described to his colleagues how he bartered for favorable publicity by having his men befriend politicians and then leak news tips to reporters. In 1927, PT&T officials exchanged notes discussing the proper amount for a monthly subsidy from the company to a newspaper in Portland, Oregon. However, such manipulation worked both ways. For instance, in 1909 the advertising manager of *Moody's* wrote to J. D. Ellsworth pointing out that the magazine was going to publish a "highly complimentary" article on AT&T and asking whether the company would be interested in buying an ad for the back cover of the issue.[31]

Telephone companies courted other influential figures. As with newspapermen, they provided free service to political candidates and officeholders. They invited clergymen to tour their facilities and then speak about what they had seen, particularly the working conditions. Bell paid for its officers to join prestigious local clubs. In 1933, AT&T headquarters asked PT&T to provide a list of precisely 1277 prominent men in its territory (the number was prorated on AT&T's estimate of the number of nationally prominent men), especially professionals, bankers, businessmen, and politicians, for inclusion on a mailing list. Bell companies eagerly made presentations on college campuses, with the goals of both recruiting managers and cultivating future leaders of the nation. The Federal Communications Commission (FCC) charged that AT&T curried influence in less aboveboard ways as well, for example by dispersing its deposits and premiums among many banks and insurance agencies, especially those run by city officials.[32]

Efforts at maintaining public relations through influential connections continued throughout the period before World War II. Meanwhile, explicit advertising for goodwill, though begun under the Fish administration, was not fully utilized by AT&T until Vail took over in 1907. It developed as a key part of his strategy for rebuilding Bell's dominance as a regulated monopoly. To secure an environment favorable to consolidation, profitable rates, and freedom of action, Vail needed the public to view AT&T as a beneficent utility instead of as a rapacious monopoly.

The themes of these advertising campaigns are illustrated by the following headlines from AT&T magazine ads:

"The Sign Board of Civilization" (the Bell symbol, 1909)

"The Pony Express: A Pioneer of the Bell System" (1912)

"Loyal to the Service" (Bell employees are dedicated, 1916)

"Owned by Those it Serves" (1922)

"Telephone Service is a Public Trust" (1928)

The ads always repeated Vail's slogan of "One Policy—One System—Universal Service." The explicit messages of such ads claimed that Bell had steadfast workers dedicated to serving the public, that its universality democratically allowed everyone to bridge the barriers of space and position, that it responded democratically to millions of shareholding owners, and that telephony was a grace of modern science. Historian Roland Marchand believes the ads also carried the deeper message of a promise that the Bell telephone was a way for users to transcend and simplify the complexities of modernity.[33] (See Photo 13.)

During the early 1910s, AT&T's advertising and other public relations efforts addressed a particular threat: government ownership. Movements for municipal telephone systems and especially the foreboding threat of nationalization so concerned Bell executives that they vigorously campaigned against government ownership by, for example, counterposing the success of American telephony with the supposed failures of European systems.[34]

Finally, the level of vigilance that had been attained by AT&T's growing corps of publicists was demonstrated in 1920, when the company frantically and apparently unsuccessfully pressured film producer Hal Roach to edit out from a Harold Lloyd movie a slapstick scene among operators at a telephone exchange.[35]

In these ways, AT&T, and to a lesser degree the independents, tried to manage customer opinion. In the first generation of the telephone, users found political pressure to be an occasionally effective counterweight to the telephone monopoly. Local office holders delayed renewing franchises for telephone wiring, placed rate caps in franchise renewals, or, until the 1910s, called in competitors. The telephone companies, especially the Bell companies, often prevailed anyway, but the political networks formed a significant obstacle. Dealing with regulators became easier when state commissions assumed regulatory authority, but the telephone men needed the state politicians to believe that the public held the companies in high esteem. Eventually, the telephone industry gained that prestige, although how much resulted from the publicity campaigns and how much from concurrent changes—the end of competitive warfare, the heroism of the telephone employees who staffed communications units in World War I, the improvements in technical quality, and so on—cannot be determined. Yet during the twentieth century AT&T certainly evolved from the "Bell Octopus" to "Ma Bell."

DISCOVERING SOCIABILITY*

In 1989, popular columnist Ellen Goodman pondered the ease with which her children ran up huge telephone bills. She blamed the telephone company:

> The marketing miracle of the age is the success of the telephone company in selling something as ephemeral as conversation. They have convinced a generation of young Americans to casually, nay whimsically, use the telephone.[36]

This comment would warm the heart of the ad writer who wrote the copy quoted earlier in this chapter. The actual history is, however, more complex.

For a half-century, the telephone industry relied on one basic rationale to motivate people to put telephones in their homes: practicality. The telephone would help people better manage household affairs and cope with emergencies. However, in the 1920s the industry started equally emphasizing another reason: sociability.** The telephone would facilitate conversation with friends and family. This shift in focus reveals much about the industry, the technology, and the users.

Sociability as a Sales Pitch

The rare early telephone ads that included the argument that the telephone might serve social ends almost always promoted calling to pass along only a brief message, such as greetings, invitations, or news of safe arrival. Telephone ads rarely suggested holding a nonbusiness conversation. After World War I, both ads for basic telephone service, where they appeared, and ads for long-distance still overwhelmingly pointed out business and practical uses. However, "visiting" with kin became a frequent theme. Bell Canada, in particular, compared to other AT&T companies, stressed family in its long-distance ads. Typical of copy from the next two decades is this ad from 1921:

> It's a weekly affair now, those fond intimate talks. Distance rolls away and for a few minutes every Thursday night the familiar voices tell the little family gossip that both are so eager to hear.[37]

*This section draws on and adds some evidence to an earlier article, C. Fischer, "'Touch Someone.'" See the article for a fuller development of the arguments.

**By *sociability,* I mean the pursuit of noninstrumental personal relations, particularly, in this regard, personal conversation.

In the 1920s, the advertising industry developed atmosphere techniques that focused less on the product and more on its consequences for the consumer (e.g., you should buy a particular stove not because it cooks quickly, but because that is what the "modern housewife" uses).[38] Signs of a similar shift appeared in this Bell advertisement in 1923:

> The Southwestern Bell Telephone Company has decided that it is selling something more vital than distance, speed or accuracy. . . . [T]he telephone . . . almost brings [people] face to face. It is the next best thing to personal contact. So the fundamental purpose of the current advertising is to sell the company's subscribers their voices at their true worth—to help them realize that "Your Voice is You," . . . to make subscribers think of the telephone whenever they think of distant friends or relatives.[39]

However, this was apparently only an early harbinger because during most of the 1920s telephone advertisers generally restricted the "friends or relatives" theme to ads encouraging long-distance calling, not ads promoting telephone subscription itself. Also, Bell salesmen spent the 1920s largely selling ancillary services, such as extension telephones, upgrades from party lines, and long-distance, to current subscribers, rather than seeking out new customers. What little advertising for new subscribers existed pressed the practical themes of earlier years. For instance, a 1928 Bell Canada sales manual stressed household practicality first and social invitations second as useful pitches for signing on new customers.[40]

Then, in the late 1920s, Bell System leaders, perhaps prodded by the embarrassment that for the first time more American families were purchasing automobiles, gas service, and electrical appliances than were subscribing to telephones, adopted a more aggressive strategy for attracting new subscribers. They built up a full-fledged sales force and sought to market the telephone as a "comfort and convenience" (i.e., as more than a practical device), drawing somewhat on the psychological and consumption-oriented themes found in automobile advertising. Bell enlarged its advertising focus to include not only upgrading the service of current subscribers, but also reaching those car owners and electricity users who lacked telephones. The social character of the telephone was to be a key ingredient in the new sales strategies. For example, a direct-mail campaign in 1929 used such pitches as: your friends are listed, your friends look for your name in the directory, and "Hail, Hail the Gang's All Here."[41]

Before the "comfort and convenience" campaign for new customers could go far, however, the Depression riveted the industry's attention on simply retaining existing subscribers. During the 1930s, marketing executives continued to advertise long-distance service, employing both business themes and the newer themes of family and friendship. But they strongly augmented the advertising of basic service, addressing it now to both nonusers and would-be disconnectors.

The first line of argument for getting or keeping household customers in the Depression was still practicality—emergency uses, in particular, and the low cost of the service—but pitches for sociable conversations were more prominent than they had been before the late 1920s. A 1934 Bell Canada ad featured a couple who had just resubscribed: "We got out of touch with all of our friends and missed the good times we have now." A 1935 advertisement asked, "Have you ever watched a person telephoning to a friend? Have you noticed how readily the lips part into smiles . . . ?" A 1937 AT&T ad reminded readers that "the telephone is vital in emergencies, but that is not the whole of its service. . . . Friendship's path often follows the trail of the telephone wire."[42] (See Photos 10 and 11.)

A survey of two northern California newspapers concurs with the impression given by industry archives that sociability themes increased after the mid-1920s. In the *Antioch Ledger,* aside from one 1911 advertisement referring to farm wives' isolation, the first sociability message, addressed to parents, appeared in 1929: "No girl wants to be a wallflower." Following it in the 1930s were ads for basic service with slogans such as "Give your friends straight access to your home," and "Call the folks now!" Sociability themes likewise increased in both the basic and long-distance ads printed in the *Marin Journal* during the late 1920s and the 1930s. Ads suggested that people "broaden the circle of friendly contact" (1927) and call grandmother (1935). One included the line, "I got my telephone for *convenience.* I never thought it would be such *fun!*" (1940).[43]

Social conversation also appeared as a selling point in guides for telephone salesmen beginning in the late 1920s. As noted earlier, a 1904 Illinois booklet presents but one paragraph addressing residential service. That paragraph's only social note is that the telephone "invites one's friends, asks them to stay away, asks them to hurry and enables them to invite in return." Conversation per se is not mentioned. By contrast, a 1931 memorandum to sales representatives entitled "Your

Telephone" is full of tips on selling residential service and encouraging its use. Its first and longest subsection begins:

> *Fosters Friendships.* Your telephone will keep your personal friendships alive and active. Real friendships are too rare and valuable to be broken when you or your friends move out of town. Correspondence will help for a time, but friendships do not flourish for long on letters alone. When you can't visit in person, telephone periodically. Telephone calls will keep up the whole intimacy remarkably well.

A 1935 manual puts practicality and emergency uses first as sales arguments but explicitly discusses the telephone's "social importance," such as saving users from being "left high and dry by friends who can't reach [them] conveniently."[44]

In sum, various sales materials confirm a pattern of change. From the beginning of telephony to roughly the mid-1920s, the industry sold telephone service as a practical business and household tool, with only occasional mention of social uses and these recommended only brief messages. Later sales arguments for both long-distance and basic service prominently featured social uses, including the suggestion that people use the telephone for conversations with friends and family.[45]

Industry Attitudes toward Sociability

Industry men, of course, did not suddenly discover in the 1910s or 1920s that the telephone could be used for chit-chat, but they had misgivings about such conversations in the nineteenth century. Some worried that the telephone permitted inappropriate or dangerous discussions, such as illicit wooing. To be sure, a few telephone men applauded and encouraged telephone sociability. Alexander Graham Bell himself wrote his wife in 1878, "When people can . . . chat comfortably with each other over some bit of gossip, every person will desire to put money in our pockets by having telephones." Yet, this remained a minority opinion during the first half-century of the telephone's existence.[46]

Instead, early telephone men often fought their residential customers over social conversations, labeling such calls as frivolous and unnecessary. For example, a company announcement from 1881 complained, "The fact that subscribers have been free to use the wires as they pleased without incurring additional expense [i.e., by using flat rates] has led to the transmission of large numbers of communications of the most trivial character." In 1909 a local manager listened in on a sample of conversations coming through a residential Seattle exchange and determined that 20 percent of the calls were orders to stores and other businesses, 20 percent were from subscribers' homes to their own businesses, 15 percent were social invitations, and 30 percent were "purely idle gossip," a rate that he claimed matched those in other cities. The manager's concern was to reduce this last, "unnecessary use." One tactic, noted in the earlier discussion of telephone etiquette, was to place time limits on calls. This was often an explicit effort to stop people who insisted on chatting when there was "business" to be conducted. Instructions on party-line etiquette in a 1900 Canadian directory read, "It is, of course, well understood that business conversations cannot be limited as to time, but 'visiting' can beneficially be confined to a reasonably short duration of time."[47]

An exceptional few in the industry, believing in a more populist telephony, tried to encourage such uses. Edward J. Hall pled for lower rates and defended "trivial" calls, arguing that they added to the total use-value of the system. However, the evident isolation of men like Hall underlines the dominant anti-conversation position of the pre-World War I industry.[48]

Official AT&T opinions moved closer to Hall's in the late 1920s when executives announced that, whereas they had previously thought of telephone service only as a practical necessity, they now realized that the telephone was a "convenience, comfort, luxury"; so-called trivial social uses contributed to its value. In 1928 new Publicity Vice President A. W. Page, who had just left the publishing business, was most explicit when he critiqued earlier views:

> There had also been the point of view [in the Bell System and among the public] about not using the telephone for frivolous conversation. This is about as commercial as if the automobile people should advertise, "Please do not take out this car unless you are going on a serious errand. . . . " We are faced, I think, with a state of public consciousness that the telephone is a necessity and not to be trifled with, certainly in the home.

Bell executives told their salesmen to sell telephone service as a "comfort and convenience," including as a conversational tool.[49]

Although this reversal was most visible in the Bell System, similar changes, especially concerning rural sales, showed up in the independents' journal, *Telephony*. Early conflict over telephone conversations was most acute in rural areas (see discussion of rural telephony in Chapter 4). Although episodic sales campaigns aimed at farmers stressed the practical advantages of the telephone, such as the ability to receive market prices, weather reports, and emergency aid, the industry addressed the social theme sooner and more often to them than to the wider public. For example, a 1911 PT&T series that focused on the telephone's utility in emergencies, staying informed, and saving money also included an ad that called it "a Blessing to the Farmer's Wife. . . . It relieves the monotony of life. She CANNOT be lonesome with the Bell Service. . . . " (See Photos 14–15.) Nevertheless, telephone professionals who dealt with farm families often fought the use of the line for nonbusiness conversations, particularly in the early years. The pages of *Telephony* overflow with complaints about farm customers, not the least of which is that they tied up the lines with their chats.

Greater appreciation of the value of telephone sociability to rural residents emerged later among salesmen. A 1932 article in the *Bell Telephone Quarterly* noted that "telephone usage for social purposes in rural areas is fundamentally important." Ironically, in 1938 an independent telephone man claimed that the social theme had been but was no longer an effective sales point because the automobile and other technologies had already reduced farmers' isolation.[50]

Parallel with the shift toward sociability in advertising themes, industry attitudes moved from irritation with conversations to approval of sociability.

Explaining the Discovery of Sociability

Why were the telephone companies at first reluctant to suggest sociable conversations as a use? Why did they eventually shift to selling sociability? There are several answers, none mutually exclusive.

The explanation that comes to mind immediately (and that is suggested by some scholars who have read this work or heard me speak about this topic) holds that telephone men switched their themes as an economically rational response to some technical development, changes in demand, prices, or other market factors. Although I do not deny that such considerations played a role, my explanation stresses the cultural changes that affected how telephone men viewed their product and their customers.*

---

*By inference, I am abandoning the assumption that all business behavior is economically optimal. There is plenty of evidence otherwise. For example, Donald Finlay Davis, in *Conspicuous Production,* argues that many of the first automobile manufacturers ran their companies out of business because they were more concerned with raising their social status than with competing for the mass market. The telephone business allowed even more room for nonmarket considerations because it was generally monopolistic.

In several ways, the telephone industry descended directly from the telegraph industry. The instruments are functionally similar; technical developments sometimes applied to both. The key men who developed, built, and marketed telephone systems were predominantly telegraph men. Theodore Vail, for example, came from a family involved in telegraphy and started his career as a telegrapher. In contrast, marketing dissidents such as E. J. Hall and A. W. Page had no connections to telegraphy (although J. I. Sabin did). Many larger telephone companies began as telegraph operations. Telephone use often directly supplanted telegraph use. Even the language of the telephone revealed its ancestry; telephone calls were for many years labelled as messages and measured in message units.[51]

It is not suprising, then, that the uses for the telephone these men first proposed and then repeated for decades largely replicated those of a printing telegraph: business communiques, orders, alarms, and calls for services. In this context, they reasonably considered telephone "visiting" to be an abuse or trivialization of the service. AT&T Vice President Page was reacting precisely against this telegraphy perspective in his 1928 defense of "frivolous" conversation. At the same conference, he also decried the psychological effect of telephone advertisements that explicitly compared the instrument to the telegraph.[52]

Industry leaders long ignored or repressed telephone sociability, for the most part, I believe, because social conversations did not fit their understanding of the technology. Feeding these attitudes, no doubt, was the common perception that women made most social calls and their conversations were not serious (see Chapter 8). That view, in turn, may have reflected a general close-mindedness toward people different from themselves. Many early telephone company officers, in correspondence or other comments, dismissed immigrants, blacks, and farmers as people who could not use or perhaps could not comprehend the telephone (see Chapter 4). The dissidents were the ones who suggested that such people, like the Chinese laundrymen solicited by John I. Sabin in San Francisco, were plausible customers.[53]

The telegraph heritage of telephony faded as the telegraph industry itself declined and as new people entered the telephone business. These new men, I suggest, were more open to reinterpreting the telephone in the ways that A. W. Page urged. They were more willing to acknowledge that people did or should use the device for sociability. After all, "visiting" had obviously occurred often enough in the past to stimulate efforts at suppression. The same Seattle manager who complained in 1909 of "idle gossip" found that 38 percent of the residential exchange's calls exceeded five minutes and that the average call was $7\frac{1}{2}$ minutes long. Such conversations drew strong reactions. Michèle Martin claims that "when women began to use the instrument for sociability, in order to break their isolation in the household, men started to cry out at the futility, frivolity of the use of the telephone in such a way, and to ridicule them in newspapers, in journals, even in books."[54]

The foregoing explanation asserts that for a generation or more there was a mismatch between the ways people actually used the telephone and how industry men imagined it would or should be used. I must next address the theory that rather than resulting from such a mismatch, the industry's original resistance to social uses and its later endorsement of them were both shrewd responses to the economic environment. There are several variations of this argument.

One is that subscribers did not begin having social conversations on the telephone until the 1920s. Thus the subsequent industry switch is a reaction to this change in behavior. Such increased sociability could, in turn, be accounted for in several ways: declining real costs, increasing subscriber rolls—which meant more friends to call—higher-quality transmissions, more comfortable instruments, automatic dialing, changes in American culture, and so on. But whether an increase in chatting occurred in the 1920s is hard to prove. Although oral histories conducted today suggest that people visited by telephone less often and more quickly in the "old days," those interviewed cannot specify rates or dates (see Chapter 8).[55] Moreover, some evidence from recollections and from contemporary discussions about the problem of telephone visiting suggests that social calls were common before the mid-1920s. An increase in the frequency of social calls may have contributed to a change in advertising strategy, but I found no direct evidence that this was how the decision-makers understood it.[56]

Another variation of the economic argument holds that the sociability theme was a response to a pricing change from flat to measured charges. On flat-rate systems, customers could call as often and for as long as they wished with no proportional return to the telephone company. Futhermore, if "visiting" filled up local lines or interexchange trunk lines, other clients might become frustrated and decide to drop the service. However, under message-rate systems, which charged subscribers per call or per minute or both, long conversations became profitable to the companies, thereby making sociability a logical sales pitch. Yet, as sensible as this explanation is, the geographical and historical patterns of flat- and measured-rate schedules do not correspond to the marketing shift of the 1920s.[57]

A third version, perhaps one industry spokesmen would have used, is that technical considerations limited "visiting" by telephone before the 1920s. Extended conversations monopolized party lines, so warnings against extended talk accompanied party-line instructions in company publications. Yet, a shift from party to individual lines cannot explain the shift toward sociability advertising. The proportion of party-line telephones varied tremendously among cities, from under 10 percent in places like Detroit and greater New York to over 60 percent in places like Minneapolis and Oakland. As late as 1930, 63 percent of Bell's residential telephones were still on party lines. That proportion had decreased modestly over the previous four years, but was actually higher than it had been before World War I.[58] Thus, the history of party lines does not correspond with that of advertising.

Although both economic and technical considerations no doubt reinforced the industry's original reservations about conversations, neither seems sufficient to explain the adoption of the sociability theme in the 1920s. Much of the explanation lies in the cultural mind-set of the telephone men. Originally slow to sell to households at all, they were surely reluctant to encourage or even endorse "idle" chatter. However, new leaders and a half-century's experience eventually led to the solicitation of social calls as profitable new business.

Other cultural changes may have also contributed to this discovery of sociability. In particular, advertising philosophy shifted in the 1920s toward softer themes and focused increasingly on women as the primary consumers. Also, telephone men believed that the automobile and other technologies had succeeded, in part, because those industries marketed their goods as luxuries. Still, both circumstantial and direct evidence suggest that the key change was the loosening of the telegraph tradition's hold on the telephone industry.

Conclusion

Today people make most of their residential calls to friends and family, often holding sociable conversations, conditions that may also have existed for two or three generations.[59] The telephone industry actively promotes such calls, encouraging people to "reach out and touch someone." For many years, it did not. Telephone salesmen from the 1880s to the 1920s praised the residential telephone for its utility in emergencies; that function is now taken for granted. They also claimed the telephone was good for shopping; that function persists ("Let your fingers do the walking"), but never seemed too important to residential subscribers (see Chapter 8). Sociability, obviously an important use of the telephone today, was ignored or resisted by the industry for almost the first half of its history.

A NOTE ON AUTOMOBILE MARKETING

Automobile salesmen, like telephone salesmen, credited themselves with having created the desire for their product, but the automobile men claimed to have accomplished that task faster. One advertising man wondered in 1906 whether "the automobile is to prove a fad like the bicycle or a lasting factor in the industry of the country." By 1914, as an automobile man wrote, "Now every normal person in this country hopes some day to own a motor car." The automobile industry's initial selling points paralleled those of its ancestor, the bicycle—a luxury for touring, joy-riding, and racing. Newspapers celebrated automobile stunts, usually endurance races, more than they had telephone demonstrations. Early advertisements and free publicity in newspapers stressed power, style, and the automobile's recreational uses, notably sightseeing through the countryside.[60]

The realization that the automobile had practical uses for transportation soon dawned on the industry. Especially after the success of the Ford Model T, automobile advertisements began to stress more mundane themes. These included utility, particularly for those models targeted at lower-income customers, and social uses such as the strength families could gain by touring together. Once the technology itself no longer had to be sold—by the late 1920s perhaps 60 percent of American families had an automobile—advertising increasingly focused on which car customers should buy rather than on selling the automobile itself.[61]

As the telephone industry did with the telegraph, automobile manufacturers initially followed a marketing strategy based on the experience of an ancestor technology. Although the automobile vendors started with luxury themes and then added more practical ones, the reverse of the telephone men, a deeper similarity existed. Both industries stressed a limited set of uses and had to be awakened, it seems, to wider and more popular uses. A historian of the automobile points out that the industry was slow to provide features that would appeal to women, especially easier starting and a closed compartment. She attributes this development process to cultural lag.[62] The automobile producers, however, apparently learned faster than the telephone men.

CONCLUSION

Looking at the role of the telephone in society through the eyes of its marketers, those most interested in knowing—or creating—that role, presents an informative case of distorted perception. Although sophisticated in their method of introducing the telephone, persistent in finding business clients, and eventually skillful at public relations, the marketers of telephone service were slow to employ as a sales tool the use that was to dominate the home telephone's future, sociable conversation.

The story of how and why the telephone industry discovered sociability provides a few lessons in the nature of technological diffusion. It suggests that the promoters of a technology do not necessarily know or decide its final uses; that they seek problems or needs for which their technology is the answer, but that consumers themselves develop new uses and ultimately decide which will predominate. The story suggests that in promoting a technology, vendors are constrained not only by its technical and economic attributes but also by an interpretation of its uses that is shaped by its and their histories, a cultural constraint that can persist over many years. The insistence of consumers on "visiting" over the telephone and the eventual adoption of the sociability theme in the industry's campaigns to "educate the public" represents a case in which a use was found and propelled by the consumers of a technology.

CHAPTER FOUR

The Telephone Spreads: National Patterns

Today, few American households lack a telephone. These few tend to lodge disadvantaged families or only a single male.[1] Three generations ago, however, it was unusual to have a telephone. In this chapter, we begin looking at who and where the early subscribers were.

Innovations generally spread from the well-off to those of modest income and from urban residents to rural people.[2] The telephone's diffusion is both typical and uncharacteristic of this pattern. After the first decades, it was not the people in the metropolitan areas but those in the hinterland who most often had telephones, and working-class families were unusually slow in getting telephones. A complete understanding of this history incorporates both the independent judgment of users—who wanted the telephone for what reasons—and the political economy within which consumers made their choices—the structure of marketing and of government regulation.

In the first part of this chapter, we examine the spread of the telephone across the Continental United States, considering what characteristics of a state encouraged or discouraged telephone development. In the second part, we focus on the adoption and abandonment of the telephone in rural America. In the third part, we explore the adoption of the telephone by the urban working class, looking at specifically who subscribed.* The automobile provides a reference point throughout.

DIFFUSION ACROSS THE STATES

Adoption of the telephone varied greatly from place to place. As late as 1927 the percentage of families with telephones varied from 22 percent to 65 percent across Bell's 100 largest exchanges. In 1930, for example, 12 percent of households in Mississippi and South Carolina had telephones, compared to over 65 percent of households in Iowa, Kansas, and Nebraska.[3] What was it about some states or their residents that encouraged telephone subscriptions at rates several times that of others? What might that tell us about the attractions of the telephone?

In the first few years, industry men rapidly seeded the continent with telephone exchanges, as Bell and its pre-1880 competitors rushed to claim territory. For example, between the third week of February and the third week of April 1878, first exchanges opened in California, New York, Delaware, and Massachusetts. The growth was so sudden that the U.S. Census Bureau confessed in 1880 that neither it "nor any statistical agency can deal in a wholly satisfactory manner with anything which is subject to rapid and violent changes."[4]

Hubbard's strategy of franchising telephone service and the press of competition required setting up exchanges everywhere, using a flock of local representatives. In addition, entrepreneurs around the country started up their own exchanges, with or without Bell franchises. The geographical distribution of exchanges in the first few years reflected Bell's fight with Western Union over the urban centers. After settling that war, Bell continued to concentrate its construction in the major cities outside the South.[5]

With the beginning of full-scale competition in 1894, thousands of places gained some form of telephone service. The number of

---

*Inferring people's purposes and uses from an analysis of who subscribed has pitfalls. Someone might get a telephone for one reason (say emergencies), but plan to use it most often for another (say, ordering goods). In turn, his or her actual use could still be different (say, chatting with friends). Finally, the consequences of subscribing may be different yet (say, irritability arising from bothersome calls by relatives). Nevertheless, knowing who subscribed is a first and necessary step in understanding why they did so and to what effect.

telephones per capita multiplied over sevenfold by 1902. The Census Bureau reported that the most rapid spread of telephony had occurred on the West Coast and in the Midwest rather than the industrial Northeast (and this despite the Bureau having missed many Midwestern rural systems in its survey.) The level of telephone subscription in the nation's largest cities by the turn of the century was only at the national average.[6]

The authors of the 1902 census report generally credited price competition with generating this regional and urban-rural pattern of distribution. Where competition flourished, they argued, it had forced down prices and stimulated aggressive marketing. Almost half of the 1000 urban centers in the country had two or more competing telephone companies; and even the mere threat of competition spurred sales efforts. Abundant competition partly explains the Midwest's advantage and the nationwide pace of expansion.[7] Most independent commercial telephone providers and most rural lines were in states such as Iowa, Kansas, and Minnesota. Midwesterners also reputedly engaged in "telephone crusades," rallies to raise the funds and the poles for farmer lines.[8] (More on rural enthusiasm below.)

The census report isolated tariffs as the single most important determinant of where the telephone spread. It credited California's cheap, measured-rate service for that state's nation-leading figure of 64 telephones per 1000 residents. Variations in prices do help explain variations in telephone subscriptions.[9] Yet, other factors played important roles. The report credits telephone diffusion in California and especially in San Francisco, which had a world-high 111 telephones per 1000 residents, to the "assiduity with which the telephone habit has been cultivated by the managers of companies." PT&T President John I. Sabin was notorious for his marketing zeal and, as we shall see, unusual in his pursuit of all customers. A PT&T telephone directory boasted in 1900:

> It is the purpose of the management to extend the use of the telephone to every town, no matter how small, throughout the Pacific States; and in all of the cities and towns, to put a telephone, at most reasonable rates, within easy and immediate reach of every person whose income is sufficient to permit him to ride on street cars.[10]

Unlike other managers, Sabin sent his salesmen to rooming houses, Chinese laundries, saloons, and working-class homes. He also developed cheap "loss-leader" services, such as residential pay telephones and 20-party "kitchen telephones" that could only call out to one number. He was determined, a colleague recalled, to " 'hang up that Blue Bell sign' in so many places that everybody would get the telephone habit."[11] This is in stark contrast to the South, where telephony started slowly and continued to lag behind. As late as the mid-1930s, small-town families in the South were less than half as likely to have telephones as families living in small towns in other parts of the country. Southern telephone managers' timidity was in part to blame for the slow rate of diffusion.[12]

Still other important characteristics included: wealth, nature of economy, population composition, and so on. Statistical analysis shows that in 1902, states tended to have more telephones per capita to the extent that they had abundant commerce, many white farmers, and prosperous farms. Taking all factors into account, the Pacific states were ahead of national averages in telephone development, and the states in the Deep South were behind (see Appendix B).

The geographical pattern of 1902 largely held up through 1940.[13] Figure 3 shows the distribution of U.S. telephones in 1907, 1917, and 1927. (We should chart residential telephones only, but the electrical census did not separately tabulate them until the 1930s.) The Plains states and the West Coast were the growth centers of telephony throughout the two decades; the Deep South remained largely unserved; and, surprisingly, the industrial Northeast lagged behind the Midwest, at least until the 1920s.*

From 1902 to 1937, telephones spread most widely in states that urbanized, grew commercially (as measured by employment in trade and transportation), added farm owners (but not farm tenants), and whose farmland appreciated in value. Prosperous farmers were critical in spurring telephone development largely in the period before World War I. (Although commercial development was important, industrialization and industrial workers' incomes seemingly were not.) A closer look shows that subscriptions to Bell telephones grew rapidly after the 1900s, in contrast to the stagnation of independent telephony. Whereas the independents relied on farm owners and residents of small towns, Bell grew in response to urbanization and increased commercial activity.[14] These numerical findings

---

*The 1937 map is similar to the 1927 map but shows slight increases in the industrial Northeast and declines in and around the Midwest.

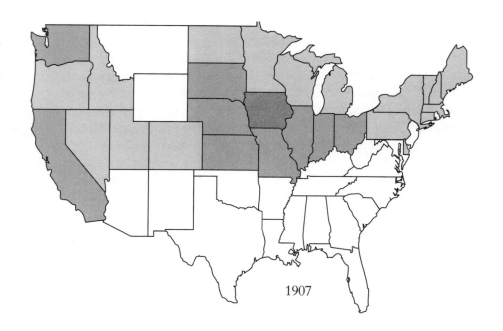

1907

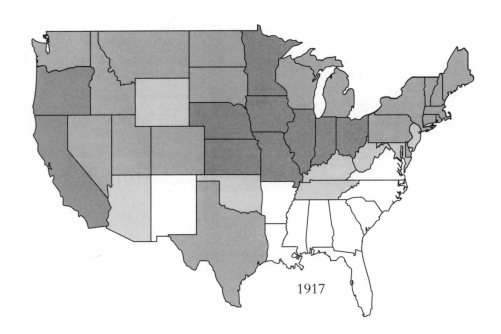

1917

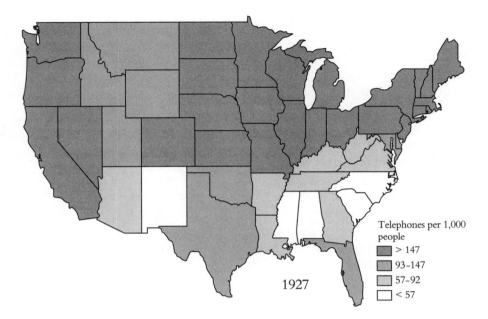

FIGURE 3. TELEPHONES PER 1000 PEOPLE IN 1907, 1917, AND 1927. The telephone spread fastest in the Midwest and Far West. Its diffusion was particularly slow in the Deep South. (Source: U.S. Bureau of the Census, *Special Reports: Telephones 1907; idem, Census of Electrical Industries 1917; idem, Census of Electrical Industries 1927.*)

are consistent with the nature of the methods each sector of the industry used to market the telephone. Bell was based in the business and commercial centers; the independents and mutuals concentrated on small-town and rural America.

In 1917 the diffusion of automobiles reached the level telephones had reached in 1902. The spatial distribution of automobiles was roughly similar to that of telephones: Western and Plains states were in advance, Southern states lagged, and the industrial Northeast was in the middle. Generally, there were more automobiles in states with more farmers, prosperous farms, and more urban wealth. After 1917, automobiles diffused rapidly, led, as were telephones, by the Midwest and the Far West. Automobile diffusion largely responded to urban commercial development and to farm conditions (see Appendix B).

A comparison of the spread of the automobile and the telephone after 1920 yields two intriguing subtleties. First, more farm tenants in a state meant more farm automobiles but fewer telephones.* Second, urban and industrialized states—industrialization measured, for example, by average wages in manufacturing—experienced greater expansion in nonfarm automobiles than in nonfarm telephones. Both findings hint that the people in the poorer classes—farm tenants and industrial workers—were more willing to buy automobiles than telephone service. (Or, alternatively, that these people were pursued more vigorously by automobile than by telephone vendors.) We will return to this possibility later in the chapter.

Overall, the geographical patterns of telephone and automobile diffusion were similar, with the Midwest and Pacific states leading the way and the Southern states trailing. Regional variations in telephone diffusion seem to be due, in part, to the nature of the telephone industry. Where competition with independent companies, lower rates, and intense marketing existed, the telephone was adopted more rapidly. But the similarity with automobile diffusion suggests that more of the variation resulted from local conditions. Where there were more well-off farmers (especially before World War I) and more commercial activity telephones diffused faster. The influence of these local factors prompts consideration of telephone subscription, first, in rural areas, and second, among urban classes.

THE RISE AND FALL OF RURAL TELEPHONY

Many of the jokes industry men told in the early years depicted the initial encounter of a country bumpkin with a telephone. The stories usually portrayed the farmer as ignorant and sometimes afraid of the instrument. An independent telephone executive claimed in 1910 that the farmer had been "so unfamiliar with the telephone as to be absolutely frightened with it."[15] (Similar jokes described immigrants as shocked by the device's ability to speak their native language.) This attitude reflected the conviction that farmers lacked interest in the telephone, even when they had some comprehension of it, as well as the conviction that farmers did not "need" the telephone sufficiently to make serving them profitable. The industry avoided the rural market until the early years of the twentieth century. Even later, arguments

---

*Other evidence shows that farm tenants were less likely than farm owners, far less likely in the South, to have telephones (U.S. Bureau of the Census, *U.S. Census of Agriculture, 1930*, volume IV, 530–39; idem., *U.S. Census of Agriculture, 1940*, volume III, 541–46).

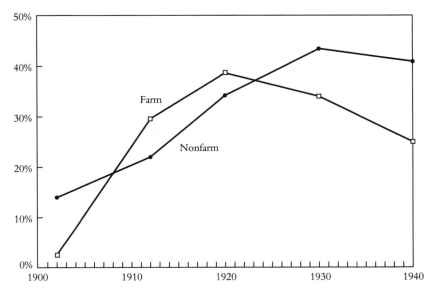

FIGURE 4. PERCENTAGES OF HOUSEHOLDS WITH TELEPHONES, 1902–1940. The distinct history of farm telephony in the U.S. is shown by its relatively rapid growth up to 1920 and its subsequent rapid decline.

for serving farmers had a defensive, beleaguered tone.[16] Yet, a remarkable event occurred around the turn of the century: Farmers, largely on their own, adopted telephones at rates surpassing those of city-dwellers. This was followed about twenty years later by a surprising turnaround in which many American farmers abandoned their telephones.

Figure 4 displays these trends. In 1902, relatively few U.S. farms had telephones (the estimates for that year are only approximate). By 1912, however, more farm than nonfarm households had telephones. That advantage persisted up to 1920, when 39 percent of farms but only about 34 percent of nonfarm households had telephones. This small difference is impressive when one considers that American farmers were poorer and faced greater difficulties in acquiring telephone service than did nonfarmers. In the 1920s the proportion of farms with telephones began to decline, dropping sharply during the 1930s. By 1940, only 25 percent of farms had telephones, compared to about 41 percent of nonfarm households. It was not until 1950 that farmers again reached the 39 percent rate of 1920.[17] Figure 4 raises two complementary questions: Why did farmers adopt the telephone so fervently? Why did they abandon it so rapidly?

The Rise: 1894–1920*

In the nineteenth century, Bell managers had little interest in serving rural areas. They were busy building and improving urban facilities and, at any rate, doubted that there was a profit to be found in the countryside. Often, small-town businessmen or farmers petitioned the nearest Bell franchise for a local exchange only to be turned down. In Colfax County, New Mexico, for example, farmers who had experimented with a barbed-wire system asked the local Bell telephone company to provide them with service. The company refused and would not allow them to connect their wires to its switchboard in the nearest town. The farmers subsequently organized a joint–stock company and started an independent exchange service in 1904. Sometimes country tinkerers built illegal lines.[18]

Within several years of the Bell patents' expiration, however, rural groups had organized over 6000 small farmer lines and mutual systems. By 1907, there were about 18,000 such operations, comprising 1.5 million rural telephones, most of them in 10 midwestern states. This accomplishment was extraordinary given the barriers farmers faced: the disinterest of major telephone companies; the individualism that "country life" experts considered endemic and damaging to rural life; the shortage of cash; and the departure of farm youth to the cities.

Typically, a leading farmer or a small-town merchant or doctor solicited members for a farm line or a telephone cooperative. Investing $15 to $50 and often labor and materials, 2 or 3 dozen families became shareholders. Annual rental fees for the telephone ran from $3 to $18 a year. If the system had a switchboard—and some did not, operating as one big party line—a farm wife or daughter served as a daytime operator. (See Photo 17.) Typically, there was no night service. Some systems connected to larger companies' switchboards in towns, enabling contact with the wider world. Most of these operations remained small, had little cash, used inferior equipment, postponed needed maintenance, and suffered from poor sound and occasional breakdowns. In some places, rural lines were literally barbed-wire affairs, running their currents over farmers' metallic fences. Managers often underestimated costs, failed to depreciate materials, and relied on subscribers to maintain their own equipment. One elderly Indiana woman recalled, "When I was young, the farmers through

---

*This section summarizes and supplements material reported in C. Fischer, "The Revolution in Rural Telephony." For further sources on rural telephony, see also Appendix A.

here organized their own telephone company.... Then that disbanded, and for a long time we didn't have a telephone. Then Mr. Schonfeld in Butlerville, who had the exchange, came out through the country and asked if we would like to have a telephone line. So he put up a line, but as he became older and wasn't able to take care of it, the line went into disuse again." Eventually, many of these companies failed, consolidated, or joined a larger system.[19]

Bell and particularly the large independents turned seriously to the rural market after 1900. Previously, most companies had only served farmers living near town, charging them mightily for extending the phone wires so far. But then the independents began aggressively soliciting farmers, offering them cheap, rural service in order to build the kind of widespread networks that would attract town customers. An independent man explained in 1906: Although entrepreneurs like him had originally ignored farmers, they now realized that farm customers were "essential" to selling in the larger exchanges. With outlying subscribers hooked up, they could offer merchants and other townsfolk access to customers, suppliers, and kinfolk in the hinterland.[20]

Bell responded accordingly. A 1908 report from PT&T to AT&T headquarters lays out the logic plainly:

> Prior to the year 1905 this company made no special effort to encourage farmers to build lines to connect with the Company's exchanges. ... In the early part of the year the fact developed that opposition could successfully secure our business by first canvassing the adjacent rural territory and, by inducing the farmers to join them, force the town people to subscribe also.
>
> To combat the opposition the Company established a fighting rate of $1.00 per year, covering exchange service and the loan of an American Bell transmitter and receiver.... *
>
> [A]t several exchanges where we have numerous farmer subscriptions ... opposition has been unable to enter.... [W]here we had but few farmers connected, our exchange was practically wiped out by the opposition.
>
> Aside from the benefit of protection, there can be no doubt that, at lower rates, we are furnishing service below cost.[21]

Theodore Vail changed Bell's strategy from direct competition to preemption. He tried to integrate rural systems into Bell's network, even at the cost of compromising technical standards. Bell companies

---

*That is, for $1 per year, farmers could lease a Bell instrument and connect, through their own mutual line, to the town exchange.

also sponsored farmer lines with technical aid and subsidies, both to preempt opposition and to save themselves the cost of construction.

The abatement of competition in the 1910s apparently sapped Bell's interest in farmers. In a 1916 system-wide conference, J. D. Ellsworth, AT&T's advertising vice president, asked his counterparts with the local companies to report on what tactics they were using to attract rural customers. He reminded them that country residents "are numerous, they are indispensable and they are politically powerful." After hearing that the local Bells were doing little to attract farmers, Ellsworth chided, "Mr. Vail has frequently boasted of our high rural development and I do not want him to be boasting without grounds."[22] Yet, in later years AT&T still shied from rural customers and sometimes sold or closed the farmer lines they operated. By the late 1930s, farmers in Bell-dominated territories were less likely to have telephones than those in areas dominated by independents.

The commercial independents, meanwhile, were more tightly entwined with farmers. One leader claimed in 1908, "It stands as an undisputed fact that the development of the rural telephone has been the potent weapon in the hands of the independents. . . . "[23] Yet it was an uneasy relationship. Managers originally doubted farmers' interest in the telephone and took credit for teaching them its value. It was a tough sell, but servicing farmers was even worse. The independents considered farmers — and their telephone cooperatives — to be hardheaded, short-sighted, and tight-fisted. They were frustrated by the mutuals' demands for toll-free connections to town, resistance to rate increases, and frequent complaints. The independents clashed with rural subscribers over maintenance, technical quality,* the monopolizing of party lines for gossip or even banjo-playing, chronic eavesdropping, and the like. Still, dealing with farmers seemed necessary, particularly in the era of heated competition. In 1903, an independent executive complained that

> farmers as a class are . . . troublesome customers to handle and are apt to have an exaggerated idea of their own rights. . . . The bumptiousness of certain farmers can be overcome only by constant efforts to educate them as to their rights and the rights of others. . . . There should certainly be some way of checking the senseless gossip that is carried to interminable lengths on some rural lines. . . . [24]

---

*Poor sound on a call from a town to a rural subscriber would often be blamed on the town company even though the problem might lie in the farmer's line.

The independents' ambivalence is captured by one manager's reassurance to his colleagues in 1910: "Do not let anyone convince you that you cannot reason with the Iowa or Nebraska farmer. . . . Do not be afraid of them!"[25]

Given that rural telephony outdistanced urban telephony through 1920, why were Bell and independent executives so reluctant and ambivalent about pursuing farmers? The simplest answer is that they saw no profit in it. Serving scattered rural customers seemed costly, with more wire to string per subscriber, and farmers seemed less able or willing to pay those costs. Certainly, farmers were poor customers for PBX's (private branch exchanges), extensions, and the like. In addition, the technical deficiencies of rural lines undercut the high-quality service that the larger companies, notably Bell, tried to market to businessmen. The assumption that serving rural customers is unprofitable is contestable. Expenses other than wire, such as land and labor, were lower in rural areas. One must also include—as the telephone men did in occasional discussions—the increased business generated in town by the ability to connect the residents to out-of-town farmers.[26] Still, industry men genuinely saw rural service as an unprofitable investment.*

Simply put, many in the industry claimed that farmer demand was limited; that rural people did not want or appreciate the telephone enough to pay the real cost of providing it. Ironically, contemporary studies suggest that farmers were willing to pay more of their income for telephone service than were comparable townsfolk; and other evidence suggests that cost was less of a barrier to subscribing than was the low quality of the rural service. Nonetheless, even in the 1940s, as they resisted federal legislation to subsidize rural mutuals, spokesmen for telephone companies argued that farmers did not really want or even need telephone service.[27]

Mixed in with these economic concerns was some measure of cultural prejudice, expressed in "rube" jokes and cartoons in the industry

---

*Bell's pricing logic seemed to shift over the years. At times, Bell considered an appropriate price to be the cost of providing service to an individual plus profit. This logic justified high rates to hard-to-reach households. At times it argued that prices should reflect the direct value to each subscriber of the service provided (value indicated, for example, by how many others the subscriber could reach through the local exchange). This would imply low rates for farmers, since they could reach fewer people for the price of a basic call. And at times, Bell argued that the telephone was a collective good; all users benefited commonly from access to one another. This logic rationalized what it considered to be cross-subsidies from, say, easy-to-serve town subscribers or businessmen to hard-to-serve farmers.

press and more subtly in industry discussions of rural service. It would have been difficult, cost calculations aside, to convince most telephone leaders that farm families—uneducated, poor, rustic, far from commerce, and prone to misuse telephones for purposes such as gossip sessions—were serious customers for a serious product. For these reasons, both hard and soft, telephone salesmen, especially Bell's, tended to disdain rural customers unless they faced competition or political pressure.

Yet, U.S. farmers, at least those outside the South, eagerly sought telephone service, even if they had to build it themselves. Most, by far, seemed not shy of the telephone but impatient for it. Furthermore, farm residents used the telephone more than did urban folk. Intriguing exceptions to this rural welcome were provided by a few retreatist sects, such as the Amish and Mennonites. Members argued over whether the telephone was a theologically acceptable device or an intolerable worldly seduction. These isolated cases, however, constitute an eccentric contrast to the overwhelmingly positive reception, even from mainline churches, that rural folk gave the telephone.[28] Why such enthusiasm?

The fee for rural service was nominally cheaper than for urban service, but since payment required cash, telephones may have been effectively more costly to farmers than to townsfolk.[29] Rural telephony sometimes required labor from the users, and the service was usually inferior—multiple-party lines, static, breakdowns, and the like. The answer to why farmers had more telephones than urban people before the mid-1920s probably lies not in its cost or quality, but rather in the uses farmers saw in telephones.

Observers, from telephone salesmen to government commissioners, agreed that the telephone had many practical uses on the farm: calling for help in emergencies, obtaining weather forecasts and crop prices,* ordering goods, recruiting temporary labor, and so on.[30] A New Mexico dairyman who was also a newspaper columnist campaigned for the Colfax County system by pointing out first its life-saving potential and then its use for efficient and profitable marketing. The catalog of functional ends to which farmers could use the telephone was longer and more serious than any comparable list for city-dwellers. (An urban list included items like ordering theater tick-

---

*It was claimed that crop price information, in particular, was important because it allowed farmers to avoid being shortchanged by traveling grain buyers.

ets.) The farm telephone was, in part, a business as well as a residential telephone. Small surveys show that farm families made half or more of their calls for business ends.[31]

Similarly, the telephone's social role was more evident for rural than for urban families. Telephone men were fully conscious of this fact. One wrote in 1904 that the "telephone takes from the farmer's family its sense of loneliness and isolation. . . . Largely through its influence will disappear [the] pathos and tragedy of the lives" of farm women. Popular magazine stories dramatized the same points. The 1907 census of telephones argued that in regions of isolated farm houses, "a sense of community life is impossible without this ready means of communication. . . . The sense of loneliness and insecurity felt by farmers' wives under former conditions disappears and an approach is made to the solidarity of a small country town." Federal investigations of the "Country Life" problem in 1909 and in 1919 echoed these appraisals. An academic claimed in 1912 that, "No one feature of modern life has done more to compensate for the isolation of the farm than the telephone. The rural mutual lines have brought the farmers' wives in touch with each other as never before." Although we lack hard evidence for the claim, there seemed to be a consensus that the telephone had become crucial to the social lives of some rural Americans.[32]

An irony of telephone history is that, just as the industry struggled mightily to create the need for the telephone among urban Americans (see Chapter 3), it largely ignored a strong, spontaneous demand for the telephone among rural Americans. For city-dwellers, a telephone in the home could be a convenience, facilitating household management and sociability, but it was not a need. Rather than call, the city-dweller could go out for groceries, or walk or take a streetcar to visit a friend. For rural Americans, lives and livelihoods, and sociability as well, came more to depend on the telephone. In retrospect, then, the wonder is not that rural demand for telephones exceeded urban demand, but that the industry was so sure that rural Americans were uninterested in telephony.

The story of rural automobility in some ways parallels that of the telephone, though at a much accelerated pace.[33] Automobile manufacturers first marketed cars as luxury goods to city-dwellers. Farmers initially reacted hostilely to the devices that scared their horses (see Chapter 5). Quite soon, though, many farmers bought low-priced automobiles, the Ford Model T in particular. By the 1910s, rural areas had disproportionately many automobiles, farmers' low incomes notwithstanding. By 1930, most farm households had automobiles. Farmers expressed considerable enthusiasm for the automobile, joining in collective road-building in many areas. Even poor sharecroppers in the South who would never consider telephones scraped together money for automobiles.[34]

Rationales for the automobile's popularity among farmers were similar to those for the telephone: that the automobile was a practical machine, surely better than a horse, as well as a social tool, especially for isolated farm wives. Farm families spent most of their automobile budgets for social ends such as family excursions and short trips by women and teenagers. The Country Life Commissions and experts applauded the car for its material and social uses at least as much as they did the telephone.[35]

Although the rural demand for automobiles in the late 1900s surprised them, the manufacturers, notably Ford, moved quickly to solicit farmers. Sometime in the mid-1920s, farm automobiles began to outnumber farm telephones. Although far more expensive than telephone service, automobiles could perform some similar and many additional tasks on the farm. Driving did not require coordination with neighbors or an operator, was subsidized by road-building taxes, and was aggressively marketed to farmers by automobile manufacturers.

Rural electrification followed yet another pattern up through the 1920s. Like telephone companies, electric utilities were reluctant to serve farmers, doubting that they formed a serious market and claiming that farmers were too ignorant. However, farmers were unable to seriously develop their own electrification systems, as they did with the telephone, at least before the 1930s. Thus, American farmers lagged far behind their northern European counterparts in the early years of eletrification.[36]

Which Farmers Subscribed: Iowa in 1924

In 1924 Iowa led the nation in the number of farm households with telephones and was, in that way, an atypical state. But an unusual directory from Dubuque County allows us to look closely at which farmers were likely to be telephone subscribers in the period and to outline the situations in which rural telephony hit its peak (see Appendix C).

Farmers who had large farms and those who owned their farms were more likely to subscribe to telephones in Dubuque County. (Owning versus tenanting made no difference when it came to having an automobile.) Aside from those households in the town of Dubuque itself, the farther out farms were located, the likelier they were to have telephones. Location was confounded, however, with the servicing telephone company. The findings strongly imply that households in areas dominated by the Bell company were considerably less likely to subscribe than those in areas served by independents, other factors being equal.

All else equal, younger families were likelier to subscribe (and to have automobiles), and families with a high proportion of adult women were more likely to subscribe than those with relatively more adult men (but gender made less difference for automobile ownership). Some of these findings, such as the gender pattern, will reappear later when we consider which urban Americans subscribed.

The Fall: 1920–1940*

In 1920 almost 2.5 million U.S. farms had telephones. By 1940, one million fewer had telephones, and the percentage with service had dropped from 39 percent to 25 percent (see Figure 4). During the Depression in the 1930s, rates of telephone subscription dropped about 2.5 percentage points among nonfarmers but 9 points among farmers. More striking still, farmers were purchasing automobiles and electrical service (see Figure 5) during this period when they were giving up telephone service.[37]

The telephone industry did not seem to notice this decline until in the late 1930s and the 1940s, when rural politicians proposed federal loans to farmer lines.[38] This proposal sparked debate over what had happened. Some argued that farmer demand for telephone service had dropped because the agricultural depression had impoverished many, because rates had increased, or because other technologies, like the automobile and radio, had made telephones less worthwhile. Others argued that the supply of rural telephones had declined because the mutual systems had collapsed or because the major companies had lost interest in farmers. Still others combined various arguments. The Institute of Independent Telephony argued in 1945, for example, that farmers had disconnected their telephones in the 1920s and early 1930s due to financial strains, then once prosperity returned, had found "too

---

*This section essentially summarizes C. Fischer, "Technology's Retreat: The Decline of Rural Telephony, 1902–1940."

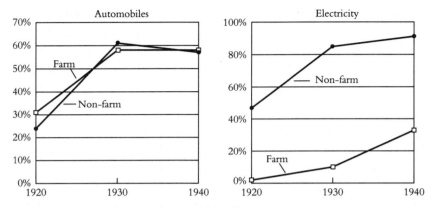

FIGURE 5. PERCENTAGES OF HOUSEHOLDS WITH AUTOMOBILES AND ELECTRICITY, 1920–1940. Although many farm families gave up or lost telephone service after 1920 (Figure 4), they continued, like other Americans, to spend on automobiles and electrical service.

many [alternate] avenues of contact with the outside world" to warrant paying for poor telephone service.[39]

The explanation that the agricultural crises of the 1920s and early 1930s forced destitute farmers to give up telephone service will not suffice. As Figure 5 shows, farmers spent money on other products. Many farmers in the 1930s who lacked telephones still had gas service, washing machines, and other consumer products. The cost of rural telephone service, while not negligible at $18 to $24 a year, represented a small fraction of an automobile's costs and an amount roughly equal to that spent by the average farm family on tobacco. Although some farmers could never afford telephones, before or after the 1920s, many had the option of choosing other spending.[40]

Telephone companies and farm representatives debated whether telephone rates had increased. The fragmentary evidence suggests that, nationally, nominal rates did not increase much between World War I and the 1930s, but the drop in farm prices meant that real costs did increase.[41] Even though these charges may have led some farmers to disconnect their telephones and angered many others, cost does not seem, for the reasons just discussed, to give a full explanation.

Industry men also defended themselves by claiming that farmers no longer wanted telephone service. One explained in 1945:

> To the snowbound or mud-bound farmer of 1920, the telephone was the only readily available link to the outside world. It brought him news, weather information, social contacts generally, and . . . the cooperation of

his neighbors. . . . Since then, with our great highway system and distribution of automobiles, and rural electrification with its radios and other
appliances . . . [the telephone] did not, admittedly, have the same indispensable quality which it had prior to 1920. . . . When the depression com-
menced to lift for the farmer—about 1936—. . . the farm telephone found
itself in competition for the farmer's dollar with a whole host of modern
farm improvements. . . .[42]

Others, including those outside the industry, made similar arguments.
This analysis, although self-serving, does provide part of the answer.

Some observers blamed the small, rural telephone companies,
which were often marginal operations. In the late 1920s they were
compelled by regulators and Bell to upgrade their technical facilities
while in the middle of a depressed farm economy. Many collapsed,
abandoning their customers. Yet, this, too, does not suffice as a general explanation. Small independent and mutual companies actually *increased* in number during the 1920s; they then diminished in the 1930s,
but at a rate far less than the drop in total rural subscribers. Also,
financially solid Bell lost almost as many farm subscribers as the independents did. Finally, if the failure of rural phone companies resulted
in an unsatisfied demand, why did new investors not come in?[43]

Other commentators pointed at AT&T. In 1944, the staff of the
FCC claimed that, between 1925 and 1934, Bell had helped independents win rate increases to shield its own high prices. Kenneth Lipartito
has suggested that regulatory commissions collaborated with Bell to
raise the quality standards for all lines, forcing some rural providers
out and forcing others to raise rates to pay for upgrading. Both Bell
and the large independents were accused of simply turning away from
farmers.[44] Although there is little direct evidence of a conspiracy, these
large companies no longer needed to compete for farmers after the
1910s. They did not mount major sales campaigns except during the
Depression. Bell and the independents also collaborated on many issues, including presentations before regulatory commissions. The industry, of course, completely rejected this collaboration argument.

Perhaps the best conclusion that can be drawn from these testimonies is that the disinterest of the large companies, the poor quality of many mutuals' service, and the increasing availability of other
means of communication set the conditions for the decline of rural
telephony, which was then triggered by decreasing farm income and
increasing real costs.

As another approach to explaining the decline of farm telephony,
I conducted statistical analyses to identify what characteristics of

particular states were associated with high losses of farm telephones.[45] In 1920 states varied greatly, from South Carolina, where 5 percent of farms had telephones, to Iowa, where 86 percent did. States with more farm telephones had more owner-operated farms, more valuable farms, and fewer nonwhite farmers and were geographically smaller. Also, states outside the Deep South and states with many independent telephone companies had more farm telephones (all other factors equal).

During the 1920s, farm telephones contracted by 14 percent. States that lost proportionately the most telephones tended to have many tenant farmers, many nonwhite farmers, and poorer farms, and were generally located in the Deep South. They were at least equally dominated by Bell companies, if not more so, as states that better sustained farm telephony. In addition, the substantial losers tended to have had greater increases in the numbers of farm automobiles. In the 1930s, farm telephones declined further by almost one–third. Again, states that lost the most tended to have somewhat more tenant-operated and somewhat poorer farms. They tended also to be more sparsely settled. The type of company, Bell or independent, made little difference. Again, the greatest losers had experienced slightly greater increases in farm automobiles.

From 1920 to 1940, then, the distribution of farm telephones became increasingly distinguished by farm tenancy and farm wealth—states with many tenant farms and poorer farms lost more farm telephones—and by region and density—with Southern states and sparsely-settled states losing more. Having local Bell companies did not help save farm telephones; it may have hurt in the 1920s. Finally, automobiles may have undercut telephone subscription.*

These statistical findings point to financial strain (farm value, tenancy) and perhaps to substitute technologies (automobiles) as explanations for the decline in rural telephony. They challenge the argument that collapsing small companies explain the disconnections; being in Bell territory was not an advantage. Some findings also suggest that marketing decisions undercut rural telephony. The cluster of three factors—nonwhite farmers, tenant farms, and location in the deep South—may have discouraged telephone marketers from seriously approaching certain rural areas.[46] The possible role of Bell provides another hint in that direction.

*A word of caution: Later data analyses have not confirmed the findings with respect to automobiles.

In 1930, an official of Southern Bell was explicit. He told a Bell conference that his region contained "tenancies of extremely low economic character and not potential telephone subscribers," as well as some districts with "high class" prospects. He suggested offering lower construction charges to would-be subscribers in neighborhoods where there were "good toll-using types,"even if legal regulations were circumvented in doing so. In other words, wealthier farmers would be offered subsidized telephone installations.[47]

I also examined changes in rural telephony on a smaller scale, comparing California counties in 1930 and 1940. The results show that counties had more farm telephones the more farm owners they had (but, unlike farm automobiles, not the more tenant farms they had), the more valuable the farms, and the more urbanized the county. (The number of farm automobiles in a county seemingly did not affect the spread of farm telephones.) Although California was atypical in that its farm telephones increased during the 1930s, the results are generally consistent with those from the analysis of states (see Appendix B).

Yet another way to analyze explanations for the decline of rural telephony is to contrast this story to those of rural automobiles and electricity. Between 1920 and 1940, rates of farm electrification (which, as already noted, has several parallels to farm telephony[48]) and farm automobile ownership increased (see Figure 5). A key distinction may be that cars in the 1920s and 1930s and electricity in the 1930s, although otherwise different technologies, were both subsidized by taxes. General taxes subsidized rural highway construction. The New Deal funded the construction of electric power lines directly to farm homes and by doing so spurred private companies to follow suit. Telephone service, on the other hand, typically had to be capitalized, without credit, by rural users. During the earlier era of competition, major telephone companies sometimes subsidized rural service, but they were less interested in doing so by the 1920s. State-directed subsidies of rural telephony, through rate caps, were not substantial until after World War II, and it was not until then that federal loans for rural telephony began. Besides, as Kenneth Lipartito has argued, state regulations often required technical upgrading, which in turn raised costs. Thus, one weakness of rural telephony compared to rural electrification and rural automobility was the requirement that its users fully capitalize the systems.

International comparisons illustrate a similar point. Europe's governmental telephone monopolies supplied many fewer farmers than did American systems. During the Depression, however, rural telephones steadily increased in Europe.[49] Government ownership may explain both contrasts. Before World War I, private competition accelerated American rural telephony far past that of Europe. In the Depression, though, European governments supported telephone expansion while the private American oligopoly could or would not.

Canada's Prairie provinces provide an even more complex case. Using various devices, the provincial authorities of Alberta, Manitoba, and Saskatchewan subsidized rural telephony. All three systems lost many farm subscribers during the Depression, although, as in the United States, Prairie farmers kept buying automobiles. Commentators variously attribute the losses to government mismanagement, unreasonably low rates which underfinanced maintenance, the enrollment of too many marginal farmers, the lack of an urban base, and the high costs of serving low-density areas. The government of Alberta claimed in 1934, when it sold its plant to rural cooperatives, that good roads and radio had reduced the farm telephone "to little more than a social convenience"; in other words, farmer demand had dropped. Eventually, rural telephony in these provinces recovered considerably sooner than did rural telephony south of the border.[50]

In sum, the economic crises of the 1920s and 1930s alone do not account for the decline of rural telephony. Among the other circumstances that undercut rural telephony in the United States, the evidence highlights two. The first is the new environment of plentiful automobiles, surfaced roads, radios, and other technologies after World War I. Some directly substituted for the telephone, and all competed for the farmer's dollar.* The second is the changing political economy of the industry. Rural telephony flourished in the era of active competition. When the industry became a mature, regulated monopoly, corporate interest in farmers slackened. Invitational rates, technical assistance to mutuals, and aggressive marketing fell away. Many remaining farm systems, of marginal quality and totally dependent on local support, were especially vulnerable when customers dropped out because of high costs or technical alternatives. These lines were absorbed by larger companies or folded. Within this technical environment and political economy, financial strains led many American farmers to abandon their telephones.

*Nonfarmers typically received better quality telephone service and were probably not as financially pressed to trade off among the goods.

Summary

A few themes emerge from the history of rural telephony. One is the active role of the consumer. Rural Americans largely discovered, demanded, and developed telephone service for themselves. While telephone companies tried to create needs for city-dwellers, farm families recognized their own practical and social uses for the telephone. To get it, they mounted unusual collective efforts. These telephone users were not mere passive recipients of sales manipulation (though industry salesmen thought so), but acted as agents for themselves. Thus, rural America led the spatial diffusion of this innovation. When, however, the burdens of rural telephony increased after 1920—higher rates, disinterest from commercial companies, tough economic times—many farmers gave it up. They now had alternatives to the often irritating telephone service, the automobile and radio in particular, to meet some of the same ends.

A second theme is the importance of the institutional, economic, and political contexts. Farmers exercised their choices under constraints. When the industry was competitive, rural users benefited from the aid and subsidies the competing companies provided. Before and after competition, however, they faced more hurdles in obtaining service than did city-dwellers. Lacking the serious governmental help available for rural telephones in other nations and for other technologies in this nation, American farmers in most periods had to support telephone service on their own.* Faced with reduced incomes and higher real costs after 1920, many disconnected their telephones.

Finally, this tale challenges simple models of technological modernization. Rural telephony did not advance inexorably, but by starts and stops. The process was not pushed ahead by the internal logic of technology, but was shaped by marketing decisions, government policies, and technical and cultural contexts.

TELEPHONES AND THE URBAN WORKING CLASS

In 1902 Bell Canada fought off municipalization of its Ottawa exchange with the argument that "of the 60,000 people in the city not more than 1200 have or require telephones. . . . Telephone service is not universal in its character" and should not be supported by tax money.[51] Besides being an ironic counterpoint to Bell's later boast of

*Users of other technologies were subsized by taxes, as with automobiles, or sponsors, as in radio.

"universal service," this claim reveals Bell leaders' sincere conviction throughout the era from the telephone's invention to 1940 that the telephone was not for the masses.

People of higher status generally adopt innovations earlier than those of lower standing for several reasons. New devices are usually costly. People with more education are more often open to experimentation. Higher-status people adopt some innovations just to differentiate themselves symbolically from lower-status people. These considerations help explain the class differences that we shall see in the adoption of the telephone.

Company strategies, however, are also a factor in these class differences. The industry's marketing practices provide considerable evidence that, for the most part, its salesmen doubted that they could sell telephones to the working class. High, monopolistic prices before 1893 precluded so many potential customers that they yielded, by some estimates, suboptimal revenues. Advertising copy had a strong upper-middle-class tone. At times, AT&T officials urged local managers to focus on selling extras such as extension telephones to existing customers rather than trying to convert nonusers into subscribers. Internal company projections at the turn of the century assumed that the working class was not a serious market. AT&T promoted residential zoning during the Progressive era because it allowed the company to concentrate its wiring in "good" neighborhoods. Men in the industry who spoke up for reaching the "small fry," specifically John I. Sabin and E. J. Hall, were drowned out by skepticism.[52]

Bell did not always ignore the working-class market. During the period of aggressive competition AT&T officials worried that the independents would become a threat by building on a foundation of "small users." Cheaper service, particularly party lines, and aggressive marketing allowed Bell to preempt the independents in some urban as well as rural places. Bell also took notice of working-class customers in the late 1920s, when its executives saw that many people "in the middle or lower strata of the market" had cars, electricity, and appliances, but not telephones. Even so, selling to these people was a minor concern.[53]

Although Bell officials discounted the working class, automobile salesmen did not. According to automotive historian James J. Flink, by "1903, the belief was common that the automobile would soon replace the horse, and mass ownership . . . seemed probable in the near future." From 1908 on, many car manufacturers produced low-priced auto-

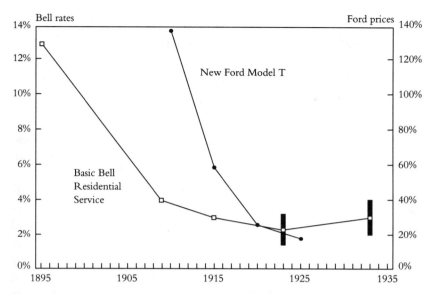

FIGURE 6. ESTIMATES OF TELEPHONE AND AUTOMOBILE COSTS AS PERCENTAGES OF MANUFACTURING WORKERS' EARNINGS. After 1915, the costs of telephone subcription levelled off, but the cost of automobile ownership kept dropping sharply. This was the period in which working-class Americans became more likely to own a car than to subscribe to telephone service.

mobiles, and by the 1910s, led by Henry Ford, they reduced prices in competition for the modest-income—though not the poor—consumer.[54]

A critical consideration for understanding the working-class market is, of course, the cost of telephone service. Charting telephone rates is extremely difficult because they were so diverse. As late as 1923, Bell had 206 different local rate schedules, varying up to threefold in the cost of the same service. Nevertheless, we can make a few approximations. Figure 6 shows some estimates for the cost of basic telephone service over the years (1923 and 1933 are represented by ranges). These estimates are crude, but are confirmed by a few others (see Chapter 2).[55]

Costs for the cheapest Bell home service dropped rapidly, both absolutely and relative to income, during the competitive era, to roughly two percent of industrial workers' wages. In the Depression, real costs rose, albeit still below antebellum levels. From about World War I on, basic residential service cost the average, urban, wage-earning or salaried family about what it spent on tobacco.[56] The rates do not reflect the increasing value subscribers received for their payments: clearer signals, freer lines, cheaper long-distance calls, quicker response time, and more subscribers to call. Still, in the postwar years the cost of basic telephone service ceased to drop.

The automobile provides a marked contrast. Figure 6 shows the relative price for a new Ford Model T Runabout, the standard for inexpensive cars (which Ford stopped making in the late 1920s). The figure even understates the true plunge in automobile prices. Before 1910, most cars cost far more than the Model T's $900. After World War I, the burgeoning used car market meant that families could have automobiles for far less than official prices. In the 1930s, low-income families bought cars for about $100. In addition, the quality, power, and weight of automobiles increased. Yet, purchase prices represented only about half the cost of owning an automobile, considering its appetite for fuel, parts, repairs, and embellishments.[57] Two points emerge from these numbers: First, automobile ownership, even at its low point, cost far more than telephone subscription—between fourfold and sevenfold (see the discussion in the next two paragraphs). Second, Americans between 1910 and 1930 experienced a sharp cut in the cost of automobile ownership, but relative stability in telephone costs.

Given the costs and their changes, how did the telephone and automobile spread through American social classes? Appendix D presents a compendium of over two dozen studies, largely surveys of how urban American families spent their money, showing which families, categorized by their class positions, had telephones or automobiles and how much they annually spent on each. The studies vary greatly in sample and method and are not strictly comparable. Also, the studies overestimate consumption in that they generally exclude broken families and many of the poorest. Still, they suggest general trends.[58]

Figure 7 summarizes four of the surveys listed in Appendix D and illustrates the connection of income to telephone subscription and automobile ownership. The first chart shows that in 1903 in San Francisco—which then led the world in telephones per capita—the more affluent families were, the likelier they were to subscribe. The second chart summarizes a 1918–1919 survey of lower white-collar and blue-collar (but not poor) families in major American cities. Again, the wealthier a family, the likelier it was to have a telephone. Subscribers' spending on telephones ranged from an average of $17 for the poorest to $23 for the richest quartile. Automobiles were still rare in this nonelite, urban population; only 10 percent of even the wealthiest had

FIGURE 7. PERCENTAGES OF FAMILIES WITH TELEPHONE AND AU-
TOMOBILE, BY INCOME, 1903–1934. Throughout the period, family income strongly determined whether people would subscribe to telephone service.
This was less so for automobile ownership, which came to exceed telephone subscription, especially among less well-off families.

cars. The third chart, from an unspecified area in Michigan surveyed in 1927, shows that car ownership exceeded telephone subscription in all classes, the margin being especially wide among the lower two strata. The last chart, for 1934–1936, is another multi-city survey of wage-earners' and clerical workers' families. Again, automobile ownership exceeded telephone subscription, especially among the less well-off families. Owners' spending for automobiles ranged from an average of $96 for the lowest quartile of families to $340 a year for the highest, compared to a range of $23 to $31 telephone subscribers spent on that service. And again, better-off families were much likelier to subscribe.

Subscription rates for working-class families underestimate their access to telephones. Many borrowed neighbors' telephones or used public telephones. An elderly man who grew up in Brooklyn around World War I recalled his role in the telephone system:

> I earned money by walking over to the corner drugstore and waiting for a telephone call: pick up the receiver, and ask who they want to speak to—which was generally confined to the block. I would run over and get the person they wanted to speak to, and usually was rewarded with a penny or two. There were four or five telephone booths. At times I would handle all five. I had competition but I was fast; I turned out to be a sprinter. I remember once earning fifty cents.[59]

We looked for systematic evidence on who had access to public telephones, but could find none. Pay telephones comprised fewer than 5 percent of all telephones in 1907, a figure which probably shrank over time,[60] so that even though subscription rates underestimate the total access of the working class to telephones, they still give an accurate indication of class differences. Another shortcoming of the statistics on subscriptions is their failure to tell us directly about class differences in how often people made calls. (Today, with near-universal subscription, people do not vary by income in how often they make local calls, but the wealthier make more toll calls.[61]) Yet, the statistics in Figure 7 and Appendix D do show which households could afford immediate, constant, and personal telephone use, an ability that varied substantially by class.

The data summarized in Figure 7 and Appendix D also show some historical trends. By 1905 or so, a majority of upper-income families had telephones. By World War I, about half of middle-income families had telephones, but few low-income families did. By the end of the 1920s and through the Depression, the telephone could be found in most middle-income homes (especially outside the South). It was still an exception in most lower-income households, appearing in perhaps one-third or less of those. Expectations that working-class people would have telephones were also low. During the 1910s, social reformers' model budgets for middle-class families included money for telephone service, but the National War Labor Board did not include any in its "minimum budget" for 1918. Official "subsistence budgets" did not include money for telephone calls until 1960. "Minimal adequacy" budgets, which are one step up from subsistence, included money for public calls starting in 1935 and basic telephone service starting in 1960.[62]

In the period of our concern, whether or not a family had a telephone depended on its income. (This criterion still holds today.[63]) Strikingly, the gap in telephone subscription by income stayed constant or widened between 1900 and 1940. In the 1920s, middle- and upper-income households apparently increased their telephone subscriptions at a faster pace than did lower-income households. In the Depression, the affluent were likelier to start or keep service than were less well-off families.

The story for automobiles is notably different. Around World War I, only the affluent among city-dwellers owned cars. Ownership subsequently spread through the class system. By the mid-1920s, almost all high-income families had automobiles. By the mid-1930s, perhaps two-thirds of middle-income families had cars and roughly one-third to one-half of modest-income families did, as well. (Remember that the truly poor are underrepresented in these surveys.) The Depression slowed but did not stall this process. Less affluent families simply kept their old cars longer. Unlike the telephone, differences by income in automobile ownership narrowed between the two world wars. Indeed, contemporary observers noted, usually with alarm, the rush of moderate- and even low-income families buying what had so recently been a plaything of the profligate. They worried about the "auto craze" and the havoc they claimed it caused to prudent family and national budgeting.[64]

The contrast between telephone and automobile diffusion is evident. Working-class families bought sooner and more often kept automobiles than they leased telephones. Between the 1918–1919 and the 1934–1936 Bureau of Labor Statistics surveys, families of modest income became a bit likelier to rent telephones, but the proportion with cars roughly tripled. From the late 1920s, working-class families possessed cars much more often than telephones, in spite of the substantial cost differences. In the 1930s, the average wage-earning family that owned a car spent seven times as much on its purchase and operation than a comparable family with a telephone did on that service. Automobile operating costs alone were annually four times that of telephone service for all classes, three times that of telephone service among the lowest-income households.[65]

By 1970, however, all but the poorest few percent of American households had telephones, more than had automobiles (see Figure 1). Rising affluence, more rate regulation, and perhaps other changes ultimately provided the working class with telephones. (The automobile had not in 1970 saturated the market as completely as the telephone, reversing its position of 1940.) Yet, the following questions arise: Why did not more working-class families subscribe to telephones during the 1920s (which is what one would have expected in a typical innovation diffusion)? Why did they disconnect so readily or fail to subscribe in the Depression? Why did they prefer automobiles even at far greater costs?

Some explanations highlight intrinsic traits of the telephone and automobile; others stress marketing differences. To these we can add the previously mentioned political economy, in which driving an automobile was subsidized in various ways but telephone service was not.

Americans may have felt that the automobile delivered more value for the dollar, although they drove cars largely for leisure, not work, during most of the period. So much more, both practical and social, could be done with a car. In the interviews we conducted (see Chapter 8), elderly respondents expressed enthusiasm when recalling their automobiles. Some described the car's role in words such as "the biggest change," and "opened a whole new way of life." Some comments suggest a sense of empowerment associated with driving. Although many respondents noted the telephone's importance in emergencies and in other uses, they expressed less excitement over it. In their minds, the telephone, rather than the car, seemed more of a luxury, a dispensable good. Also, since they could use pay telphones and neighbors' lines, people may have felt less need to have personal telephones. (The industry continually campaigned against telephone borrowers whom it labelled "deadheads.")

The marketing explanations include the familiar critique of the automobile industry, that its aggressive advertising, easy credit, and annual model changes enticed Americans to buy automobiles, perhaps foolishly.[66] In contrast, telephone men were shy about such broad sales campaigns. In 1927 a Bell sales executive confessed that he and his colleagues were unfamiliar with sales drives, having just recently undertaken them and regarding them in part as public relations devices. Another admitted that Bell's rates were probably too high: "[I]n many cities a substantial proportion of the population are not subscribers. . . . [I]n many instances the minimum rate is what may be regarded as relatively high . . . ," there being few minimum-service options. A third official argued, in 1930, that the company should canvass for customers among families earning over $1300 per year, but let others "come to us voluntarily." He thus wrote off perhaps one-fourth to one-third of American families. This reluctance to broadly market the telephone affected not only rates and sales campaigns but also whether service was even available. Using zoning, as noted earlier, Bell often directed its construction to avoid what it deemed to be unpromising neighborhoods.[67]

The spread of the telephone stalled among the American working class in the 1920s and 1930s. Combined, this phenomenon and the loss of rural telephony account for the slowdown in the diffusion of the telephone shown in Figure 1. The slowdown may indicate a flaw in the telephone as a consumer good, in how its vendors operated, or in the surrounding political economy.* The fact that telephone subscription eventually exceeded automobile ownership tends to direct us toward marketing and politics rather than intrinsic explanations.

Nonmonetary Considerations

Families may have heavily weighed their incomes in deciding whether to subscribe, but it was not their only consideration. Professionals, businessmen, and other white-collar workers were certainly more likely to have telephones at home than were blue-collar workers with similar household incomes (about twice as likely in the 1934–1936 Bureau of Labor Statistics national survey in Figure 7). This difference

---

*The importance of the political context is further illustrated by the Bolsheviks' decrees concerning telephones when they moved the revolutionary government to Moscow in 1918: "With the exception of personal phones belonging to high government officials, doctors and midwives, telephones in private flats were placed at the disposal of 'house committees,' to be made available for general use free of charge. Houses without telephones were entitled to free use of the communal phone of a neighboring house." This system, however, did not last long (Solnick, "Revolution, Reform and the Soviet Telephone System, 1917–1927," 162).

by occupation, however, did not exist for automobile ownership, at least in the mid-1930s. Wage-earner families were just about as likely as clerical, business, and professional families of the same income to own automobiles.[68] Robert Pike, studying Hamilton, Ontario, for the years up to 1911, suspects that many white-collar families used their home telephones for business purposes.[69] Several of our interviewees recalled that their fathers used the family telephone to conduct their jobs as doctors, investors, and the like. White-collar workers were not the only ones to do so. One interviewee's father worked for the gas company and was required to have a telephone. A woman recalled that her divorced mother depended on being telephoned for offers of maid work. Yet the telephone was almost surely more often a job requirement for white-collar workers.

Subtler aspects of the "collar" difference also could have played some role. Perhaps white-collar families more often had status needs, such as keeping up with church and club activities, that required a telephone. Perhaps white-collar families with modest incomes nevertheless had affluent kin and friends who subscribed, so they felt that need to have telephones as well. Perhaps white-collar families lived in neighborhoods targeted by telephone companies for marketing, whereas similarly well-off blue-collar families lived in areas the companies ignored. Perhaps the geographic dispersal of white-collar families through suburbanization made telephones more necessary for them than for the blue-collar families who typically remained in the center cities.[70] And, perhaps white-collar families were culturally more open to innovations. Unfortunately, many of these explanations involve factors that are hard for people to perceive and to recall (e.g., that their parents were status-conscious), so we cannot easily confirm or disconfirm them as explanations for the collar difference. Most suggestions on this list also imply a collar difference in car ownership, which we did not observe. Only the explanations of job needs and industry marketing seem more distinct to the telephone. The basic point is that, when income differences are discounted, white-collar families were likelier to subscribe to telephones (but not to have cars) than blue-collar families.

Homeowners were likelier to subscribe to telephone service than renters, a condition due only partly to the formers' greater wealth. At middle-income levels in 1934 to 1936, for example, owners were 1.5 to 2 times as likely to have telephones as were tenants. (Here is an instance where looking at telephone access rather than subscription might yield a different conclusion. If many apartment buildings had common telephones, then the owner-renter gap in actual use may have been narrower.)[71]

Telephone men assumed for most of the period that recent immigrants and blacks were unlikely customers. This may have become a self-fulfilling prophecy when the industry skirted minority neighborhoods in sales and in construction. The only statistical evidence on ethnic differences in subscrition is the 1934-1936 Bureau of Labor Statistics spending study. It shows that urban blacks were considerably less likely than whites at comparable levels of disposable income to subscribe. Blacks were also considerably less likely than whites to own automobiles. (Both comparisons, however, may be contaminated by regional differences.)[72]

Working-Class Families in 1918–1919

We took a closer look at the connection between class and the telephone by analyzing the original data from the 1918-1919 national survey displayed in Figure 7 and listed in Appendix D. To estimate the cost of living in 1918 and 1919, the U.S. Bureau of Labor Statistics interviewed the wives of over 12,000 blue-collar and lower white-collar workers in cities around the United States, asking for detailed accounts of their spending in the prior 12 months. We drew a random subsample of about 2600 of these households for analysis. This survey has several peculiarities and is only a one-time "snapshot," but it is a rare picture of the working class before the 1920s. (Appendix E presents the details of the analysis.)

We tried to determine what distinguished the 20 percent of families that had telephones from those that did not. (Six percent of the sample had cars, and 40 percent had electricity in the home.) Money and living expenses clearly did. The more household income, the more valuable the home, and the fewer the young children families had, the more often they had telephones. One discovery is that income earned by adults other than the head of the household—usually these others were adult sons and daughters—contributed about as much as did the head's income to having a telephone but was not a factor in getting an automobile or electricity. It is possible that extra adults in the home, notably daughters, specifically applied their independent incomes to telephone service, either because their jobs or because their social interests demanded it. (This gender effect, which reminds one of a roughly similar result in the analysis of Iowa farmers, comes up again in Chapter 5.)

The occupation of the household head was also significant. Only 15 percent of families with manual, production-worker fathers had telephones, compared to 37 percent of those white-collar fathers. (See the foregoing discussion about differences by occupation.)

Differences in location were important. High rates of telephone subscription were reported by Westerners (34 percent) and Midwest-erners (26 percent). The former might be credited to aggressive pro-motion by Pacific Bell—the heritage of John I. Sabin—and the latter to the low prices and heavy marketing of telephones in the Midwest by Bell's independent competitors. Another pattern is greater subscrip-tion by residents of smaller cities. Those in cities with populations under 50,000 had a subscription rate of 30 percent (versus a rate of 13 percent for those in cities over 500,000). The advantage of smaller communities may partly be a result of lower rates and greater com-petition but may also indicate a greater demand for telephones in less densely populated places. Compatible with the second speculation are the higher subscription rates of people who lived in detached homes (28 percent versus 11 percent for all others). Such neighborhoods tend to be more sprawling.

Summary

That income initially determined whether urban Americans sub-scribed to telephones is no surprise. That it apparently continued to do so about as strongly over 40 years is somewhat surprising. Diffusion of the telephone down the class system seemed to stagnate during the first decades of the twentieth century. This contrasts with the diffusion of the automobile; there working-class Americans rapidly shortened the ownership gap with middle-class Americans during the 1920s. I have already mentioned several explanations for the telephone's rel-atively languid diffusion. Some address the technologies themselves: The telephone may have been a less cost-effective (and less exciting) purchase than, for example, the automobile. Some address market-ing: Telephone industry skepticism about the working-class market, as with farmers, may have caused them to miss sales opportunities. This contention, however, faces a chicken-and-egg conundrum: Did industry skepticism retard diffusion or did sluggish sales discourage the industry? There is probably enough evidence to suggest that both effects operated, but the role of marketing decisions should not be underemphasized. Finally, these forces operated in a political economy, specifically one of government subsidies, that favored mass automobility over mass telephony.

CONCLUSION

This chapter has outlined the national pattern of telephone diffusion in the early twentieth century, presenting the spread of the automobile as a counterpoint. Although originally targeted for the urban North, telephones diffused most rapidly in the Midwest and West. Farmers were likelier to subscribe than city-dwellers, at least in the first two decades of the century, but many subsequently gave up their telephones. Automobiles followed the same geographic pattern, but swept ahead of telephones. Farmers not only bought cars but also kept them.

Residential telephones spread from the elite to the (non-Southern) middle class by the 1920s, but their diffusion into the urban working class stalled. To be sure, even low-income urbanites used telephones, those located in drug-stores, bars, and neighbors' homes. However, working-class households did not subscribe in the numbers one might have expected, though they spent far more on other consumer goods, especially the automobile. The latter quickly overtook the telephone in working-class households.

In understanding these national patterns, a recurrent theme has been the counterposition of marketing and consumer autonomy. On the one hand, some features of telephone diffusion apparently reflect marketing decisions. The weakness of telephone development in the South and its strength in the West, the persistent reluctance to deal with farmers before and after the era of heavy competition, the particular weakness of farm telephony in Bell-dominated regions, the disregard for working-class customers, and so on—these decisions shaped the patterns of diffusion. Industry leaders preferred to believe that their sales policies simply reflected the market. No doubt that was often so, but in many cases it was not. Marketing policies, that could have been chosen differently, had a strong influence on the diffusion process.

The grass-roots telephone revolution that farmers mobilized on the other hand, points to the autonomy of the consumer, to the role of the user in the "user heuristic" presented in Chapter 1. The story told in Chapter 3 about telephone sociability gives a similar indication. In these accounts, people act independently of marketing manipulation to shape the history of the technology. These two seemingly contradictory arguments can be reconciled by understanding how choice operated within constraints.

Although the industry was not effective at creating needs and shaping use, it did set the structure within which consumers could exercise choice. Before the era of competition, of course, Bell served whomever it wished and charged virtually whatever it wished. The monopolies that local telephone companies held after about 1915 renewed customer vulnerability to marketing decisions. Companies laid lines in certain areas of cities and along certain rural routes. Typically, they solicited subscribers in those areas and ignored potential business elsewhere. The regulated oligopoly kept prices up. In these ways, the industry impeded adoption of the telephone. By contrast, automobile selling was competitive almost everywhere all the time. Customers even had the option, with the used car market, to buy at almost any price.

The government also provided a structure that limited consumer choice. By guaranteeing local monopolies and not subsidizing lower-income users for most of the period, it left such potential telephone subscribers with higher costs to bear than comparable users in other nations. In the later years, state intervention in rate setting did begin to attract some otherwise excluded users in some places. By contrast, American governments allowed a wide-open market in automobiles and even subsidized their use by applying tax monies to roads, safety services, and the like.

In these ways, the marketing and politico-economic environment placed limits on consumer choice. At the same time, Americans pushed those limits by determining whether, when, and how they would use the technologies, given that the devices were practically available.

At the outset of this chapter, I wrote that knowing who subscribed to the telephone can suggest why they subscribed and how they used it. This chapter has looked at the "who" question broadly, by analyzing telephone diffusion across geographical and class categories. We found that farmers eagerly adopted telephones for reasons of practical and social utility, but later abandoned them, perhaps in part because other devices could serve those same ends. We found that the urban working class, in contrast, did not rush to subscribe. Aside from having limited income, these families shied away from the telephone perhaps because they had fewer job-related needs for it, perhaps because they could use public or borrow neighbors' telephones, but also perhaps in part because the telephone industry ignored them. Moreover, the telephone's utility to such families, living in the denser cities, may not have been as great as it was for families living in smaller places. At the end of the chapter, we looked more closely at which working-class households subscribed around World War I. The next chapter will focus still more closely on the question of who subscribed and why by looking at the diffusion of the telephone and automobile over a 40-year period at the local level, specifically in three California towns, and among specific households in those towns.

CHAPTER FIVE

The Telephone Spreads: Local Patterns

In 1893 citizens of Antioch, California, gathered together at the Methodist Episcopal Church and heard what a local journalist called "the most unique and novel entertainment ever given in Antioch": nine musical numbers performed three blocks away and brought to them by telephone wire.[1] The impressario of the event, the recently arrived Dr. W. S. George, was starting his rise to local celebrityhood by organizing this fund-raiser. He eventually became physician to several lodges, councilman, church leader, and winner of the prize for the best-decorated automobile in the 1908 Fourth of July parade. (He also later moved on to the more liberal, higher-status Congregational Church.)

Although this episode is reminiscent of the stunts that Alexander Graham Bell used to introduce his device to naive audiences, the telephone had arrived in Antioch nine years earlier, with a single line into the town drugstore. The telephone spread with varying speed from place to place, but American communities such as Antioch first encountered the telephone in roughly similar ways.[2] In this chapter, we use as case studies over the years from 1880 to 1940 three towns of small to medium size located in northern California—Antioch, Palo Alto, and San Rafael. We also explore which residents of the towns subscribed to the telephone in which years.

The first few telephones in many nineteenth-century towns appeared at the railroad station, the druggist, a major landowner's home, or the sawmill, and were connected by copper wire to a switchboard in a larger town. When the regional Bell company decided that the town's growth justified a local exchange, it franchised a native businessman or, in a later era, posted a salaried agent to solicit subscribers and build the switching station. Residents often viewed such agents suspiciously, as representatives of a foreign "trust."[*] After the expiration of the Bell monopoly in 1893, hundreds of local entrepreneurs tried to set up their own independent exchanges. Whether Bell or independent, a telephone enterprise required 15 to 25 subscribers to survive. Many towns lacked telephone service altogether, or had only a few telephones at the end of a toll line to a neighboring town.[3]

Telephone men needed town governments' permission to conduct business, plant poles, and to string wire across streets. In return, the companies often provided free telephone service for City Hall, agreed to put caps on rates, and sometimes paid bribes. As local merchants became subscribers, they occasionally used their political clout to restrain rate increases. At times, subscribers protested over issues such as the conversion from flat- to measured-rate service or the companies' elimination of flat-rate public telephones in favor of "nickel-in-the-slot" machines. In the early twentieth century, some towns were host to fierce battles between competing telephone companies. This competition gave town officers new leverage in suppressing rates, extracting concessions such as underground wiring, and soliciting "contributions."[4]

Industry-wide developments in the 1910s (described in Chapter 2) calmed the local politics of telephony. By then, almost all towns had local exchanges. Head-to-head competition decreased, and regulatory reforms moved oversight authority from municipalities to state commissions. Telephone service became more regular, predictable, professional, and controllable.

We will explore the details of this history through our three case studies. A case study permits one to detail the process of diffusion—to identify the specific actors, trace local reactions, and see what issues arose. It also enables the researcher to rise above the colorful anecdotes plentiful in the telephone literature. Using three cases helps to ease the

---

[*]Such suspicion was perhaps with reason: In the 1890s one Bell manager in Abilene, Kansas, embezzled funds, and another skipped town leaving behind bad debts (M. L. Beveridge, "A History of the Development of the Telephone in Abilene, Kansas," Manuscript, n.d., in MIT).

tension between generality and depth. A single case is unique; more cases yield greater generality. However, given a researcher's finite time and energy, more cases can become more superficial. I chose to look at three different towns to attain some, albeit limited, variation. Still, of course, these are only three towns in one region, so there are limits to the generalizations.*

I sought three communities that were located reasonably close to our research center in Berkeley, yet far enough away to be distinguishable from San Francisco; were of modest size and thus more easily graspable than a large city; had continuous boundaries (although that was only partly achieved); had coterminous telephone exchanges run by PT&T; had local newspapers with all or most of the numbers from 1900 to 1940 available; and that varied from one another socially and economically. From about a dozen possibilities, I chose three: Palo Alto, which had an affluent and well-educated population; San Rafael, which, although it included some elite San Francisco commuters, was largely a working-class and ethnic community; and Antioch, which was notably poorer, rooted in agriculture and later a residential community of factory laborers.

The chapter proceeds first by sketching the histories of the three communities, second, by recounting how the telephone came to the towns, and third, by comparing those chronicles to the local histories of the automobile. We will see how the telephone quickly attained an uncontroversial and mundane quality in contrast to the automobile's continuing notoriety. The fourth section describes which sorts of families in each town were likely to have telephones between 1900 and 1936. We will see how adoption of the telephone depended on social class but also on more personal aspects of the household.

THE THREE TOWNS

Palo Alto

In 1894, railroad baron, Governor, and U.S. Senator Leland Stanford created Palo Alto—about 30 miles south of San Francisco on the peninsula—to serve the university he had founded in his late son's

*The primary archival research was done almost exclusively by John Chan on San Rafael, Steve Derné on Palo Alto, and Barbara Loomis on Antioch. Oral histories (discussed in more detail in Chapter 8) were conducted by John Chan for San Rafael, Lisa Rhode for Palo Alto, and Laura Weide for Antioch.

name.[5] He also required it to remain free of liquor and otherwise worthy of the university. California historian Kevin Starr notes:

> ... Palo Alto could escape the burden of California history. It would not have to go through the vulgar frontier stage, but could appear from the first a community of achieved maturity. The near-by town of Mayfield, with its false fronts, saloons, unpaved streets, and atmosphere of shiftlessness, violence, and shabbiness, provided a perfect example of what Palo Alto would transcend. ... One Palo Alto pilgrim of the 1890's remembered the community as "horribly puritan."[6]

Palo Alto grew rapidly and prosperously. (See Figure 8 for comparisons among the towns.[7]) From about 1000 residents and a few hundred Stanford students in the year of its incorporation, it reached almost 6000 in 1920, and through growth and annexation reached almost 16,000 (plus 5000 Stanford students) in 1940.

Palo Alto's population began and remained affluent, educated, largely Protestant, and American-born. From the 1890s, many professionals lived in town and commuted to San Francisco by train. Administrators, clerical workers, and faculty from Stanford, as well as local school teachers, formed much of the community. Merchants served the students and the surrounding agricultural region. Although craftsmen, construction workers, and service providers, such as gardeners, also lived there, few laborers did, at least until Palo Alto annexed the neighboring town of Mayfield in 1926.

Palo Alto was a center of enlightened high Protestantism and Progressivism. It blended, to quote Starr, "elitist talent with middle-class democracy," and also with moralism and then-respectable racism. Many residents were active in reform movements, including feminism and pacifism. The town even practiced a modest socialism, municipalizing some utilities. Churches and women's organizations, aided by Stanford University, enforced moral codes, including restrictions on dance halls, Sunday baseball, and unseemly movies, and sustained an unrelenting prohibition. One long-time resident told us: "They couldn't screen everybody who came to Palo Alto to live, but ... there were some pretty stringent qualifications that were dropped on you ... [that] we expect a certain kind of lifestyle." In 1913 the *Palo Alto Times* commented:

> In a town as free from flagrant evils as Palo Alto it must sometimes puzzle the militant Christian of the old type to know what there is for the church man or church woman to do here aside from ambling along to meeting.

There is no saloon to fight, no gambling to fight and no red-light district to fight.*

At times, rambunctious Stanford students upset this gentility, although they usually did so in other towns.[8] The Progressive-era racism that accompanied the town's gentility expressed itself in efforts to undercut immigrant votes (which partly fueled the women's suffrage movement) and in occasional efforts to ban racial minorities from parts of the town.[9]

For those who could live there, Palo Alto was (and remains) a remarkably comfortable community, well-endowed with public services, well ordered, and composed of an educated, successful, and dutiful citizenry.

San Rafael

An old mission town 15 miles north of San Francisco across the Golden Gate, San Rafael grew in the 1850s and 1860s into a simple marketing center, serving the workers of Marin County's main industries, lumber, dairy, beef processing, and quarrying.[10] It was a rough place. In the early 1870s, rail and ferry improvements made San Rafael more accessible to San Francisco. The town became, in turn, a favorite destination for city day-trippers, a summer community for San Francisco high society, and finally a bedroom community for businessmen seeking impressive suburban homes. The new commuting gentry led San Rafael into incorporation, sponsored civic improvements, and tamed its frontier culture.** One major town leader was the wealthy merchant William Coleman, who had garnered fame as a leader of the San Francisco vigilantes.

The real estate boom of the 1870s and 1880s stalled, however, and population stagnated in the 1890s. It did grow in the 1900s, in part thanks to an influx of refugees from the 1906 San Francisco earthquake, many of whom later returned home. In 50 years of ups and downs, San Rafael's net growth rate was a mere 1 percent per year, far less than either Antioch or Palo Alto. The opening of the Golden Gate

---

*These social evils, however, could be found in neighboring places such as Mayfield and Menlo Park.

**For example, the town fathers toned down and then abolished the rowdy and besotted annual "San Rafael Day."

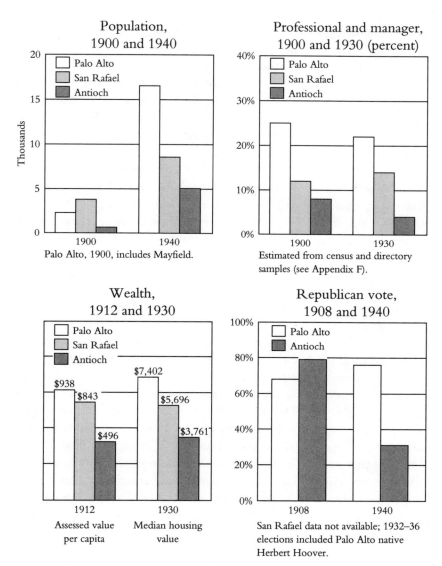

Population, 1900 and 1940

Thousands

Palo Alto
San Rafael
Antioch

1900   1940

Palo Alto, 1900, includes Mayfield.

Professional and manager, 1900 and 1930 (percent)

Palo Alto
San Rafael
Antioch

1900   1930

Estimated from census and directory samples (see Appendix F).

Wealth, 1912 and 1930

Palo Alto
San Rafael
Antioch

$938
$843
$496
$7,402
$5,696
$3,761

1912   1930
Assessed value   Median housing
per capita   value

Republican vote, 1908 and 1940

Palo Alto
Antioch

1908   1940

San Rafael data not available; 1932–36 elections included Palo Alto native Herbert Hoover.

FIGURE 8. SELECTED CHARACTERISTICS FOR PALO ALTO, SAN RAFAEL, AND ANTIOCH. The three towns in our study varied widely in growth, employment patterns, wealth, and political ideology. Palo Alto thrived and remained faithful to the Republican Party. At the other extreme, Antioch became even more of a working-class town and shifted sharply toward the New Deal.

Bridge to San Francisco in 1937 and the postwar suburbanization of Marin County, however, stimulated later long-term growth.

San Rafael did not become the wealthy suburban enclave that Palo Alto was, so the town fathers searched for other sources of economic growth. A few factories and creameries moved into town, as did many automobile-related businesses. By 1940, San Rafael's economy had settled largely into providing commerce and services for Marin County. The occupational profile of town residents was more diverse and balanced than either of the other two communities.

The population of the town reflected the makeup of Marin County, with a heavy representation of Catholics—Swiss-Italian and Portuguese vintners and dairymen, as well as Italians in petty commerce. Germans and Irish were also common. A few Chinese lived in town; others were exiled to a peripheral community. The proportion of residents who were either immigrants or first-generation Americans never dropped below about 60 percent before World War II, higher than either Antioch or Palo Alto.

The culture of San Rafael reflected its social diversity. On the one hand, its frontier history, Catholic immigrants, and tourist business encouraged tolerance of liquor, gambling, and other entertainments. During Prohibition, city police regularly arrested bootleggers, fined them, and released them. Police Chief O'Brien supposedly admitted, "Yes, that's the way we do business. Every so often we raid the boys, and they pay a fine. The money helps to keep up the police and fire departments and keeps the tax rate down."[11] Two of our interviewees used the refrain, "Everyone was a bootlegger," to scoff at Prohibition. On the other hand, The Protestant elite and segments of the business community resisted San Rafael's saloon and beer garden culture, trying to regulate it and make it more respectable. The tension between Protestant propriety and Catholic permissiveness lasted through much of this period.

Antioch

Antioch is on the southern shore of the Sacramento River Delta, up 3 bays and 50 miles northeast of San Francisco, a 2-hour train ride away in 1910.[12] A group of families from Maine founded the town in 1851 and, emulating their ancestors on the *Arbella,* listed their reasons for coming to "the golden shores of California, the Paradise of America, where summer reigns perpetually. . . . " One motivation was to escape the "despotism of fashion" in New England. "[L]ike our Puritan fathers, in order to preserve our integrity, we flee into the wilds of the far west."[13] However, the search for a Californian "City on a Hill" was difficult. Many original settlers moved away and were replaced by ex-miners returning from the gold country.

Because it lay near the confluence of the Sacramento and San Joaquin rivers, and along two railroads, the Southern Pacific (laid in 1875) and the Santa Fe (laid in 1900), Antioch was a transshipment point downriver toward the Golden Gate and eastward to the Midwest. In the mid-nineteenth century, Antiochians loaded grain from nearby farms onto barges. Later, they loaded coal from local mines, but that industry, like the local wheat farms, succumbed to competition, the last mine closing in 1902. Increasingly, Antiochians turned to canning and shipping asparagus, celery, and other vegetable produce of the Delta area. These industries, in addition to services for farmers, formed the town's economic base.

A pair of major paper manufacturers provided some factory employment, but the largest plants came to the neighboring city of Pittsburg. From 1910 on, new steel, rubber, chemical, and other mills in Pittsburg attracted laborers, many of whom found homes in Antioch. In 1910, fewer than 20 percent of Antioch's workers were industrial laborers, but by 1937 over half were.[14] Throughout the first decades of the twentieth century, Antioch was a working-class community.

Antioch grew steadily after 1900, doubling its population during the 1920s and reaching a figure of about 5000 in 1940. Antioch was always ethnically diverse (with proportionately more residents of foreign stock than Palo Alto but fewer than San Rafael), including mostly Irish and Germans in the nineteenth century, then adding southern Europeans, especially Italians, in the twentieth. Many of these later immigrants came to take the new industrial jobs. During the Depression, there was also an influx of Southern "Oakies." Race tensions revolved largely around anti-Asian agitation, such as the movement to ban Japanese ownership of farmland.

Lacking a sizeable elite (many successful Antiochians moved to Oakland or San Francisco), Antiochians struggled to make modest civic improvements and to resist the improprieties of a working-class town. The clientele of farmworkers, rivermen, and laborers supported a lively vice industry; some such establishments floated in offshore houseboats. In 1877 Antioch had one saloon for every 40 residents. Fifty years later, during Prohibition, an editor of the *Antioch Ledger* wrote:

> Antioch is not the quiet little community around midnight hours it seems to be from the daytime appearance. Rather is it a little forty-niner camp, liquor apparently being easy to procure, and its drinking being easier or more effective than would be imagined. . . . [15]

One elderly informant reported 53 bars and 4 brothels in town during his youth.

Although its economic base changed during the early twentieth century, Antioch continued to be largely a working-class town composed of farmers, farm laborers, boaters, miners, canners, and factory laborers. It also functioned as the center of a small agricultural world that was a long train trip from the metropole. Today, the San Francisco-Oakland suburban tide has reached Antioch, and the community is in part a bedroom suburb for the middle class.

The three towns of Palo Alto, San Rafael, and Antioch, although all part of the larger Bay Area, differed in many respects. Any parallels in the course of telephone and automobile history among them ought therefore to be generalizable. Any variations we find may provide clues to the contingencies that influenced the histories of those technologies.

THE TOWNS ASSIMILATE THE TELEPHONE

Figure 9 shows how quickly household telephones diffused through the three towns. [16] (The apparent sag in Palo Alto's telephones circa 1910 is an artifact. The true trend line would not dip. See Appendix F.) In the 1910s the three towns began diverging widely in their levels of household telephone subscription, both absolutely and per capita. The basic explanation for this divergence appears to lie in the differing social class compositions of the towns.

Palo Alto

Merchant John F. Parkinson put Palo Alto's first telephone in his lumber and hardware store in 1892, connecting it to a switchboard in neighboring Menlo Park. (See Photo 19.) In 1893 Bell's Sunset

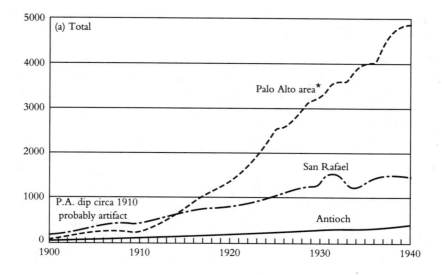

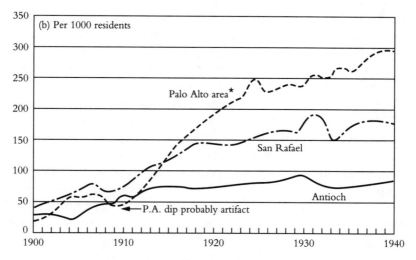

* Includes Palo Alto, Mayfield, and Stanford telephones divided by Palo Alto and Mayfield population estimates.

FIGURE 9. ESTIMATED NUMBER OF RESIDENTIAL TELEPHONES (a) TOTAL AND (b) PER 1000 RESIDENTS, 1900–1940. These estimates, based on a few sources, of the number and proportion of telephones show the distinctly different patterns of diffusion in the three towns.

Telephone Company installed a switchboard in his store. The company's announcements read:

> The Bell Telephone Co. is now putting in a switchboard and Telephone exchange at the Parkinson Lumber and Hardware Company's Office. Those wishing Phones in their homes can have the same at a rate of $4 per month.
> On [the lower] floor the company has also fitted up a room for the public telephone where conversations may be carried on privately.[17]

Parkinson also housed the telegraph and post offices. He was on his way to becoming what the *Palo Alto Times* called "the pioneer settler and businessman in Palo Alto," a major entrepreneur, builder of water works and streetcar lines, Progressive Republican activist, mayor, and leader of the volunteer firemen, as well as winner of the July 4, 1901, "fat man's 50-yard dash." Back in 1893, he took telephone calls and often delivered messages personally or dropped them in residents' mailboxes. Only two other subscribers were on the line then, a realtor and the butcher.[18] Late in 1893, the exchange moved to Hall's Pharmacy. (See Photo 21.) An 1896 newspaper ad for the drugstore noted, "Sunset Tell-Phone. New rates. Sound-proof conversation room. Special operator."[19]

Before 1896, only isolated items about telephones appeared in the *Palo Alto Times*. The newspaper reported a few dramatic uses of the technology, such as doctors called to emergencies and the family of a Stanford student called with news of their son's suicide. Some advertisements for out-of-town services, such as those by a doctor in Menlo Park and an undertaker in San Jose, appeared with telephone numbers. Yet, probably fewer than ten Palo Altonians subscribed. In 1896, however, the Merchants Association discussed starting an independent company. Although they later decided that it would be too costly to proceed, this discussion suggests that the businessmen had some complaint with Bell. About the same time, the town trustees began to regulate telephones. They permitted Bell's Sunset Telephone Company to erect poles in town, but in the years that followed, the trustees repeatedly sparred with Sunset over poles, rates, and free service for the city.

In 1897 the Peninsula Lighting Company announced that it would offer cheaper telephone service to Palo Alto. Almost instantly, a representative of Sunset Telephone Company appeared in Palo Alto offering service at $1.50 per month—a price cut of 62 percent from 1893 rates—with free toll connections to Stanford, Menlo Park, and Mayfield. By the end of the year the Bell company had installed about 20 telephones, including instruments in the homes of John F. Parkinson, two doctors, and the editors of the two local newspapers.[20]

Palo Alto's telephone exchange grew rapidly over the next eight years. New offices opened above Smith's Cyclery, with three daytime operators and a Stanford student on call at night. The switchboards joined over 500 telephones, including ones in Menlo Park, where the original exchange closed. Disgruntlement, however, grew as well. Rates rose to $2.50 per month by 1900. The town council repeatedly argued with Sunset over the size and color of poles, pressuring it to place wires underground. (The company agreed to do so in 1905, an unusual concession to a city as small as Palo Alto.) Would-be competitors to Bell prowled the territory. Explicit criticism appeared in the press, including complaints about operators breaking into conversations, listening in, or prematurely terminating calls. This displeasure culminated during a town meeting in 1905, called over the issue of unsightly poles, but which spilled into general dissatisfaction. At the meeting, the local Bell manager reportedly admitted that the service was "rotten," as subscribers claimed. "The notorious inefficiency of the service, he explained, was due to the town's having outgrown its trowsers," the *Times* reported. "He promised that the company would do better in the future."[21]

It did. Following its reorganization in 1907, PT&T, Sunset's corporate descendent, spent $75,000 on plant improvement, moving to new offices in Palo Alto, laying metallic wiring underground, and providing new telephones to subscribers. Unfortunately the company did not move fast enough to avert an operators' strike in April over increased hours and workload. Nevertheless, these changes, together with a new manager and chief operator in 1908, quelled the complaints.[22]

Telephone subscriptions increased over the next several years, from a few hundred residential listings to about one thousand in 1917 (see Figure 9). Expectations that businesses and their customers would have telephones grew as well. A hardware store suggested in 1908 that customers "shop by mail or telephone"; Bixbie & Lillie Grocers boasted in 1912 of having two telephones "for better service"; and the Stanford Deli suggested that Palo Altonians order "home cooking" by telephone. In 1913, the *Times* began running an advertising section listing businesses' telephone numbers. By 1914, subscribers could summon a taxi by telephone.

Still, some political controversy dogged the local telephone manager. In 1910 the city council voted unanimously to set telephone rates at $1.00 a month for four-party service and banned any deposit for installation, but apparently nothing came of it. The establishment of the California Public Utilities Commission in 1911 soon usurped these regulatory powers from the town, thereby defusing most local controversies about telephony.[23]

Subscription expanded further during the 1910s, into outlying areas of Palo Alto and onto the Stanford campus. The first coast-to-coast call to Palo Alto arrived in 1915.* In 1918 the Army opened Camp Fremont nearby, and among PT&T's responses to the subsequent demand for service was the laying of a 57-circuit cable to San Francisco. In 1919, during the waning days of the federal government's temporary ownership of American telephony, Bay Area telephone operators went on strike. The Palo Alto merchants, the *Times,* and apparently the local citizenry (who, though Republican, sympathized with unions) supported the strikers, both morally and monetarily, against "the niggardly policy of the telephone company."[24]

The 1919 strike may have been the last controversy over telephony in Palo Alto. Certainly, telephone news became scarce after that. In 1922 the police chief urged members of the Business and Professional Women's Club to call in suspicious behavior: "Get what the boys call the 'hit-the-phone habit.'" In 1929 PT&T transferred the Palo Alto system to automatic-dial service.[25] Otherwise, telephones seemingly faded into the background of Palo Alto life.

San Rafael

In 1879 a Western Union subsidiary helped place the first telephone line in San Rafael, running between Thompson's Drug Store and doctors' offices. A long-distance line also ran from the construction site of Mt. Tamalpais Cemetery via underwater cable across the Golden Gate to San Francisco. In 1882 Iverson's Wood and Coal Yard advertised its telephone line to Gieske's Grocery, inviting customers to phone in their coal orders when shopping at Gieske's. Western Union built the first local switchboard in 1884, with about 30 subscribers wired to it. Other interests purchased the exchange from Western Union in 1885

---

*Mrs. B. S. Mitchell received a call from her son-in-law in New York telling of her daughter's operation. "Telegrams are not always so satisfactory where relatives are most anxious," read what was no doubt a company press release, "and that was why R. J. McClelland wished to talk direct to his wife's mother to reassure her" (*Palo Alto Times,* 9 February 1915).

and then sold it to Bell's Sunset Company five years later for $900. Sunset's first manager, H. B. Armstrong, was also the local Wells Fargo agent, and his daughter was the switchboard operator. E. E. Bogle, who later became a successful local businessman, took over the exchange management in 1894 and held it for 14 years. His tasks included setting up poles, stringing wire, billing, and being the relief operator.[26]

Beyond these notes, unfortunately, we know little specifically about the telephone business in San Rafael. The Bell company no doubt had some disputes with the local town council, including a case in 1908 when the council insisted that its $50 monthly bill must be an error. PT&T upgraded its system and put wires underground about the same time as it did in Palo Alto, in 1908 and 1909. In 1917, the town renewed PT&T's franchise in return for 2 percent of receipts. Generally, the telephone company did not advertise its service much, placing newspaper ads around 1911 and again in the late 1920s and through the Depression. For the most part, the development of telephony in San Rafael seems not to have been particularly noteworthy to the local press.[27]

Antioch

In 1884 J. Rio Baker—later to become a banker and city councilman—installed Antioch's first telephone instrument in his drug store cum post and telegraph office (much like John F. Parkinson in Palo Alto). Within five years, the Antioch Drug Store, the Empire Mine, and Dr. Wemple also had telephones, all connected to a switchboard in the county seat, Martinez, about 20 miles away. In 1892 the Sunset Telephone Company established the first exchange in Antioch, with six subscribers. The next year, Dr. George brought the telephone its first recorded notoriety in town with his church concert.

By 1903, telephones were sufficiently common in the county to inspire fulmination by the editor of the *Contra Costa Gazette:* "With a new invention springs up a new crop of fools. A boar is always a boar even at the other end of a telephone line." In 1904 Antiochians demanded all-night telephone service. In 1907 the *Antioch Ledger,* which was generally supportive of telephony, encouraged farmers to the east to organize a system, saying that the "telephone is a necessity, not a luxury." Also in 1907 a new manager arrived and installed an enlarged switchboard. Growth in subscribers required the adoption of telephone numbers for the first time in the spring of 1908. In the same year, the local PT&T manager helped farmers organize a system; rural subscribers purchased their instruments for about $20 and then paid 25 cents per month in charges. (A few of our interviewees from Antioch noted that it was the farmers who seemed most interested in the telephone.)[28]

The Antioch Board of Trustees began regulating telephones for the first time in 1909, requiring permits for putting up poles and charging PT&T $60 per year and two free official telephones for its franchise. In 1910 the telephone exchange moved into its first brick building and employed five operators and enough linemen to fill out a baseball team. Also in 1910, town merchants purchased the wire for the new farmer system in the Lone Tree district, south of Antioch. According to anecdote, the new country lines allowed Dr. George to dispense with his previous means of maintaining contact with rural patients, homing pigeons.[29]

Merchants' advertisements referring to the telephone appeared in the *Ledger* in this period, notably one placed by Antioch's first wholesale liquor business recommending that customers call for home delivery. Liquor dealers were quick to advertise this use of the telephone, a practice that bedeviled temperance efforts nationwide.[30] Advertisements by PT&T itself soliciting subscribers were rare until 1911, when the company ran a series in the *Ledger* urging farmers to subscribe.

Progressive-era reforms changed the regulatory climate, first eliminating City Hall's free telephone service and then subjecting rate increases to municipal review. In 1912 PT&T's efforts to raise rates and require a deposit for installation raised howls from the usually friendly Chamber of Commerce and the *Ledger*—"Talk about nerve!" the editor wrote on April 20—so PT&T backed down. In 1913 the town ceded regulatory power to the state. During the rest of the 1910s, telephony made steady progress in and around Antioch, with yet another numbering system, yet another new office, new farmer cooperatives (a few of the older ones asked PT&T to absorb them), and automatic switching.[31] By 1920, 363 business and residential telephones connected to the Antioch exchange. About 300 of those were in town, or 15 per 100 Antiochians.[32]

How common the telephone was depended on one's sector of Antioch society. An elderly informant, who had been an automobile mechanic in the 1920s, did not have a home telephone until after World War II. "It was a luxury for our family. We were lucky to have enough to eat." Yet, a member of a leading Antioch family recalled that "telephones were pretty universal."

In the 1920s and beyond, little appeared about the telephone in the *Ledger* or elsewhere. An occasional controversy arose. In 1939 residents in one neighborhood objected to new telephone poles.[33] In general, however, telephones were simply not newsworthy, especially in contrast to the automobile, for which even individual car purchases were worth publishing (see the next section).

In all three towns, news stories about the telephone quickly diminished. In the nineteenth century, the local press included items about the wonder of telephony, drawing largely on out-of-town stories (which were probably planted). The first local news about telephones, in the 1880s and 1890s, concerned the placement of a lone station in a centrally located store, followed in a year or two by the first switchboard exchange. By the turn of the century, sometimes bitter controversies arose pitting the Bell telephone company against its subscribers and the town government, in arguments over service, operators, and unsightly wires, but mostly over rates and fees. These controversies would occasionally draw the ire of the local newspaper, a departure from the friendly relations Bell generally cultivated with the press. Subscribers' concerns about rates encouraged efforts to establish cheaper competitors, at least in Palo Alto. PT&T responded to the threat by improving service and by sharply cutting its monopoly prices, although it sometimes raised them again later. By the late 1910s, controversies had largely disappeared, in part because the state had taken over regulation and perhaps also because the threat of competition had disappeared. In subsequent years, the press was largely silent about the telephone. What appeared were typically press releases by PT&T announcing open houses, lectures, plant improvements, and the like. (If active telephone competition had been present in any of these towns, there would surely have been more news.) Note that throughout the controversies, there was little evidence that critics resisted the telephone. When they objected, it was to Bell and its delivery of the technology, not to the technology itself.

Adoption of the telephone was slow in the nineteenth century, restricted to a few functionaries and a few members of the elite (such as Parkinson in Palo Alto, Baker and George in Antioch). Some, doctors and merchants especially, saw advantages to subscribing by the early twentieth century. They advertised their availability by telephone* and, in Antioch, encouraged farmer telephone lines. The telephone soon became commonplace to the affluent and familiar even to those who were not wealthy. A woman who had worked the An- tioch switchboard while in high school told us that even for nonsub- scribers in the late 1910s, "It wasn't like you never saw a telephone. They had one at the store or at the neighbors."

The telephone soon lost its newsworthiness. The quiet atmosphere that came to surround it probably reflected the removal of political control from the towns to the state, the emerging dominance of Bell, and perhaps the way in which the telephone literally became part of the unremarkable furniture in the homes of middle-class people.

It was not so with the automobile.

THE AUTOMOBILE ARRIVES

The first reports of automobile sightings were filled with amazement. Wonder soon mixed with outrage, however, as touring automobilists upset local buggy drivers, made horses bolt, kicked up dust, and caused accidents. An Antioch old-timer recalled to us that "when the vehicle came, horses got scared, kids screamed, and mothers pulled their children off the road." A 1906 issue of the *Antioch Ledger* blared:

MANY DEATHS DUE TO "DEVIL WAGON!"
Dead and Unconscious Forms Left in Wake of Speeding
Automobiles Operated by Reckless Chauffeurs[34]

Such experiences with "scorchers" from the cities led locals to impose restrictions on drivers. For example, in 1903 angry residents passed around petitions to ban automobiles in San Rafael's Marin County. The president of the Auto Club of America was able, however, to per- suade the Marin County Supervisors to drop the ordinance by inviting them for rides in members' cars.[35]

As elsewhere, acceptance of the automobile soon won out over hostility, in part because community notables bought cars. In 1907 Antioch's state senator, C. M. Belshaw, purchased a Packard, and, in the personal jesting common to stories about the elite, the April 13 *Ledger* claimed that he was "negotiating for a large tract of level land, free from fences or ditches, in order that he may learn to oper-

---

*This indicator of diffusion is pursued more fully in the next chapter.

ate the beastless vehicle without endangering his friends or destroying property." (Ironically, Belshaw eventually died in an automobile accident.)[36] People—and horses—became more accustomed to seeing the new technology. The *Ledger,* in June 1907, defended the automobile: "[T]here is no denying the fact that the machines are practical, popular, and have come to stay. . . . " To be sure, there were still motorists who loved to scare horses, but those were not local folk, the editor claimed. The *Marin Journal,* which had earlier listed the nuisances of the automobile, concluded in 1908 that "[t]hey are good things and here to stay."[37]

Automobiles were eventually accepted, but they were not taken for granted. Well into the 1920s, newspapers reported new car purchases from friendly local dealers,* and organizations used automobile caravans to drum up attendance at events. In 1913, for example, the Sisters of the Antioch Congregational Church offered automobile rides as a fund-raising entertainment. "The auto 'rubber-neck' tours were popular pastimes and the cars . . . were filled every trip with a merry crowd."[38]

The news, of course, included serious matters. The press described automobile accidents fully, even luridly.** Citizens argued among themselves over the regulation of motorists, particularly speeders. In the Depression, for example, Palo Alto residents pressed the police to crack down on speeders, but the merchants worried that heavy enforcement would discourage business. Furthermore, as time passed, auto-related and auto-assisted crimes—auto theft, drunk driving, burglary, as well as driving violations—increasingly occupied police officers' attention. Auto-related arrests for drunkenness, for example, rose in Palo Alto from 8 in 1915 to 657 in 1925.[39]

Controversies over road building and improvement that predated the automobile swelled after its arrival. No matter how large the investment, the condition of the streets remained a perpetual scandal to town boosters. In 1907 Palo Alto lumberman John Dudfield complained, "Certainly nothing can be expected of Palo Alto in the way

---

*E.g., "The first new Studebaker to be received in Antioch was delivered this week to Dr. and Mrs. E. W. Bell by W. A. Christiansen. The car is a five-passenger, Duplex-Phaeton of the Special Six class. It is the latest perfection. . . ." (*Antioch Ledger,* 2 October 1924).

**For the record, complaints about horse accidents and about livery drivers racing down the street were common in the years before the automobile (e.g., *Palo Alto Times,* 15 November 1895, and 1 November 1901).

of speedy growth as long as her streets are allowed to remain in the present impassable and disgraceful conditions." Twelve years later, according to a memoir, most Palo Alto streets were still "dustbowls in the summertime and mudholes in the winter."[40]

Town leaders pressed for increasingly better-paved streets, but their efforts met with resistance. Originally, owners of facing property initiated and paid for street improvements. By the 1910s, however, town councils more commonly funded the work out of city revenues, floated bonds, or contracted for road work and then billed the property owners for it. Many owners balked at paying for projects implemented without their consent. The most heated battles occurred in San Rafael. In the 1920s Mayor Bowman imposed repaving—and its costs—on property owners, including homeowners in the poorer south side. A challenger subsequently won the 1927 mayoralty election by running against these "confiscatory burdens." Although he, in turn, lost the next election, future road work proceeded with more consideration and financial help to south siders. By 1935, the San Rafael city council itself fought against state plans to widen the city's main street.[41]

Business leaders—again, especially in San Rafael—also pushed for highway construction around town. As early as 1910, they claimed that highway development through San Rafael to the north coast would bring badly needed tourist dollars. The most dramatic campaign was to span the Golden Gate. A bridge was debated in the 1920s, supported by officials, business interests, and labor, and opposed by lumber companies, ferry lines, and environmentalists.* It was eventually completed in 1937.[42]

What business the automobile can bring, it can also take away. Merchants worried about local customers who drove to the bigger cities to do their shopping. Similarly, San Rafael's luxury hotel of the nineteenth century, the Hotel Raphael, closed in 1909, a victim of that same automobile, according to some: "Rich people no longer wanted to spend a whole summer month in one spot when they could, with convenience, travel to many different locations."[43] For many businessmen, the automobile was a more direct threat. Blacksmiths and livery operators largely sank without a trace. The struggles of other, larger enterprises were more visible. The water transporters around Antioch lost freight business to trucks, especially after a deckhand strike

---

*One environmentalist wrote in 1924, "When you have one of the most romantic approaches in all geography, why spoil it? Let the landowners of lovely Marin County stew in their own juice . . . but in the interest of our dear San Francisco, do not bridge the Golden Gate. Leave that kind of gesture to Los Angeles . . . ."

in 1919, and responded by seeking regulation against wildcat jitneys in the 1920s. The boatmen eventually succumbed, and the last firms consolidated their holdings into two luxury steamers.[44]

The automobile itself became a major business, spawning dealerships, service stations, repair shops, and so forth. The number of garages on San Rafael's Fourth Street grew so rapidly that the city council started denying permits in 1925. Homeowners resisted new garages in residential neighborhoods by the 1930s, if not sooner.[45]

News coverage of the automobile encouraged the impression that "everyone" had one, though this was far from true, even in affluent Palo Alto. Yet, in the 1920s the *Palo Alto Times,* arguing against subsidies for the failing public transit system, claimed that "[t]he majority of taxpayers have access to automobiles, you know."* Whenever the editors of the *Times* commented on a quiet Fourth of July, they typically explained that almost "everyone" had driven to the mountains or shore. A former Antioch automobile mechanic who had described the *telephone* as a luxury told us that "everyone" had cars by the mid-1920s.

The story of the automobile in our three communities is similar to that of many towns in the United States. Commonly, the first serious notice of the automobile came when wealthy "scorchers" raced through, frightening horses, scattering pedestrians and livestock, tearing up gravel roads, and littering the countryside. Reaction was quick and hostile.[46] Soon, however, town notables were tooling about in the latest models. Leading doctors were particularly quick to purchase cars.[47] Several subsequent developments, from about 1910 to 1930 — falling prices, improving performance, better roads, credit financing, advertising — accelerated the diffusion of the automobile described in Chapter 4.

Extensive public notoriety and controversy accompanied the automobile. The outrages of the first years gathered much attention, most notably the bloody accidents. Later, accidents became commonplace items in the newspaper and no longer served as a justification

---

*The "majority" quotation is from 15 January 1925. An earlier debate on public parks drew the following letter to the editor from a reader who took the pen name Motorless:

> Being one of a large number of families that I know of that have no automobile, I disagree with "Member of Chamber of Commerce," who said in the Forum yesterday that every family has a motorcar and, therefore, we don't need parks .... Anyone who actually thinks that everybody has an automobile is apparently quite out of touch. (*Palo Alto Times,* 24 February 1924)

for restricting automobile use.[48] Discussion, debate, and diatribes over traffic, congestion, and parking became staples of the local press, commonly pitting local merchants trying to ease shoppers' access to their stores, against taxpayers, urban managers, and commuters. In Los Angeles, for example, the city council tried to ban parking downtown during midday to help speed up the streetcars. Merchants fought back when they saw that parking restrictions discouraged shopping motorists from coming to the center city.[49] Behind the headlines of accidents and the arguments over traffic jams were concerns of a deeper sort. Experts worried that average Americans were driving into extravagance and hopeless debt.[50] Others blamed the car for teenage sexual promiscuity, family breakdown, and the decline of the church (see Chapter 8).

The Silence of the Telephone

The attention paid to the automobile far outweighed that received by the telephone. Except for a few small controversies, the three communities of our study seemed to have assimilated the telephone into daily life with little notice. In contrast, residents seemed constantly fascinated by and concerned about the automobile. (This distinction appears in the recollections of our elderly informants, as well.) Note that the fascination included, at first, rejection of the automobile itself and, later, hostility to many of its correlates, such as accidents and congestion.

There are many reasons for the automobile's notoriety. It was public, noisy, and glaring in ways the telephone was not. It had a considerable and direct economic role. The machine was costly, and its purchase generated other spending. (The monies also sustained publicists and advertisers whose job it was to keep the automobile visible.) Driving helped bring commerce to town—and helped take it away. Use of the automobile entailed drama—speed, contest, romance, danger, and death. Finally, its use raised a profusion of policy issues, such as road building, policing, taxation, and urban planning, that kept the automobile near the top of the political agenda.

In contrast to all this, the diffusion of the telephone, once it passed the technical marvel stage in the nineteenth century, proceeded with little visibility. This distinction places the telephone among the technologies of modern life that are taken for granted, along with technologies of food preservation and personal hygiene, for example. Soon after their invention and initial diffusion, these kinds of devices become commonplace in two senses: They are nearly universal and they are mundane. In contrast are technologies that, even long after they are common, remain noticed and controversial. Examples include television and the automobile. What makes a technology either mundane or remarkable is not evident. The scale of its consequences may make but a modest contribution to its notoriety. One could, for example, credit food canning, refrigeration, and sewage treatment with significantly extending Americans' lives and yet these technologies go almost unmentioned in general discussions of modernization.

The automobile was (and is) much more noticeable and notorious than the telephone. In economic, political, and even mythic ways, the automobile was, by all appearances, more important. What is less evident, however, are the implications of the technologies in people's personal lives—how the use of the automobile and telephone altered, if it did, social action, community, and individual consciousness. It may be true that the telephone was a part of the backroom furniture, whereas the automobile was proudly displayed in the front of the house, but we are still left to wonder what personal difference, if any, each technology made. That topic will be the subject of the later chapters. First, we must understand which residents of Palo Alto, San Rafael, and Antioch had telephones.

WHO HAD THE TELEPHONE WHEN?

We systematically examined which households in our three towns subscribed to telephone service in which periods. (See Appendix F for details.) Drawing on census lists and other compilations, we selected random samples of households from each town for 1900, 1910, 1920, 1930, and, to study the Depression, 1936. We assembled as much information as we could about the households and then looked up which ones were listed in the telephone directories.* Censuses of this era are incomplete, missing perhaps 20 percent of the foreign-born,[51] and other sources are even less comprehensive, but, as various researchers have found, they nevertheless yield useful cross-sections of community residents. For each of the five years, we randomly selected an average of 110 households from each town. In addition, we picked

---

*This research required intensive work collecting census and directory data and coding it. That was performed by several assistants: Melanie Archer led the task, aided by Barbara Loomis, William Barnett, Barry Goetz, and Kinuthia Macharia.

a special set of 20 households that included people whose names had appeared in the local press — town *notables*.

The telephone directories are essentially complete sources (we do not need to worry about unlisted numbers). Some people, however, had personal access to telephones not listed in their names, such as the live-in superintendent of the San Rafael Municipal Baths, or apartment-dwellers who shared a building telephone. So, our counts slightly underestimate access to telephones.

The samples give us another estimate of telephone diffusion in the towns (see Figure 10; cf. Figure 9). In 1900, only 3 percent of households had residential telephones. But by 1910 about 25 percent of the households in each town had telephones. Afterward, brisk growth in subscriptions continued in Palo Alto and somewhat less sharply in San Rafael, so that most households in both towns had telephones in 1930. By contrast, telephone diffusion to 1930 was slow in Antioch. Telephone subscriptions declined in all three towns in the Depression. The patterns in Figure 10 resemble those displayed in Figure 9(b), except that the household samples show a greater loss in Palo Alto during the Depression.

Who had telephones in 1900? A total of only 3 percent of households subscribed. Some were headed by town leaders whom we have already encountered, such as Dr. W. S. George in Antioch. Some were people with business reasons for subscribing, such as attorney R. B. Bell of Palo Alto. A few surprises, such as the telephone listed for 76-year-old Antioch bricklayer Wesley Dunnigan, can be explained by knowing who else shared their homes. Mr. Dunnigan, for example, lived with a son who was a druggist and a daughter who was a telephone operator, each of whom had business reasons for subscribing.

Besides those with business reasons, it was the social elite who subscribed in 1900. Families whom we defined as *notables* were six times more likely to subscribe than were the households we randomly selected, 18 percent versus 3 percent.*[52] Of the 106 San Rafael residents listed in the 1899 Bay Area *Blue Book,* 44 percent had telephones.[53] Statistical analysis shows that the strongest indications that a household had a telephone were being a *notable* household, having a servant, or being headed by a man who worked as a manager. Beyond that, little distinguished subscribing from nonsubscribing households. (House-

---

*Among the notables were some people who were not town elite, such as outstanding athletes. Households with these sorts of notables were *not* likelier than other households to have telephones.

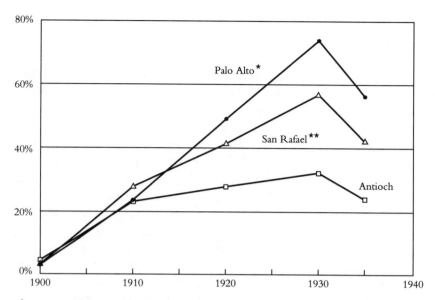

* Palo Alto in 1910 is probably underestimated.
** San Rafael from 1920 to 1936 is probably overestimated.

FIGURE 10. PERCENTAGES OF SAMPLE HOUSEHOLDS WITH TELE-
PHONE LISTINGS, 1900–1936. These estimates, based on randomly sampled names traced in telephone directories, show a similar pattern to that of
Figure 9(b): Over time, the three towns' levels of residential telephone diffusion,
which largely reflected their social class composition, diverged.

holds with a married head and with a middle-aged head were slightly
likelier than others to have a telephone.)

During the rest of the 1900s, telephone subscription escalated to
roughly 25 percent of all households. Having a telephone increasingly
reflected occupation and household composition. In 1910, households
headed by blue-collar workers were about 15 percent less likely—all
else being equal—than others to have telephones, whereas doctors'
and managers' families were substantially more likely to subscribe. A
Palo Alto resident born in 1892 recalled that around 1905, "Very few
people had phones. They were too expensive. Mostly businessmen
had them." Others remembered that, in later years, their families
had telephones because their fathers needed it for work. The fathers
of these respondents included doctors, a contractor, an investor, a
lineman for the gas company, and a school principal. Again, having
servants sharply distinguished subscribing households; over 70 percent of those with a servant had telephones. Almost half of the notable
households subscribed. (But unlike 1900, being a notable, by itself,

did not distinguish subscribers; it merely reflected economic standing.)
Nevertheless, telephones were no longer exclusive to the elite. More
than 1 in 10 blue-collar families now had a telephone.

Also by 1910, subscribing reflected a household's gender composition. Female-headed households were less likely to have telephones
than male-headed ones (21 percent versus 38 percent). However, the
more adult women living in male-headed homes—a wife plus adult
daughters—the higher the chances of telephone subscription. Having adult sons at home, on the other hand, made little difference.
That is, the likelihood of subscribing increased with the number of
adult women at home, other household traits being equal. This finding echoes a gender effect hinted at in our analysis of the 1918–1919
national survey (Appendix E) and of the 1924 Iowa farm directory
(Appendix C). These three sets of results imply that women had more
interest in the telephone than did men (see Chapter 8).

Comparing telephone subscription between 1910 and the later years
is difficult, because for 1920 on we had to use city directories and voter
registration lists rather than census lists. Nevertheless, subscriptions
increased steadily, if not spectacularly, from 1910 to 1920. White-collar
workers, who were probably most accurately recorded in the sources,
raised their rates of telephone subscription from 35 percent to almost 50
percent. During the 1920s, telephone diffusion advanced rapidly, such
that by 1930, almost three-fourths of white-collar families had telephones. Petty merchants, salesmen, factory foremen, and craftsmen
(especially in Palo Alto) increasingly subscribed to telephone service.
The Depression, however, reversed these trends. By 1936, telephone
subscription had dropped to roughly the level of 1920. Again, about
half of white-collar workers had telephones.

We also traced our households in sources published four years after
the initial listings (e.g., we looked for the 1900 families in 1904 city
and telephone directories) to see how commonly subscribers kept
their telephones. After World War I, subscribers renewed at high and
constant rates, which implies that the fluctuations shown in Figure 10
reflect historical changes in the chances that *new* subscribers would get
telephones.[54] The decline in subscriptions during the Depression occurred not because subscribers canceled the service in greater numbers,
but mostly because hardly any new subscribers signed up. One likely
explanation of the drop in new subscriptions is the financial strain of
the Depression. Another is that telephone diffusion may have topped
out. That is, as more people subscribed in the 1920s, those who did not

were increasingly "hard-core" nonsubscribers. Or, adapting the argument of Chapter 4 concerning farmers, perhaps by the 1930s, people who had not yet subscribed looked at their automobiles and radios, checked their shriveled pocketbooks, and decided that a personal telephone was not that important. Those already with telephones, however, generally kept them.

Although the level of telephone subscription fluctuated widely between 1920 and 1936, the profile of telephone subscribers did not.[55] Households headed by married couples or widows were likelier to have telephones than were single people. The more related adults (usually, grown children) living with a couple, the likelier the household was to subscribe. As in 1910, it apparently was not the extra adults' earning power, but simply their presence that affected subscription. We can assume, based on the earlier years (although, unfortunately, we did not code the gender of adults from 1920 on), that these adults were mostly women. The younger informants whom we interviewed often gave social reasons for their parents' decision to subscribe to telephone service around the 1920s. One woman in San Rafael recalled that her parents got a telephone after World War I only because the children wanted it to talk to their friends. Another San Rafaelian explained that his father subscribed because they were too isolated—both for emergencies and for the youngsters' dating. (More of these recollections appear in Chapter 8.)

Families headed by white-collar workers were likelier to subscribe than those headed by blue-collar workers. Professionals were especially likely and laborers were especially unlikely to have telephones.[56] (Between 1930 and 1934, laborers were especially likely to cancel their telephone service.) Complicating this class pattern were town differences; blue-collar workers in Palo Alto subscribed as often as their white-collar neighbors did, but few of Antioch's blue-collar workers subscribed. This contrast between the towns probably reflects the nature of blue-collar work in the three communities. Antioch blue-collar workers were more commonly unskilled laborers and, by the 1930s, factory workers; those in Palo Alto were more often craftsmen, frequently self-employed; and the San Rafael ones were a mixed variety of workers. In addition, there may have been a local climate of opinion encouraging subscription among Palo Alto's working classes and one discouraging it in Antioch.

To recap, in 1900 residential telephone subscription was an elite indulgence or a practical necessity for a few, such as major business managers. By 1910, professional and other upper-white-collar house-holds generally subscribed. There was also some sign that subscribing reflected household size, especially the number of women in the home. Through 1930, telephones increasingly became a typical item included in middle-class furnishings, as well as a more common possession of blue-collar households, perhaps most notably those of craftsmen. Through these years, again, the more adults in a household, the more likely it was to have a telephone. The Depression brought a drop in subscription rates across the classes, though most sharply among blue-collar workers and in Antioch. The decline resulted largely from a severe drop in the number of households subscribing for the first time.

Occupational and Household Patterns

Two primary characteristics distinguished households with telephones between 1900 and 1936: economic position, as indicated largely by the head's occupation and by the presence of servants; and household structure. Figure 11 displays, by year, the subscription rates for four major occupational categories, holding constant the specific town and household composition.[57]

Differences among occupational groups in 1900 were essentially moot; only the elite had telephones. Differences widened in 1910 as many professionals and, to a lesser extent, other white-collar house-holds obtained residential service. By 1920 and then 1930, most white-collar homes had a telephone. Blue-collar households had started to subscribe, as well. (The 1920 estimates for the lower-status groups are probably inflated because of missing data in Antioch; see Appendix F.) In 1930 over a third of even unskilled workers' homes had telephones. Up to this point, the familiar diffusion-of-innovation picture appears: Adoption proceeds earliest and fastest with higher-status groups, later and more slowly with lower-status groups. The Depression then in-tervened, however, setting back telephone diffusion, especially among the lower-status groups.

Occupation affected telephone subscription through its relation to household income, of course, but also perhaps in other ways, as suggested in the earlier discussion of occupations and telephones (see Chapter 4). The noteworthy point here is that between 1910 and 1930 the gap in telephone subscription between the white-collar groups and the lowest blue-collar workers widened slightly rather than shrunk. (This is also what we saw in the national surveys discussed in Chap-

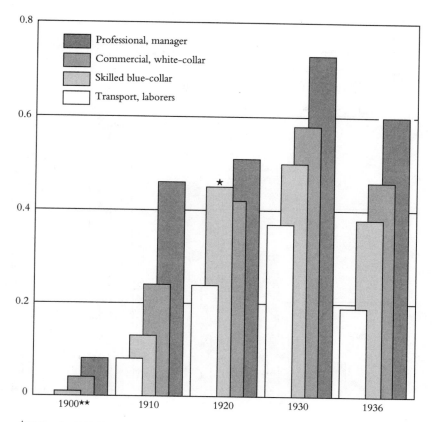

* Values for 1920 blue-collar households are probably overestimates due to missing data in Antioch.

** In 1900 the category of transport and laborers equals zero.

FIGURE 11. PROBABILITY OF SUBSCRIPTION, BY HOUSEHOLD HEAD'S OCCUPATION, 1900–1936. The chances that a household subscribed (holding constant family composition and town differences) varied by the occupational position of its head—and increasingly so over time.

ter 4.) It would take another 30 years, until the 1960s, to complete the spread of the telephone down the class hierarchy.

Figure 12 summarizes adoption of the telephone by household structure, holding constant specific town and occupation. Household types are distinguished by the number of family adults in them (i.e., excluding boarders or servants).[58]

The figure reinforces a few observations made earlier: The more adults in a household, the likelier it was to have a telephone. Women made a greater difference in this regard than did men. Single female heads more often had telephones than did single males. Those in the "other(s)" category are mostly women. Finally, as noted earlier, it is

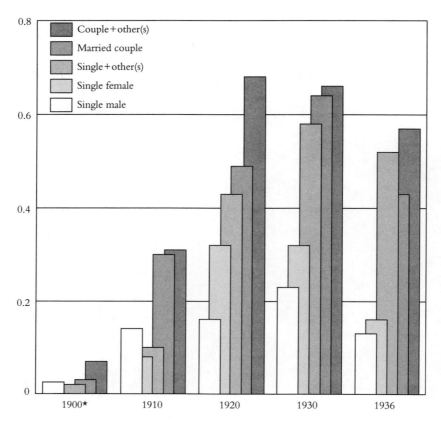

* In 1900 the category of single females equals zero.

FIGURE 12. PROBABILITY OF SUBSCRIPTION, BY HOUSEHOLD ADULT COMPOSITION, 1900–1936.   Except in 1900, the odds of a household having a telephone (holding constant town differences and occupational position) increased if it was headed by a married couple and if there were other adults there, too. Those other adults who supported telephone subscription were disproportionately women.

adults, not adult earners, who seemed to matter.[59] (See Appendix E for a discussion of similar results in the national 1918–1919 sample.)

At least two important implications follow. First, the finding about adults implies that the more people in a household, the more reason to call. This simple point suggests, in turn, that telephone diffusion was driven by demand, rather than solely by marketing. Much of the literature on technology diffusion treats the end consumer as passive and the decision to buy as dictated by external circumstances, such as advertising and cost. A device is presented, a culture is receptive, and people adopt. The "consumption junction," in Ruth Schwartz Cowan's

terms, or the instance and the dynamics of the purchase, is largely ignored in these discussions.[60] When we turn our attention to the buyers
we must also consider their varied "tastes" for the technology.[61] This
leads to the second implication of this section, that the adults whose
demand mattered most were women, which in turn suggests that such
a taste for telephony was greater among women. This point will be
expanded upon in Chapter 8.

Geography and Telephone Diffusion

The telephone is a space-transcending instrument; demand for it
should naturally be related to space. We have seen indications of this
relationship in the demand of farmers living in rural areas for telephone service. Within our three towns, households whose locations
were spatially most isolated should have been most interested in the
telephone. Those people living closest to the centers had easier access
to services, emergency aid, kin, and companions, and thus may have
been less motivated to subscribe than those living on the outskirts of
town. A few of our interviewees' comments suggested as much. Most
explicitly, an Antioch woman born in 1903 recalled that the people "in
the country" used the telephone more than townsfolk did. Her family
did not subscribe because the town was so small, "so what would you
need with a telephone?" Such differences by location, however, ought
not to have been major; these were small towns, after all.

I classified the addresses of most households according to central
versus peripheral location. (See Appendix F for details.) Overall, centrally located households were equally likely to have telephones as
peripheral households. This is, actually, some evidence supporting a
locational effect. Neighborhoods on the outskirts of towns were, we
can presume from general telephone history, less often and less fully
wired than central neighborhoods. Equal rates of subscription probably suggest greater demand in outlying neighborhoods. Moreover,
closer examination shows a consistent difference by location among
households headed by white-collar workers. Those white-collar families living outside the town centers were usually more likely to have
telephones, other household traits held constant, than those living
in the center. The differences between central and peripheral rates of
subscription were minor in 1900, substantial in 1910, and apparently
narrowed from then on—narrowing, perhaps, because what was peripheral in the early years became increasingly central as the towns
grew outward.[62]

One explanation for a locational difference among only white-collar households combines perceived need and resources. All households farther from the center felt a greater need for a telephone, as implied by the woman quoted above. However, blue-collar households on the periphery could not afford to satisfy that need (or perhaps lived in specific neighborhoods unserved by the telephone company). White-collar households on the periphery possessed adequate resources to satisfy their demand, so they subscribed slightly more often than did otherwise similar households in the town centers.*

CONCLUSION

In the 1900s the home telephone was a business tool for some and an indulgence for a few residents of Palo Alto, San Rafael, and Antioch. By the late 1920s, it was becoming a normal, though not yet universal, feature of middle-class life. However it had still not appeared in many working-class homes. (By contrast, the automobile, as we saw in Chapter 4, spread rapidly to the working class nationwide.**) The Depression sharply set back telephone diffusion, even among some middle-class households.

In seeking to understand what factors, besides income, influenced families in these towns to subscribe to the telephone, we found indications pointing to job requirements (especially for doctors and managers), the number of adults in the household (especially women), and peripheral location (for the middle class). These results are consistent with those from Chapter 4.

The findings imply, as did those in the last chapter, that telephone diffusion was heavily influenced by the character of consumer demand. Here, we get some glimpses at the local level of how the users approached the purchase of telephone service. Household characteristics shaped the choices made at the "consumption junction," but within the constraints of income and the services that the market provided. People made their decisions in response not only to business needs and

---

*Social historian Ewa Morawska reports that immigrant families in the early twentieth century, upon moving into the suburbs of Johnstown, Pennsylvania, would subscribe to telephone service in order to stay in touch with kin and friends in the old neighborhood (personal communication). Another explanation, however, could be that residents on the outskirts were more often culturally "modern" (suggested by Mark Rose in personal communication).

**We tried to see if automobile ownership could be traced in our towns, but the only data we could find were personal property tax records in Antioch. They were too incomplete and inconsistent to yield meaningful results.

resources, but also, apparently, to more diffuse considerations, such as the presence of adult women in the household and peripheral location in town. Thus we find, in the cases of Antioch, San Rafael, and Palo Alto, more indications of consumer autonomy.

Although the telephone was a device for speaking, it existed, compared to the automobile, in silence. Except for its early years as a marvel and rare periods of controversy around rates and regulation, there was little public debate about the telephone. The melodramatic reconstructions by later essayists aside (see Chapter 1), Americans, at least after the turn of the century, took this wonder of modern science in stride. Unlike the automobile, which generated fierce disputes, even after almost everyone admitted that it "was here to stay," the telephone engendered only occasional argument. The Bell companies helped still the political controversies and, for the most part, avoided aggressive sales campaigns that might have made telephones newsworthy (see Chapter 3). For these reasons and perhaps due to the nature of the technology itself, telephones became, for the middle class at least, part of the commonplace material culture of daily life, seemingly taken for granted. This, however, does not mean that users were passive consumers. Based on budget considerations, practical needs, and personal tastes, households actively chose to adopt or reject this new technology. Whether the freedom to say "no" to the telephone diminished as it became a near-universal tool of ordinary life in later years is another issue.

PHOTO 1. This AT&T model displays the common wall telephone of the late nineteenth century. The use of similar models continued in rural areas for decades. A small crank, not visible in the picture, extends from the right side of the upper box. The caller listened into the receiver hanging from the left side and spoke into the middle box. The lowest box contained a battery. The next major technical development eliminated the battery box, instead providing current from a common battery located at the central office. (Courtesy of Telephone Pioneer Communications Museum, Don T. Thrall, Archivist and Historian.)

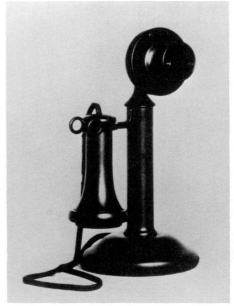

PHOTO 2. This desk set was introduced by AT&T in 1910. It and a few immediate predecessors made calling a more flexible and comfortable activity. Much of the apparatus, including the bell, was attached to a box on the wall and is not pictured. (Courtesy of Telephone Pioneer Communications Museum, Don T. Thrall, Archivist and Historian.)

PHOTO 3. This AT&T "300" series desk set, introduced in 1937, exhibits a few of the changes that the Bell System had made over the preceding 27 years: the dial allowed subscribers to directly reach other telephones automatically, bypassing operators for local calls; the "French" handset combined receiver and transmitter in one hand; and the bell was moved into the base of the instrument. (Courtesy of Telephone Pioneer Communications Museum, Don T. Thrall, Archivist and Historian.)

PHOTO 4. Like other central offices in large cities, the Sutter Exchange in San Francisco, 1914, employed many young women as operators and some as ever-present supervisors. Automatic dialing, phased in during the 1920s, eliminated many of these jobs. According to one casual estimate, if automatic switching had not been instituted, the growth of telephones would have led to the employment of all adult American women as operators. (Courtesy of Telephone Pioneer Communications Museum, Don T. Thrall, Archivist and Historian.)

PHOTO 5. The expansion of telephone systems and telephone wires around the turn of the century led to many scenes such as this one in an unidentified California town. Besides being unsightly, mass wiring could be dangerous. Town governments often argued with telephone companies over when and how the companies would bury the wires underground. (Courtesy of Telephone Pioneer Communications Museum, Don T. Thrall, Archivist and Historian.)

PHOTO 6. This tableau celebrated the completion of the first transcontinental call to San Francisco in 1915. Seated in the center are the city's mayor and Thomas Watson. Watching over the event are the two saints of telephony, Alexander Graham Bell and Theodore N. Vail. (Courtesy of California Historical Society, North Blake Library, San Francisco, FN 28803.)

THE Modern Way is to save one's time and temper by telephoning. House-keeping is such a simple matter when all the ordering is done over the wires. A morning's tiresome shopping can be done in ten minutes, in the comfort of one's own boudoir. There is so much more time left for pleasure and recreation and things that are worth while.

The modern woman finds emergencies robbed of their terror by the telephone. She knows she can summon her physician, or if need be, call the police or fire department in less time than it ordinarily takes to ring for a servant.

You can have a telephone in your residence for about $2.00 per month. There is no charge for installation.

We will be happy to furnish you any information desired.

DELAWARE & ATLANTIC TEL-EGRAPH AND TELEPHONE CO.
12th and Filbert Streets, Philadelphia
B. W. TRAFFORD, General Contract Agent

PHOTO 7. This 1905 advertisement typifies telephone companies' early arguments for having a household telephone: that the lady of the house could more easily manage her domestic sphere with a telephone. "Housekeeping is such a simple matter when all the ordering is done over the wires ... [and] emergencies are robbed of their terror...." (Courtesy of AT&T Archives.)

U NEXPECTED happenings often detain the business man at his office.

With a Bell telephone on his desk and one in his home, he can reach his family in a moment.   A few words relieve all anxiety.

The Bell telephone system is daily bringing comforting assurances to millions of people in all parts of the country by means of Local and Long Distance Service.

Are you a subscriber?

NAME OF ASSOCIATED COMPANY

Universal Service. Reasonable Rates.   14

PHOTO 8.   This 1910 advertising copy, prepared by AT&T in New York and sent to affiliated companies for their use, repeats the practicality theme. In particular, like many other ads of the era, it gave businessmen reasons to purchase telephone service for their wives. (Courtesy of AT&T Archives.)

Why
TELEPHONE SERVICE IS AN ECONOMY

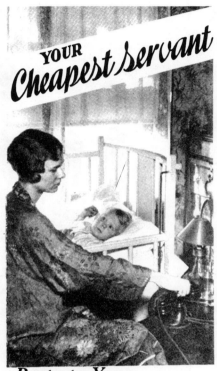

YOUR *Cheapest Servant*

Any investment that saves more than it costs is an economy. It costs no more to have a telephone for 24 hours than it does to drive an automobile about two miles, and a telephone will save that much driving easily. Besides, your time and your comfort are worth more than the few cents a telephone costs.

Sickness? .... it is economy to have a means to reach your doctor quickly. Fire? ... a few minutes delay will cause great damage. Accident or thieves? .... nothing will bring you help as quickly as the telephone!

Just one little· extra job obtained by telephone will frequently pay for your telephone service for a whole month. Then, without cost, use your telephone the other 29 days to secure additional work.

Really, how much value *can* you place on something that makes life easier and safer?

Protects Your
Dearest Possessions

PHOTO 9.   This 1932 advertisement, prepared for the United Telephone Company, contains other practicality themes: thrift and emergencies. (Courtesy of Museum of Independent Telephony.)

PHOTO 10. The first common advertisements emphasizing motives of sociability for telephoning resembled this 1915 ad. They stressed greetings, rather than conversations, and most often sold long-distance calling, rather than subscribing itself. Grandmothers were omnipresent—grandfathers occasionally appeared in the background. (Courtesy of Telephone Pioneer Communications Museum, Don T. Thrall, Archivist and Historian.)

"Do come over!"

FRIENDS who are
linked by telephone
have good times

The Pacific Telephone and Telegraph Company

Business Office: 440 Railroad Street.   Telephone Pittsburg 440.

PHOTO 11.   By the 1930s, social reasons for having a telephone were much more common in telephone advertising. This ad, which appeared in the *Antioch Ledger,* August 17, 1932, exemplifies the emphasis on friendship, not just family; on fun, not just function; and on sociability as a reason for subscribing in the first place, not just as a reason for calling long-distance. (Courtesy of *Antioch Ledger.*)

Speaking Directly Into The Transmitter

THE transmitter of the telephone is the result of years of study and experimentation by telephone engineers. It is of delicate adjustment and its fullest efficiency can only be obtained through proper use.

The lips should not be more than an inch from the transmitter, and the voice should be clear, not loud.

Speak distinctly and directly into the mouthpiece. This will mean your satisfaction and that of the person with whom you are talking.

 THE PACIFIC TELEPHONE AND TELEGRAPH COMPANY

PHOTO 12. Telephone companies also tried to educate users about the mechanics and etiquette of using telephones. (Courtesy of Telephone Pioneer Communications Museum, Don T. Thrall, Archivist and Historian.)

The Telephone Unites the Nation

AT this time, our country looms large on the world horizon as an example of the popular faith in the underlying principles of the republic.

We are truly one people in all that the forefathers, in their most exalted moments, meant by that phrase.

In making us a homogeneous people, the railroad, the telegraph and the telephone have been important factors. They have facilitated communication and intervisiting, bringing us closer together, giving us a better understanding and promoting more intimate relations.

The telephone has played its part as the situation has required. That it should have been planned for its present usefulness is as wonderful as

that the vision of the forefathers should have beheld the nation as it is today.

At first, the telephone was the voice of the community. As the population increased and its interests grew more varied, the larger task of the telephone was to connect the communities and keep all the people in touch, regardless of local conditions or distance.

The need that the service should be universal was just as great as that there should be a common language. This need defined the duty of the Bell System.

Inspired by this need and repeatedly aided by new inventions and improvements, the Bell System has become the welder of the nation. It has made the continent a community.

AMERICAN TELEPHONE AND TELEGRAPH COMPANY
AND ASSOCIATED COMPANIES

One Policy   *One System*   *Universal Service*

PHOTO 13.   Beginning in the late 1900s, AT&T placed many national magazine advertisements that were designed to increase its standing with the American public, not directly to sell its service. This text, circa 1915, credits the telephone with helping to bring "us closer together, giving us a better understanding and promoting more intimate relations" and applauds the Bell System for being "the welder of the nation," for having "made the continent a community." (Courtesy of Bell Telephone, Canada.)

Your Home
DESERVES PROTECTION

A telephone on a farm is the greatest obstacle to rural thieves. A telephone can head off the theft of your chickens, hogs, harness and gasoline---and warn folks down the road of the crooked peddler and the vicious tramp. The farm without a telephone is isolated from outside assistance. Quick reporting of crime demands a telephone to save precious minutes. You need your telephone to give your family and property protection they deserve.

(INSERT NAME OF YOUR COMPANY)

THE MODERN FARM HOME NEEDS A TELEPHONE

PHOTO 14. This ad, prepared by the Illinois Telephone Association, was among a set of advertisements stressing the practical uses of the telephone targeted to farmers in the 1930s. All the others in the series, except one (see photo 15), listed uses such as making money, checking market prices, calling veterinarians, calling doctors, and the like. (Courtesy of Museum of Independent Telephony.)

MAKING FARM LIFE ENJOYABLE

The old time isolation and lonesomeness of farm life is a thing of the past. Modern communication has increased the activities and broadened the social life of the rural family. The telephone plays a necessary part in neighborhood affairs, such as arranging social and church gatherings--planning trips and reunions--promoting community meetings. And, of course, the telephone is especially valuable in exchanging information and local news. You need your telephone to keep in touch with the rest of the world as well as your neighbors.

(INSERT NAME OF YOUR COMPANY)

THE MODERN FARM HOME NEEDS A TELEPHONE

PHOTO 15. This ad was the exception in the Illinois series, addressing "old time isolation and lonesomeness." (Courtesy of Museum of Independent Telephony.)

PHOTO 16. The crude humor in this set of cards produced by a Canadian company in 1877 clues us into some of the social uses—or perceived misuses—of the early telephone. Someone is made a fool because they cannot see what is happening at the other end of the line. In most instances, the humor involves romance or interrupted romance, as in the case where a mother intercepts a young man's query of her daughter, "When will you be alone love?" (Courtesy of California Historical Society, North Baker Library, San Francisco, FN 05332.)

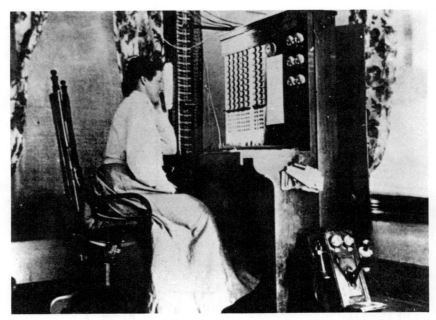

PHOTO 17. This woman probably typified operators on the larger farm systems of the early 1900s. In all likelihood, she operated the switchboard out of her home and her husband ran the company. (Courtesy of Museum of Independent Telephony.)

PHOTO 18. This 1930s scene, taken on the salt flats east of Wendover, Utah, shows telephone company personnel inspecting the lines that crossed the vast plains of rural America. (Courtesy of Museum of Independent Telephony.)

PHOTO 19. Parkinson's lumber yard, 1892, was Palo Alto's first post, telegraph, and telephone office. John F. Parkinson himself does not appear in the photo, but Alice Kelley, his assistant postmaster, is in the doorway. (Courtesy of Palo Alto Historical Association.)

PHOTO 20. This photograph depicts San Rafael's central office in the 1920s. The woman at the far right is accepting payment at the counter. (Courtesy of Telephone Pioneer Communications Museum.)

PHOTO 21. Palo Alto's University Avenue, circa 1895, was one of the muddy streets that were the bane of town boosters, horses, and bicycles (probably Stanford students' bicycles). Also visible are telephone poles and wires and a Bell System placard. B. F. Hall, whose sign adorns the roof, had taken over the town's telephone service around 1893. In 1896, he advertised Vaseline at 10 cents and "a sound-proof conversation room" for making telephone calls. In 1901, the first switchboard in Palo Alto was installed in this same drug store. (Courtesy of Palo Alto Historical Association.)

PHOTO 22. The mud, the horses, and the bicycles had vanished from Palo Alto's University Avenue by 1940, as had the telephone poles. But the automobiles were omnipresent. The visibility of the car and the relative invisibility of the telephone on the street mimic their respective positions in townsfolk consciousness. A similar contrast can be found in photographs of San Rafael's main drag, Fourth Street. (Courtesy of Palo Alto Historical Association.)

PHOTO 23. The Tonso family of Antioch's lower-income Prosserville section displays its new 1921 Ford. Many working-class families had cars but, like the Tonsos, did not have telephones. (Courtesy of Earl F. Hohlmayer.)

PHOTO 24. Antioch's volunteer firemen's truck and hose team pose during the Fourth-of-July parade in 1906. Participating in volunteer fire companies was a favorite activity for young sports in the nineteenth century, and the Fourth-of-July parade was a popular form of entertainment for everyone. As the new century aged, Independence Day festivities declined. Increasingly, town residents entertained themselves privately, often by automobile excursion. (Courtesy of Earl F. Hohlmayer.)

PHOTO 25. These San Rafael Young Men's Institute (Catholic) boys are posed before a bus that will take them to a sports competition in another town. Sport was one of the few leisure activities that clearly became more typically an out-of-town pursuit between 1900 and 1940. (Courtesy of Marin County Historical Society.)

CHAPTER SIX

Becoming Commonplace

The absence of press coverage of the telephone in the three northern California towns by the 1930s suggests that it had become an unremarkable part of middle-class American life. The telephone was on its way to becoming an "anonymous object," "so imbedded in the daily routine as to have become undifferentiated from the rest of the immediate landscape."[1] This chapter investigates how this came to be. We will look at how and when people came to take the telephone for granted in practical matters, such as buying and selling, and in social activities, such as the coordination of voluntary organizations. We also will consider the ways in which the telephone, through becoming commonplace, was transformed from an option into a requirement. To document the achievement of ordinariness, which by its definition lacks drama, we must rely largely on indirect evidence. We will focus on the display of telephone numbers in advertisements and at the prescriptions for telephone use in etiquette books.

THE TELEPHONE IN PRACTICAL MATTERS

As described in Chapter 3, the telephone companies defined a set of practical uses for their product as they began to market it. Many of these functions quickly became routine for many Americans. It appears that people turned to telephones, whether their own or that of a neighbor, so habitually in cases of emergencies, that its use soon came to be expected in crises. Sidney Aronson has described how physicians quickly integrated the telephone into their practices. Doctors "gradually but steadily assumed the moral obligation to be reachable by telephone at all times."[2]

In 1893 a Menlo Park doctor advertised that he would answer calls made to a Palo Alto number. A decade later, Palo Alto physicians had become sufficiently dependent on the telephone to complain that the town drugstore did not answer calls at night. Soon operators and nurses were calling doctors on their rounds. Telephone consultations increased after the turn of the century; doctors could now advise patients at a distance and screen would-be visits. Medical telephony was not, however, without its complications. Some experts worried that telephone consultations led patients to postpone needed examinations and to demand over-the-phone care. Others were concerned about the safety of such advice and the possibility of garbled drug prescriptions. Some doctors also worried about whether and how they might charge fees for care rendered over the telephone. Little direct evidence supports a claim made in 1924 that telephones had lowered urban death rates, but it seems that doctors and middle-class patients had fully integrated the telephone into medical care by that time.[3]

As reviewed in Chapter 4, discussions of rural telephony stressed the many different crises for which farm families could use the device: accidents, ill animals, tornadoes, floods, recruiting farm hands, and so on. *Telephony* claimed in 1934 that the decline in rural telephones had increased fire losses in Nebraska. Stories about operators, particularly in small towns, often described how they coordinated emergency help. Similar, if less extensive, claims can be made for telephones in towns. In 1901 the *Palo Alto Times* published a list of public telephones to be used for calling in fire alarms. The *Antioch Ledger* warned in 1913 that people calling in fire alarms were too often giving wrong addresses. Articles about victims and officials using the telephone to deter or catch criminals appeared frequently starting from the turn of the century. Police encouraged the "telephone habit," and it seemed to become just that. For example, in 1932 the *Antioch Ledger* asked people to call the police if they suspected that the notorious "pants burglar" was entering their homes.[4]

Yet, these emergency uses, as important as they are and as indicative as they are of the telephone's integration into American life, do not capture the manner in which the telephone attained its prosaic quality. Several indirect indicators of this subtle process of becoming mundane might be consulted. One possibility would be to identify when key organizations such as libraries, offices, and schools started routinely asking people for their telephone numbers. Another would be to track magazine advertisements, charting the migration of telephones from featured foreground positions into the backgrounds of the layouts.[5] Historians have noted the use of the telephone in plays, prose, and song during its early years; comparisons of such cultural products might show when it lost its theatrical quality.* For this study, we analyzed the inclusion of telephone numbers in newspaper advertisements.

References to the telephone and numbers in advertisements show, of course, that the parties who placed them had telephones. The telephone companies encouraged merchants to publicize their numbers, and even more, to urge their customers to shop by telephone. Many purchases were made over the phone, but not in the volume the telephone industry sought or boasted of, and it did not come close to supplanting in-person shopping (see Chapters 3 and 8). Still, by listing their numbers businesses testified to their belief that some customers had telephones and would want to call to make purchases or ask about prices, hours, and so forth. The references also indicate that it was considered a normal practice for an establishment to list its number. We take these ads, then, as reflections of the degree to which telephones were routine in our three towns.

Advertisements including telephone numbers appeared in the earliest issues of the *Palo Alto Times*. On 1 December 1893, three out-of-town men—a San Jose architect, a San Jose undertaker, and a Santa Clara contractor—listed their services and telephone numbers. In the early twentieth century, explicit references increased. The California Market, in Palo Alto, placed an ad reading, "Hello Central! Give me Main 61. Fresh Oysters." Solares Cleaners told its customers to insist on calling even if the operator said the line was busy, because they were having telephone problems. Starting in 1913 and running into the 1930s, the *Times* ran a special section of advertisements, first called the "Bell Telephone Classified Directory," later changed to "Information, Sponsored by Pacific Telephone," and still later simply entitled "In-

---

*Robert Collins claimed that over 650 telephone songs were published between 1877 and 1937, with titles such as "Hello, Central, Give Me Heaven," "Hello, Is This Heaven; Is Grandpa There?," "Love by Telephone," and " The Bell Went Ting-a-ling" (*A Voice from Afar,* 141–42).

formation." This section contained small classified ads for businesses and featured their telephone numbers. It had as many as 29 listings in the mid-1920s but shrank during the Depression. The name changes suggest that the telephone feature of the ads was no longer remarkable.

The newspapers of the other two towns also contained increasing numbers of ads with telephone references. In the *Antioch Ledger* of 1906, Manuel Viera, a leading Portuguese-American vintner, advertised, "Live Oak Wood for Sale... Inquire William Bullock... Leave money with Wm. Bullock... Oakley." Ten years later, however, after he had become an early subscriber to a farm line, Viera advertised, "Good Wine Cheap... Telephone Farmers 4x2 or call or write M. Viera. . . . "[6]

To see when and where telephone references became common, we systematically surveyed ads printed in the *Antioch Ledger,* the *Palo Alto Times,* and the *Marin Journal.* We examined, where available, two issues for each even-numbered year from 1890 through 1940. Overall, we coded roughly 8600 distinct advertisements from 134 issues of the newspapers (see Appendix G).*

The total number of ads did not change systematically over the years. They increased in Palo Alto, decreased in San Rafael, and went up and then down in Antioch (where a few major stores closed down). The proportion of ads that referred to the telephone, however, did change systematically. In two of our towns, telephone references shot up quickly. The percentage of ads that alluded to the telephone increased from almost none in 1892 to about 40 percent around 1908 in both the *Palo Alto Times* and San Rafael's *Marin Journal.* After that, the increase in telephone references was much slower. By 1930, about two-thirds of the coded ads in those two newspapers included telephone references. Antioch lagged; by the 1930s, only about half of the ads in the *Ledger* mentioned telephone numbers.

Figures 13 and 14 display the patterns more precisely, showing the percentages of different types of advertisements that included a reference to the telephone. Figure 13 displays the data for advertisements featuring consumer products, such as clothing, appliances, and food, or consumer services, such as liveries, movies, and laundries. (These trends are much like those for all coded ads pooled together.) Telephone references increased in a classic diffusion pattern in Palo Alto,

---

*We also coded newspaper ads according to whether they featured automobiles or automobile products, or referred to the automobile in any way. The data were too erratic, however, to suggest any pattern.

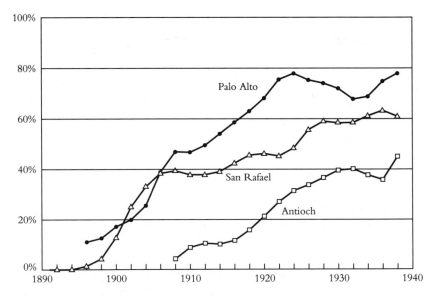

Note: The graph shows smoothed data, with the endpoints dropped.

FIGURE 13. PERCENTAGE OF CONSUMER ADS MENTIONING TELE-
PHONES, 1890–1940. As town residents increasingly placed telephones in
their homes, town businesses, in tandem, placed telephone numbers and allusions
in their ads (cf. Figure 9).

sagging a bit in the Depression, but reaching 80 percent of consumer
ads by 1938. The growth pattern was similar in Antioch, but at a much
lower level, and telephone references dropped more sharply during
the Depression. In San Rafael, the expansion of telephone references
stalled between 1905 and 1920, for reasons that are unclear. The lag
probably reflects the general economic stagnation of San Rafael during that era.[7] Put another way, by 1910 over half the advertisements
for consumer items in the *Palo Alto Times* included telephone numbers. San Rafael's *Marin Journal* did not reach that point until 1925;
the *Antioch Ledger,* not until about 1940.

These patterns echo those shown in Figure 9 (p. 131) for the number
and proportion of residential telephones in the three towns, implying
that the two different indicators track the telephone's penetration of the
communities in tandem. Statistical analysis suggests that fluctuations
in the percentage of consumer ads with telephone numbers followed
fluctuations in the number of residential telephones (Appendix G). It
appears, then, that advertisers of consumer goods and services reacted

to, or in some other way reflected, the expansions and contractions of telephone subscriptions in their communities.*

This phenomenon may be the result of a class-differentiated market: Retailers such as florists, who catered to the telephone-using middle class, included numbers in their ads, whereas those serving a working-class clientele, such as corner groceries, did not. Some of the latter may themselves not have had telephones.[8] Thus the differences in ads among the three towns may, in turn, reflect their class compositions, so that merchants in each town made different assumptions about telephones. By 1920, advertisers who placed most of the ads in Palo Alto presumed that the telephone number was useful to the readers, an accurate assessment given the general diffusion of the telephone in newspaper readers' homes. In San Rafael and more so in Antioch, this presumption was slower to develop. Only in the late 1920s did major advertisers in San Rafael consider the telephone number useful, and in Antioch this did not occur until after World War II. In other words, the telephone was almost commonplace in Palo Alto in the 1920s and on its way to being so in San Rafael, but was not at all commonplace in Antioch during this period. For the middle class of these towns the assimilation of the telephone was no doubt several years ahead of the average. Only a closer study of the advertisements and the businesses can corroborate these speculations.

Somewhat different patterns appear when we look at the two other major categories of advertisements, professional cards and classified notices (Figure 14). For professional cards (Figure 14[a]) — in effect, printed business cards for doctors, lawyers, music teachers, contractors, carpenters, and so forth — references to the telephone increased linearly and in parallel in all three towns, reaching roughly 50 percent around 1908 in Palo Alto, 1912 in Antioch, and the early 1920s in San Rafael. By the 1930s, almost all professionals and craftsmen in all three towns who placed cards also listed telephone numbers. (Why San Rafael had lower percentages than Antioch is not clear.) For classified advertisements, there were sharper town differences (Figure 14[b]). The percentage listing telephone numbers increased steadily in the Palo Alto classifieds, from none in the mid-1900s to almost all at the end of the 1930s (and one large subcategory, for-rent ads,

---

*It is not intuitively obvious that this should be so. Another logical scenario would be that once telephones were generally available, all advertisements would include numbers. If this had been the case, we would have seen rapid growth in Figure 13, leveling off closer to 100%, with similar percentages for all three towns.

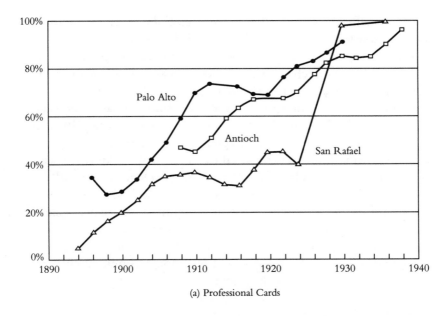

(a) Professional Cards

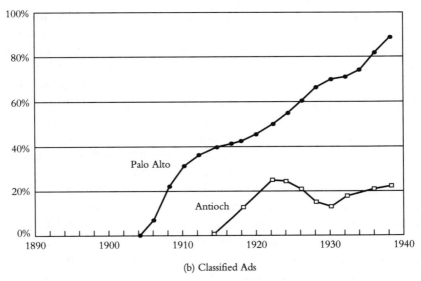

(b) Classified Ads

Note: Both graphs show smoothed data, with the endpoints dropped. Points based on under 10 coded cards or ads are also not shown.

FIGURE 14. PERCENTAGE OF (a) PROFESSIONAL CARDS AND (b) CLASSIFIED ADS MENTIONING TELEPHONES, 1890–1940. The proportion of professionals who listed their telephone numbers in their notices increased steadily to nearly 100% in all three towns. The proportion of persons placing classified ads who listed numbers, however, increased more slowly and reflected differences among the towns.

shows virtually the same pattern). San Rafael data are not shown because there were few years with 10 or more classifieds coded, but the pattern accelerates like that of Palo Alto, only tapering and leveling off sooner. For Antioch, however, there is a very shallow growth in telephone references for classified ads, leveling off at about 20 percent.

A straightforward interpretation of the results for cards and classifieds refers us back to the differences in telephone diffusion by social class and by town reported in Chapter 5. For professionals and businessmen, the predominate users of the cards, telephone diffusion came sooner, faster, and was largely complete by the Depression in all three towns. Almost all such affluent people had telephones by 1930, so they thought it worthwhile to list their numbers. Statistical analysis shows that increases in the listing of telephone numbers on professional cards followed the growth of business telephones in the towns rather than the growth of residential telephones (unlike increases in consumer ad references to the telephone—see Appendix G). Many, perhaps most, of the people who placed classified ads, however, were of lower social standing. These were people offering rooms to rent, selling used household goods, looking for jobs, or searching for lost items. In Antioch, these advertisers rarely included telephone numbers until after World War I, and even then the proportion never exceeded one-fourth. In Palo Alto (and probably San Rafael, too), telephone diffusion among the placers of classified ads, although trailing that of card-placers, proceeded steadily, with about one-third listing numbers in 1910 and almost two-thirds in 1930. Antioch was simply a less affluent and less fully telephoned community.

What do these data say about the assimilation of the telephone into the everyday lives of northern Californians from 1890 to 1940? They suggest that the telephone had not yet become a universal commonplace—else virtually all ads would have featured telephone numbers. Yet, if we take our cue from the contrast between Palo Alto and Antioch, they indicate that the telephone was approaching a commonplace status among the middle class in these years and—as other evidence suggests— its advance had stalled among the working classes.

THE TELEPHONE IN SOCIAL MATTERS

Over the years, using the telephone for social purposes became more conventional. In Chapter 3, we saw how the telephone industry initially considered sociable conversations an affront and resisted such exchanges, embracing them only in the 1920s. In this section, we shall consider the emerging acceptance of the telephone as a social device from other vantage points.

The Telephone and Proper Conduct

Historians and sociologists have often tracked changes in customs by examining formal prescriptions for behavior. Etiquette manuals are a prime source. Although experts in manners usually wrote for the elite, giving instructions, for example, on how to dine at the White House; although the canons sometimes clashed with practice, which accounts for the occasional scolding tone in etiquette books; and although the authors responded to market forces, such that some later books addressed the middle-class readers of women's magazines rather than just the upper crust—still these manuals describe in some measure the dominant standards of behavior. I examined a sample of 21 etiquette books written by and for women and published between 1891 and 1955 to see what light their admonitions might shed on the assimilation of the telephone into American social life.[9]

One clear development was the growing, if grudging, acceptance of the use of the telephone to issue invitations to dinners, parties, and similar events. In the early 1890s the major etiquette quandary concerned the appropriateness of mailing invitations rather than sending them by messenger. By the turn of the century, etiquette books had accepted the U.S. Post Office, albeit hesitantly, but the telephone was another matter. The authors either ignored it—although most elite families had one—or strongly cautioned against inviting by telephone. Annie Randall White wrote in 1900 that an invitation by telephone "is never excusable, save among very intimate friends" or in an emergency, but even then it required an apology for so public a behest. A 1906 volume cautioned that "[o]nly to an impromptu dinner can guests be invited by telephone, telegraph message, or verbal request."[10]

By around 1910, it seems that practice was challenging this admonition. "A Woman of Fashion" in 1909 still denounced the telephoned solicitation: "Verbal invitations and invitations by telephone should not be administered except to the most intimate friends. . . . [F]or most social matters the use of the telephone is questionable at best." Yet, after noting that the mail was now acceptable, the author made some concession to the telephone, too: "Invitation by telephone is one of these modern innovations . . . which shocks elderly, conventional persons still. The convenience of the telephone for quickness and prompt response appeals, however, to so many persons, that it is hopeless to inveigh against it." That same year, Ellye Glover noted that in a large city telephoning invitations might be a necessity, although it should be followed up by a note. In 1914 Florence Hall admitted that issuing invitations by telephone was extremely popular but still advised against it: "The person invited, being suddenly held up at the point of a gun, as it were, is likely to forget some other engagement" or feel constrained to accept. Several authors worried about the latter, that people who received requests by telephone would be hard put to decline graciously. The telephone companies surely promoted the idea that telephoning was consistent with elegant living. Yet, their few social advertisements during this period conformed to proper etiquette. Two 1910 advertisements prepared by AT&T for its affiliates carefully specified that the call was to be made for "informal" invitations:

> *For Social Arrangements:* The informal invitation which comes over the telephone is generally the most welcome.
> The Bell service makes it possible to arrange delightful social affairs at the last moment. . . .
> *For Impromptu Invitations:* The easiest way to get up an informal party, quickly, is by telephone. [The picture shows a young woman, dancing couples behind her, speaking over the telephone to two young men.][11]

During the 1920s, tastemakers' opinions seemed to change, making peace with practice. In the traditional vein, a 1921 book prescribed written invitations to all events, a 1923 volume stated that invitations were not to be issued by telephone unless they were "entirely informal," and in 1924 the *New York Times*'s Washington society editor wrote that "[t]here are persons who should know better who give invitations by telephone, and others who accept or decline in the same casual method. In both instances there is a deplorable lack of form. . . . " In 1923, however, Emily Post stated this injunction with a warmer tone: "Custom which has altered many ways and manners has taken away all opprobrium from the message by telephone and . . . all informal invitations are sent and answered by telephone. Such messages, however, follow a prescribed form."[12]

One of our elderly interviewees, the daughter of a Stanford professor, recalled the ambivalence about using the telephone for social communication: "We used the phone for social calls to a certain extent, but that was really cutting the corners. You should really write a note. I remember Mother sending notes to people in the neighborhood."

By the end of the 1930s, the invitation battle seems to have been largely settled. In 1938 Margaret Fishback announced that "[t]he telephone, too, is becoming more and more the harbinger of invitations to dinner, lunch, cocktails, and week-end parties. It's quick and simple, unless the list of guests is too long." In 1947 Margarey Wilson wrote, "[i]n this incredible world of ours, invitations are frequently casual. More and more, the telephone is coming into use as a means of inviting friends to do this or that." In 1948 Millicent Fenwick prescribed the correct language for a telephoned invitation to dinner. Some guardians of good taste still resisted, however. In 1942 Mrs. Oliver Harriman, who felt that "the 'telephone habit' is an evil," labeled the phoned invitation as the "greatest departure" from traditional etiquette. Nevertheless, she provided some tips on its proper execution, advising, for example, that the invitee should be informed of the appropriate level of dress and that the guest of honor should be mentioned.[13]

If we read these injunctions with appropriate skepticism as rearguard defenses of outdated convention,* they indicate that by the 1910s hostesses in proper society often used the telephone for invitations and that by the mid-1920s they did so with little guilt. For those hostesses who lived in the middle rungs of society and who probably did not regularly send engraved invitations to lunch and kaffeeklatches, it seems reasonable to presume that those with telephones called in most invitations.

The volume of advice concerning telephone manners increased vastly over the years. At the turn of the century, Annie White was an exception in devoting a section of her book to telephone etiquette, "now that the novelty has worn off, and its use has settled down to a plain matter of fact." She warned against occupying the line for long periods of time, being rude, and calling at inappropriate hours. Yet, she reassured her readers that a lady may talk to her husband, family, business agent, doctor, grocer, or butcher "with perfect propriety." In 1909, "A Woman of Fashion" also included a chapter on telephoning, which "has assumed such social proportions as to require some rules for its proper conduct." Again, the author cautioned against monopolizing the line but most strongly admonished people who had their servants call someone for them, thereby forcing the called party to hold the line while the servant brought the caller to the instrument. These two books, however, were exceptions; others

*Arthur Schlesinger, in his study of etiquette books, noted that "they scored chiefly in fostering tendencies already in the making. . . . [T]hey led the procession by following it." (*Learning How to Behave*, 66)

of the period attended little to telephone manners. By the 1920s, telephones received more attention. The early Emily Post, published in 1923, contained brief passages on telephone invitations and the proper way to instruct servants to transmit telephone messages to their employers. In 1938 Margaret Fishback cautioned single girls not to pester boys by telephone (or mail or telegraph) and warned single men never to call someone and "babble 'Guess who this is.'" The 1942 Emily Post contained many entries on the subject—referring to oneself and others on the telephone, using the telephone to forewarn people of a visit, paying for calls made on hosts' telephones, giving others in the room privacy while they spoke, and using an office telephone. The 1955 Emily Post included an entire chapter on telephone courtesy, with additional topics such as wrong numbers, children (they should not answer the telephone), and party lines.[14]

These etiquette books seemed to follow practice, rather than lead it, lagging far behind. In the late 1940s answering a call with "Hello"— the battle against which AT&T had long ago surrendered—was still controversial to society advice-givers. Millicent Fenwick judged that "Hello" was not, according to tradition, proper form for servants, but was acceptable among equals. Margarey Wilson wrote, "When answering the telephone it is perfectly correct to say 'Hello.' Some people seem to find that it is undignified . . . [b]ut experiment shows that any other words sound funnier still."[15]

These books also implied, at least at first, that respectable women treat the telephone aloofly. The earlier ones discussed the instrument largely in terms of servants' use and formal messages. They, as well as the telephone companies, also cautioned women against occupying the lines for "passing nonsense."[16] Only the later editions took common and casual use of the telephone for granted, recommending that people call before a visit, advising ways to bypass servants, and discussing problems of personal conversations, such as excessive duration or maintaining privacy.

Thus these codes of conduct certified the telephone—after the fact—as an acceptable device for high-status women to use in initiating and carrying on sociability. Perhaps the key period in the transition from exceptional to ordinary use (for all but the shockable elderly), was roughly 1915 to 1925.* It is worth noting that these books showed

*The "cake of custom" may have been harder to break in Britain. A few observers described the English as slower to adopt telephones because it was harder to "work them into the elaborate etiquette of the late Victorian and Edwardian years." Their use seemed "synonymous with a deterioration in standards and form" (Perry, "The British Experience," 78, 79).

little concern for an issue that greatly exercised Miss Manners in 1989, unwanted calls, although a few of them did admonish readers to call at convenient hours.[17]

As the telephone nearly disappeared from the pages of the daily press (see Chapter 5), it took up increasingly greater space in the etiquette manuals. Both trends demonstrate the commonplace quality the telephone attained. When it was a novelty, the newspapers covered telephone developments, controversies, and dramas, but the arbiters of upper-class taste largely ignored it. As a merely utilitarian contraption of brief pedigree, the telephone was banished to the backrooms of noble homes with cleaning supplies, kitchen sinks, and servants. Subsequently, as people of the middle class used the telephone more frequently, not with a rush of excitement but with casual familiarity, newspaper editors found it less newsworthy. The ladies of good counsel, however, were increasingly forced to confront and control its use, as they had earlier done with eating utensils.

Organizational Activities

From early on, news accounts and organization reports appeared detailing the use of the telephone in associational activities, especially by women. Accounts of farm families often noted that women arranged organizational affairs, church functions in particular, by phone. In Lana Rakow's oral histories, rural women recalled using the telephone to set up church socials, coordinate meetings, and so on.[18] Minutes of the Palo Alto Business and Professional Women's Club from 1929 to 1933 frequently mentioned telephoning. Excerpts include:

> If you can help [a member's trip arrangements], call 21745.
> Last Saturday we had a food sale. . . . Mrs. Baldwin did the telephoning.
> If you wish to go to the show with us next week please phone Jean Andrus at once.
> If this is your first notice that you are already named on the committee, it is because you don't answer the telephone when a busy B.P.W. can call you.[19]

Telephone references in charitable solicitations were especially prominent during the Depression. For example, the *Antioch Ledger*—in which notes about the telephone were not that common—asked readers in 1930 to call a caseworker if they knew of families in need of furniture and during "Relief Week" in 1931 to call with donations of clothing. Similarly, the *Palo Alto Times* in 1932 announced that the

Palo Alto Benevolence Association "begs, pleads and implores for the kindly help and cooperation of every citizen of Palo Alto" in reporting any inefficiencies in relief operations. "If you hear of families in distress send or telephone us the name and address with all the data you can collect. . . . If you have old clothing telephone to the Womens' Service League." By 1934, the editor of the *Times* complained that charitable solicitations over the telephone were becoming a nuisance and should be stopped.[20]

Election campaigning by telephone appeared in Palo Alto by at least 1922, when the Anti-Saloon League announced that it would call nonvoters to persuade them to cast ballots for the latest prohibition referendum. The earliest campaign reference that we found in Antioch was a 1939 notice to voters seeking rides to the polls that listed the telephone numbers of both sides of the "Ham-n-Eggs" initiative campaign. Telephone canvassing, like other activities, was less common in Antioch than in the other two towns.[21] The telephone occasionally came up in other political contexts. In 1931 the *Antioch Ledger* attributed bank uncertainties to radical sabotage by telephone: "The United States Attorney's information shows drives in which the telephone is used to advise people that such and such a financial institution is in trouble. . . . Every citizen owes it to himself, no less than to the rest of the public, to report immediately any person heard disparaging the credit of any financial institution." In the mid-1930s sheriffs resisting organizers' efforts to unionize farm workers living near Antioch called one another with the license plates of the organizers' cars.[22]

In Chapter 3 we saw, through the reactions of the telephone industry, that subscribers were increasingly using the telephone for everyday sociability in the early years of the century. In Chapter 8 we will read accounts of this situation from the recollections of elderly Antioch, Palo Alto, and San Rafael residents. Their descriptions were not dramatic, compared to those they gave of the automobile. The telephone became an "anonymous object" in social life—at least for the middle classes—gradually and without fanfare. We can find only indirect reflections of this process in unusual places such as etiquette books and club minutes. Eventually, the telephone became more than commonplace, however; it became a virtual necessity.

FROM OPTION TO REQUIREMENT

People rely on some technologies so frequently that their use is woven into daily life and they come to be regarded as necessities. One indication of such changing norms appears in the prescriptive budgets developed by social welfare agencies. Over the course of the twentieth century, welfare departments have added electricity, indoor tubs, refrigerators, and the like to "bare subsistence" budgets.[23] New material standards arise from at least two understandings of subsistence, one practical and the other cultural. In the first instance, agencies recognize that daily life is insufferably difficult without the no-longer-new technology. For example, in an era without vegetable peddlers, icemen, and inexpensive food delivery, a refrigerator is a necessity for a family. In the second instance, as cultural definitions of what constitutes a "normal" lifestyle change, it becomes humiliating for families to lack certain consumer products. A television, for example, is not required to sustain life, but in a nation where 99 percent of households have one, to be without is to be culturally isolated and conspicuous.

Certainly, the telephone today ranks among these virtual necessities of American life. Regulatory agencies have required telephone companies to provide minimal "lifeline" services to the poor. Indeed, most people do seem to view it as a lifeline. In a survey of Manhattan residents who had, because of a switching office fire in 1975, lost the use of their home telephones for over three weeks, two-thirds answered that they had felt uneasy and isolated.[24] The telephone did not attain this universal, requisite status during the historical era covered by this study; that probably occurred during the 1960s. Yet, we can trace its movement in that direction among the American urban middle class during the interwar years.

The budget studies reviewed in Chapter 4 (and Appendix D) suggest that a telephone became an expected item in middle-class homes around World War I. A prescribed budget for a "typical" middle-class family in 1911 allocated 1 percent of total spending for the telephone; a 1915 budget for an upper-middle-class household set aside about 2 percent; and a 1921 budget recommended to *Ladies' Home Journal* readers also included telephone charges.[25] Jessica Peixotto's 1927 study of Berkeley faculty led her to pronounce that "[c]ustom has now made a telephone a routine necessity in most homes." In contrast, budgets for the poor treated the telephone as a luxury. A "minimal adequacy" budget in 1935 allotted funds for electric lights, a radio, an iron, and public telephone booth calls only; a 1960 budget added basic monthly telephone service (as well as a refrigerator, a toaster, and a vacuum cleaner).[26]

Writing in 1933, Malcolm Willey and Stuart Rice noted that telephone subscriptions had been increasing faster than actual calls, implying that Americans had been (at least until the Depression) installing telephones even where the instruments would be infrequently used. They concluded that the telephone had been transformed from a luxury to a necessity. "To be without a telephone or a telephone listing is to suffer a curious isolation in the telephonic age," they wrote—with some exaggeration, given that fewer than two-fifths of American homes had a telephone in 1933.[27] The Depression further undercut working-class subscriptions, keeping the telephone a largely middle-class item for three more decades.

The automobile moved through a similar process of becoming a middle-class necessity perhaps a decade behind the telephone. The Lynds, who were critical of the automobile's role in Middletown, commented—with some hyperbole—that "As, at the turn of the century, business-class people began to feel apologetic if they did not have a telephone, so ownership of an automobile has now [in the mid-1920s] reached the point of being an accepted essential of normal living." About the same time, Peixotto wrote of the automobile among college professors: "Though obviously for most professors primarily an instrument of recreation... [it is] a relatively new type of expenditure that custom is rapidly ranging in the class of necessities though comfortable conservatives still regard it darkly as a luxury." In her 1929 study of typographical workers, Peixotto found that only a fifth of them had automobiles, but those who did saw it as an essential possession rather than as an indulgence.[28]

Michael Berger illustrated the process of the shift from option to requirement in his description of the car in rural America:

> What had really changed was the *structure* of rural life. So vast was the transition that, in 1929, the person who did not own a car was at a definite disadvantage in rural America. Fewer support facilities were available for his or her way of life. There was a smaller number of blacksmiths and livery stables, the railroad ran less frequently [both true as well for our three towns], and some things that might have been better accomplished by horse power, such as getting a doctor through a three-foot snow drift, were less likely to be attempted.[29]

This was, perhaps, a slight exaggeration. Even in 1929, over 40 percent of American farms lacked automobiles, so some alternative services existed.[30] Nevertheless, the basic point is valid: It became more inconvenient not to have an automobile.

Today, almost everywhere in America, the car is effectively a necessity and is generally regarded as such. In a recent national survey, about

90 percent of Americans reported that they thought of an automobile as a necessity. (Clothes washers were second on the list at about 85 percent.)[31] In a suburbanized society with little or mediocre public transit beyond the densest cores, the material structure of the society requires an automobile. For example, two of our Palo Alto interviewees stated that their mothers started to drive around 1920 because the grocery stores stopped delivering to their homes (and perhaps the stores stopped delivery because so many customers were coming in person by car). The culture, as well as the social structure, of America, seems to require an automobile as an emblem of full citizenship.

The telephone lacks the symbolic power of the automobile, but it, too, has become a structural, if not an emblematic, necessity. (The national survey did not ask about telephones, perhaps because they seem so necessary.) At the turn of the century, it became possible for middle-class Americans to rely on their home telephones for emergencies, calling a doctor or the police, for example, instead of sending a messenger or sounding a public alarm. As the decades passed, first middle-class people and then nearly all Americans became not only able but also expected to call for help. Eventually, it may become difficult to get help without the telephone. Police patrolling, for instance, now relies on telephoned requests for aid, with the beat officer rendered largely a relic of the past.

Roughly a century ago, businesses started accepting telephone orders. Within a few decades, middle-class Americans treated telephone shopping as an ordinary option. Today, consumers without telephones are at a disadvantage because they are less able, for example, to order from discount marketers and because they encounter difficulties in paying by check or gaining credit if they lack a telephone number (or a driver's license, for that matter).

A century ago, in "good society," the telephone was too vulgar to have a place in proper social life. Before the mid-century mark, it had attained an acceptable role and had even acquired its own rules of etiquette. Today, although there are still a few domains from which the telephone is barred—wedding invitations, for instance—it is hard to imagine arranging social events without it. Survey data suggest that it is an integral part of most friendships and family ties (see Chapter 8).

The telephone passed from miraculous in the nineteenth century to mundane in the mid-twentieth century to mandatory by the end of the twentieth century. The period upon which we have focused witnessed its attainment of commonplace status in middle-class, but not working-class, America. Businesses catering to the middle class presumed that customers had telephones, formal manners made room for it, and organizations built their activities around calling. Unlike that of the automobile, the assimilation of the telephone was unspectacular. Yet, it was also profound, eventually becoming a basic constituent in the material culture of modern life.

CHAPTER SEVEN

Local Attachment, 1890–1940

Sociologist Charles Horton Cooley wrote in 1912 that "[i]n our own life the intimacy of the neighborhood has been broken up by the growth of an intricate mesh of wider contacts which leaves us strangers to people who live in the same house . . . diminishing our economic and spiritual community with our neighbors."[1] Via streetcars, mass media, and telephones, people at the turn of the century, Cooley and others argued, increasingly turned to the wider world beyond their immediate communities and let fall their ties to the locality.

Cooley's contention has been repeated to the point of becoming a convention. For example, a 1987 college textbook states baldly that the locality has "lost its relevance." Many scholars use this theme to structure their narratives of American social history, attempting to pinpoint the period when American towns changed from cohesive and parochial communities into dissipated and ill-defined "localities of limited liability."[2] Historian Robert Wiebe has described the turn of the century as the era when America's "island communities" lost their insularity and were absorbed into the mass culture. Others challenge this claim, however. Sociologist Peter Rossi has argued, for instance, that the "world has become increasingly cosmopolitan, but the daily lives of most people are contained within local communities."[3]

The thesis positing the decline of local community has served scores of scholars for decades both as a guide and as a straw man. It asserts that the cohesion of American towns fragmented (among other changes), and the outlook of residents turned from parochial localism toward rootless cosmopolitanism. Although many studies have challenged or qualified these claims,[4] the general theory that modernization has undermined community remains a major interpretation of our history.[5]

This chapter will explore one aspect of the wider community debate: the relation between technology and breakdown of localism. Localism is defined here as the extent to which the locality bounds, delimits, or sets apart residents' lives, including their work, personal relations, political involvement, and identity. To the degree, for example, that outsiders take control of a town's economy, that a town's values change to conform to those of the metropoles, or that its residents turn their attention outside the community, we may speak of a loss of localism. The assertion that telephones—and automobiles, radios, movies, and so on—contributed to this loss is usually part of the story.

One assumption usually found in the argument is that the various aspects of a locality—economy, personal relations, public attention, and so on—all cohere. Another assumption is that people must trade off local involvements against extralocal ones. Both assumptions, however, are questionable. On the one hand, localism did decline in some obvious ways during the twentieth century. Today, baseball fans can watch a game played 3000 miles away more easily than they can watch one in town, and local governments pay for much of their construction with money sent from Washington, under strictures set by Washington. On the other hand, localism seemingly increased in other ways. People today live in the same neighborhood longer than their ancestors did.[6] In addition, neighborhood organizations defend local interests at least as vigorously as in the past.[7] Localism has many dimensions. For instance, Michael Frisch found that Springfield, Massachusetts, became a stronger, more active community in the nineteenth century, but also lost some emotional commitment from its residents.[8] Although localism is complex and contradictory, discussions about it tend to be global and one-dimensional. Moreover, it is conceivable that residents could increase *both* their local and extralocal activities over time.

According to the standard argument, many aspects of modernization have undermined localism: national consumer markets, as epit-
omized by brand-name products and advertising; an active central
government, including its increasing financial and regulatory roles;
and, of course, mass media. Also implicated in many explanations—
Cooley's, for example—are personal, space-transcending technologies: rail lines, bicycles, automobiles, and telephones. As early as 1891
a telephone official suggested that the technology would bring an
"epoch of neighborship without propinquity." Over the years, other
observers suggested that the telephone would permit intimacy beyond
the locality. Some current writers, looking backward, conclude that
the telephone contributed to the weakening of local boundaries.[9] Similar claims have been made for the automobile. For example, economist
Robert Heilbroner wrote that "[t]he quintessential" contribution of the
automobile is "nothing less than the unshackling of age-old bonds of
locality. . . . " The automobile allowed people to leave their towns and
more easily import new ideas into their communities.[10]

A significant dissent to the claim linking telephones and automobiles to decreasing localism appeared in 1933. As noted in Chapter
1, Malcolm Willey and Stuart Rice distinguished between broadcast
media and "point-to-point" media such as the telephone and automobile. The former, they suggested, flattened local cultural variations
by introducing national ideas. The latter, however, might have actually enhanced local cultures because people most often telephoned and
drove locally, thereby deepening their ties with people nearby. "The
intensification of local contacts [through telephone and automobile]
may act to preserve and even enhance local patterns of habit, attitude
and behavior, and serve as an inhibitor of the process of cultural leveling which is so commonly assumed as an outstanding and unopposed
tendency of contemporary life."[11]

In this chapter, we consider whether and how localism may have
changed in our three towns during the first half of the twentieth century, and then we weigh the role of the telephone and automobile in
any such transformations. Of all the community-level changes that
might be attributed to new communications technologies, a weakening of residents' ties to their localities would seem most likely. (Other
aspects of historical changes in localism in these towns are examined
in separate reports.[12]) We consider various indications of localism in
four general domains: commercial activity, social life (leisure in particular), interest in the community, and politics. What characteristics

of these domains, if any, changed between 1890 and 1940, and how did such changes occur? Did space-transcending devices—specifically the telephone or automobile—play a role in such changes?

COMMERCE AND LOCAL PATRIOTISM

Where and with whom people do their business is one aspect of localism.[13] To the extent that people shop out of town, one local tie weakens. Merchants' worries over losing their town customers presumably reflect the attraction that shopping elsewhere holds for townsfolk. The geographer Norman Moline found that residents of Oregon, Illinois, increasingly shopped in larger cities as the automobile spread between 1900 and 1930, and the local merchants increasingly voiced their opposition to this practice.[14] Merchants in our three towns also complained.

From the 1890s on, Palo Alto businessmen griped that residents too often made their purchases from outsiders. In 1893, for example, the *Palo Alto Times* featured an ad blaring "Stop her! Don't let her waste money in car fare going to San Francisco. That woman can buy what she wants at equally low prices if she trades with us." In 1895, the *Times* editorialized, "Mrs. [Leland] Stanford buys here, why not you?"[15] Several times, merchants proposed licensing fees and other restrictions on transient peddlers and on outside businesses that delivered in town, such as laundries in San Jose. When voters had the opportunity to voice their opinion, however, they voted down the restrictions.[16] Still, the city councils did succeed in passing some discriminatory legislation. The merchants also pressed the *Times* itself. In 1920 the newspaper insisted on breaking with its previous commitment not to run ads placed by out-of-town businesses, claiming that 1600 copies of San Francisco and San Jose newspapers arrived in town daily, so that "every family" saw outside ads anyway. The *Times* prevailed against the local retailers' pressure. Of course, Palo Alto businessmen worked to lure other towns' residents into their stores, raising money in 1905 for example, to build a bridge across a nearby creek, and initiating Christmas season festivals in 1922.[17]

The situation in San Rafael was similar. At the turn of the century, even grocery stores in San Francisco advertised in the local press, forcing San Rafael merchants to lower their prices. The town newspapers spoke up for buying "at home." For example, the *Marin Journal* editorialized in 1908, "The citizen who buys his household goods by mail. . . . That man's THE TOWNBUSTER."[18] The newspapers promoted several campaigns such as the "Dollar Spent at Home" drive in 1927 targeted against shopping in San Francisco. Town officials harassed itinerant peddlers, one of whom, Jacob Albert, later became a major San Rafael businessman. In the 1930s, merchants also faced the invasion of chain stores such as J.C. Penney's and Woolworth. In 1935 about 200 county businesses joined a "Home-Owned" campaign, appealing to buyers' local patriotism. The next year, however, the voters of the county rejected a state proposition to assess chain stores a special $500 licensing fee.[19]

Merchants in Antioch likewise waged campaigns against the variety and low prices that big-city stores and mail-order merchandisers provided, appealing to local sentiment. In a 1906 issue of the *Ledger* they asked, "Who sympathized with you when your little girl was sick? Was it Sears & Roebuck? Who carried you last spring when you were out of a job and had no money? Was it Montgomery Ward & Co., or was it your home merchant?" In 1925 residents were urged, "[F]or the sake of the town, for the sake of the people living in it — and FOR YOUR SAKE: Trade in Antioch" (1925).[20] Local retailers "hailed with delight" an increase in railroad fares in 1908, in hopes that it would keep shoppers home, but many big-city stores rebated part of customers' train or trolley costs. In the 1920s the competition may have hit Antioch particularly hard; several major stores — Belshaw's, which had been in business for 35 years, the Toggery, and Meyer's, among others — closed down.[21]

If we assume that worries about townsfolk shopping elsewhere reflected, in part, actual customer practice, then out-of-town shopping — be it by car, rail, or mail — was pervasive throughout the 50 years of our study. (The rise of chain stores in the 1930s is a different matter.) In Antioch, economically the weakest town and the one most "shadowed"* by a neighboring community (Pittsburg), the 1920s seem to have severely hurt local merchants. This downturn in sales could be attributed to increased automobility.[22] Otherwise — and unlike Moline's findings in Illinois — the consistency of these complaints imply no serious change in local versus extralocal shopping.

A related indicator of commercial localism is the balance of advertising by in-town versus out-of-town businesses. If we assume that the

---

*"Shadowing" refers to a neighboring large city's depressive effect on the business of a town.

businessmen who bought such ads had reason to believe they would attract townsfolk, then the frequency of out-of-town ads reflects local customers' defections to out-of-town stores. Moline reported an increase in such ads in Oregon, Illinois, between 1900 and 1930.[23] Our impression is that, over the years, national and regional companies took more space in the three newspapers we traced. Large ads for corporations such as Chevrolet or Pacific Gas and Electric became more noticeable. This trend may reflect an increasing nationalization of the towns' commerce. At the same time, however, the number of small ads placed by out-of-town mail-order merchandisers declined.[24]

Figure 15 draws on the coding and counting of newspaper advertisements previously analyzed in Chapter 6 (and Appendix G). It shows, by year, the percentage of the ads printed in the town newspapers that were placed by businesses or individuals located neither in nor adjacent to the town. (Dividing the advertisers right at the town lines does not notably alter the results.) Two types of ads are counted: those for consumer products or services, and cards placed by lawyers or doctors. Also, for Palo Alto only, the figure shows classified listings of rental units. (See Appendix H for more detailed analyses.)

Although the patterns are erratic, none shows any convincing increase in extralocal advertising. The figure for Palo Alto suggests a rise in extralocal ads after 1920, when the *Times* formally dropped its ban on outside merchants, but the increase was neither large nor long-lived. In general, distant merchants contributed fewer than 10 percent of the retail consumer ads directed to town residents.[25] The percentage of out-of-town lawyers' and doctors' cards dropped very sharply, although it rose again in Antioch at the end of the period.[26] Finally, out-of-town rental notices in Palo Alto rose to about 10 percent of such ads around 1930, but then sank back to zero. The trends for the absolute numbers of extralocal ads are even more consistently downward.

Statistical analyses show that the visibility of outside retailers in the town press generally declined. (See Appendix H.) One explanation could be the same one the *Times* gave Palo Alto merchants, that town residents increasingly subscribed to San Francisco newspapers and could read big-city ads there, but this seems, however, not to be the case. (We will look at newspaper subscriptions later in this chapter.) Instead, the decline may indicate the development of the towns into more self-sufficient commercial centers, into towns with enough local outlets to discourage outside competition.

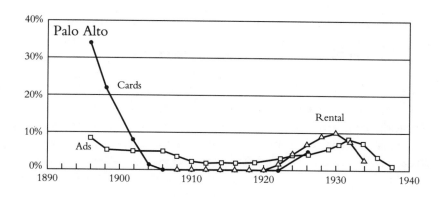

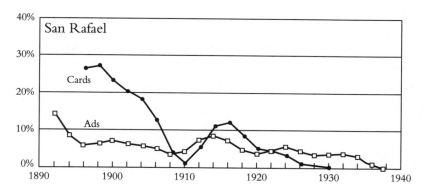

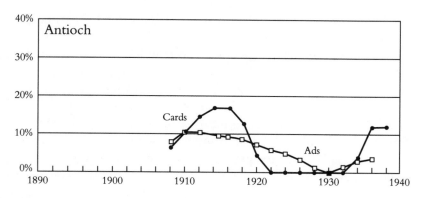

Note: The data are smoothed, with the endpoints dropped. Palo Alto also gives percentages for out-of-town classified listings of rental units.

FIGURE 15. PERCENTAGE OF CONSUMER ADS AND LEGAL-MEDICAL CARDS FROM OUT-OF-TOWN, 1890–1940. Although the trends are irregular, advertisements placed by either out-of-town professionals or out-of-town retailers tended to decline as a proportion of all ads in the local newspapers.

How do these conclusions square with the common impression that, especially because of the automobile, residents of small towns abandoned local merchants for big-city ones? Some studies suggest that rural people used their new automobiles to shop in larger towns, bypassing crossroads and village stores. This was not, however, a uniform or dramatic change.[27] For larger communities such as the three we studied, the effects of the automobile may have been yet weaker. One reason is that town businesses were already in competition with larger centers due to rail and streetcar connections. Also, population growth in the towns may have counterbalanced the increasing competition. Perhaps the undermining of commercial localism was limited to the smaller places—rural hamlets, Moline's Oregon, Illinois, and perhaps our Antioch.

Insofar as merchants' concerns about out-of-town competitors or the actual volume of those competitors' local advertising can be taken as evidence, we do not see a decline in localism between 1890 and 1940. If telephones or automobiles encouraged such a decline, then other concurrent changes must have counterbalanced those influences.

LOCAL SOCIAL LIFE

Did the residents of Palo Alto, San Rafael, and Antioch increasingly reach out beyond their town limits for social life? Editors of the *Palo Alto Times* thought so in 1921 when a story headlined "Disappearing Barriers" cited the formation of a regional drama club as evidence that diverse communities on the peninsula were uniting.[28] Commentators credited new technologies with breaking down geographical obstacles. One scholar wrote in 1933, for example, "[t]hrough the automobile and improved roads, rural social contacts have multiplied many fold, and are now based in increasing measure upon age, sex, and common interests rather than upon kinship and common residence, as was formerly the case." Ordinary people, too, such as residents studied in *Middletown* and *Plainville,* believed that their neighbors were forsaking local activities and ties for distant ones. Similarly, according to Norman Moline, citizens of Oregon, Illinois, felt in the 1920s that too many young people were driving out of town for their fun.[29]

We cannot easily measure the spatial range of people's activities and their social relations decades ago, but we can consider a few indirect indicators. The activities of formal organizations, for example, are relatively visible.

Most of the voluntary associations in Palo Alto were tied to higher, outside offices. The Civic League, formed in 1912 and later to become the League of Women Voters, was part of a statewide structure. Outside organizers stimulated the founding of the Palo Alto Women's Business and Professional Club in 1924. In turn, Palo Alto club members held positions at the statewide and national levels. These vertical links, however, do not appear to have been any greater in the later years than in the earlier ones. In 1901, for example, outsiders organized the Palo Alto Carpenters' Union, and as early as 1904 and 1913 the regional Building and Trades Council called strikes in town.

Local clubs often held social events in conjunction with branches elsewhere. This practice, too, dates from before the turn of the century. The Palo Alto Native Sons of the Golden West, for example, went to Redwood City for a Grand Parlor meeting in 1897. On the Fourth of July, 1904, they entertained Native Sons from out-of-town. In 1908 they sent a delegation to a meeting in distant Lodi. They played baseball against the San Francisco chapter in 1920. And so on. In 1910 the women's Order of the Eastern Star entertained 50 guests who arrived in three special train cars. In 1914, 450 out-of-town visitors came to the Rebecca installation of officers, by both train and automobile. And so forth. These extralocal activities did not seem to increase after World War I and may have even declined.[30]

Palo Altonians also took outings to nature spots throughout the period. In 1894, for example, over two dozen young people celebrated the Fourth of July in the mountains, ending with a stop in Redwood City to see the fireworks. In 1903 the Southern Pacific Railroad offered special rates for excursions to coastal resorts. The manner in which outings probably did change over the years, thanks to the automobile, was an increase in the frequency of private vacations at distant sites. In 1919 the *Times* noted that more Palo Altonians had registered at Yosemite Valley's Camp Curry than ever before. A few of our interviewees commented on the excitement of driving to Yosemite in the late 1910s and the 1920s.

Some evidence suggests that as the years passed members of Palo Alto's ethnic minorities increasingly spent time with their kindred outside town. Associations of Italian-, Japanese-, Filipino-, Portuguese-, and African-Americans appear more often in the city directories and in the press after the mid-1910s. News reports indicate that the Palo Alto organizations were the centers on the peninsula for these groups (this despite many other Palo Altonians' efforts to exclude minorities).

As early as 1905, the Japanese Association brought a crowd of nearly 200 from 10 miles around to Palo Alto to celebrate the accession of the emperor. The spatial expansion of ethnic social life over the 50 years might have been noteworthy but was still modest.

In San Rafael, organizational life seemed to undergo change in two ways. On the one hand, more county-level associations formed, focused, for example, around sports or ethnicity and headquartered in San Rafael, the county seat—precisely the sort of shift to association based on common interest rather than common residence asserted above. On the other hand, many suburban members of San Rafael organizations split off to form their own branches. In 1903, for example, Masons from Mill Valley founded their own unit. In 1926 the Marin Golf and Country Club lost many of its members to a new club in suburban Fairfax. Because of such fracturing, San Rafael organizations became more *local* both in their enrollments and activities.[31]

Club members in Antioch, like those in Palo Alto, regularly mingled with colleagues in neighboring towns. Yet, perhaps because of Antioch's small size, it is the extent of the regional ties maintained by its ethnic minorities that seems notable. The few Jewish merchants of Antioch spent much time in San Francisco. Merchant Leopold Meyer's widow moved to "The City," for example, and the Jacobs sent their daughter to high school there, where she subsequently married a Guggenheim. Asian farmworkers traveled to Stockton, Oakland, or San Francisco for leisure. Portuguese organizations were active both in town and regionally. Manuel Azevedo, Antioch barber and assistant fire chief, became a statewide president of one. Local Italians affiliated, as did Antiochians of other national backgrounds, with regional ethnic associations. Holy Rosary Church, founded in 1866, served Catholics over a wide span of the region.[32]

In sum, news accounts do not show any serious shift in organizational contacts with outside groups or toward outside activities between 1900 and 1940. Perhaps minorities were better able to gather with their compatriots as time went on – either due to easier communications or increasing numbers. And more regional organizations may have formed over time. In all, however, these accounts do not confirm a major decline in organizational localism. Perhaps the volume of both parochial and regional activities expanded.

We also examined social activities numerically, using newspaper reports of leisure and recreational events. (The details of the data and the analysis are reported in Appendix H.) We* recorded the location where each of nearly 10,000 events took place: in town, in an adjoining or nearby town, or further away. We defined these events broadly to include activities such as dances, theatrical programs, card parties, service club meetings, athletic contests, and so on. Figure 16 displays the findings for all leisure events combined. (Some disaggregated counts are discussed in the next paragraph.) Reports of in-town events predominated throughout; reports of both in-town and nearby events grew over the years; and reports of events further away grew more slowly or shrank. In all three towns, reports of "far" events at first outnumbered those in the "vicinity," but were fewer by the end. Proportionately, however, in-town events generally declined from about 70 percent to about 60 percent of all events during the period.

Figure 17 distinguishes two specific recreations, organized sports and paid performances (theater, movies, and shows charging admission fees), contrasting in-town versus out-of-town locations. The Antioch and Palo Alto newspapers originally reported more organized sports events occuring in-town than out-of-town, but reported more out-of-town sports by the 1920s. (The San Rafael pattern is confusing but not contradictory.) Accounts of in-town paid performances generally outnumbered those from out-of-town, and the gap fluctuated over the years. We also counted reports of participatory performances (shows presented by musical clubs, pageants by women's groups, school shows, and so on). They, too, displayed a slight increase in the proportion held out-of-town.[33]

The number of meetings by clubs (not shown) reported from both in-town and out-of-town grew roughly apace. There was therefore no net variation in the proportion of such meetings held within city limits. This finding reinforces the impression from the newspaper accounts, that the localism of club life changed little between 1890 and 1940. Finally, both in-town and out-of-town socials (dances, teas, etc.; not shown) increased over the years, with the balance changing from 83 percent local in the 1910s to 70 percent in the 1930s. (See Appendix H.)

When Norman Moline traced club meetings and recreation in Oregon, Illinois, between 1900 and 1930, he noted the widening territorial range of these events, which he attributed to the automobile.[34] Our evidence suggests that even though a similar territorial expansion

*Specifically, Barry Goetz.

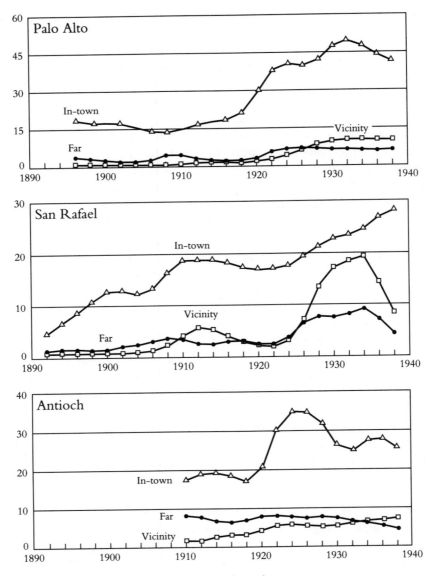

Note: The data are smoothed, with the endpoints dropped.

FIGURE 16. NUMBER OF LEISURE EVENTS PER NEWSPAPER ISSUE, BY LOCATION, 1890–1940. Throughout the period, the local press reported far more leisure events that occurred in town than either in adjacent communities ("vicinity") or farther away. Over time, the number of local events listed roughly doubled in each town, but the number of nearby events expanded at the fastest rate. Proportionately, then, in-town leisure declined slightly in favor of activities in neighboring towns.

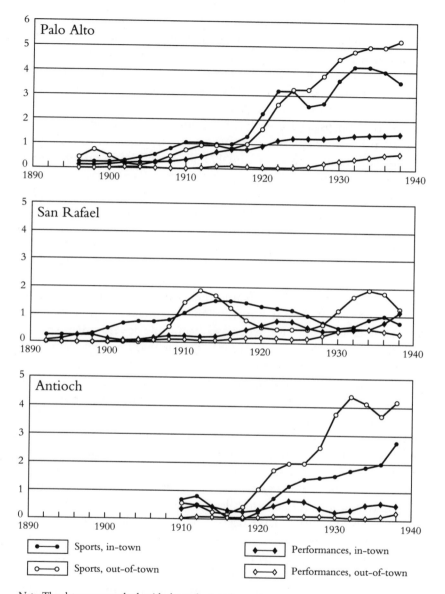

Note: The data are smoothed, with the endpoints dropped.

FIGURE 17. NUMBER OF SPORTS AND PERFORMANCE EVENTS PER NEWSPAPER ISSUE, IN-TOWN VERSUS OUT-OF-TOWN, 1890–1940. Organized sport was the leisure activity that most dramatically adopted extralocal venues. Reported out-of-town games and meets came to outnumber in-town ones in both Palo Alto and Antioch. Paid performances (e.g., movies and shows) were typical of leisure activities more generally: Local events outnumbered extralocal ones throughout, although the difference narrowed slightly over three or four decades.

occurred in our towns, it was accompanied by a great increase in local activities. The only major proportional change we found was among organized sports events (74 percent of the reported matches were in-town in the 1900s, 35 percent in the 1930s). That dramatic drop perhaps reflected the growth of away games for local amateur and school teams. (See Photo 25.) Generally, however, the volume of in-town events grew almost apace with that of out-of-town events.

What role did telephone and automobile use have in fostering the patterns shown in Figures 16 and 17? The standard argument previously discussed attributes the dispersal of social activities to the new technologies. Others have suggested, as did Willey and Rice, that these technologies intensify local activities more than extralocal ones. Evidence in support of the Willey and Rice position includes the fact that residential telephoning, even today, is largely local.[35] In our data, it is difficult to disentangle the expansion of telephony from that of automobility, since both technologies spread almost steadily in the half-century, and to separate both from other historical changes. There is a hint, merely a hint, in our statistical analysis that telephone development spurred local activity, whereas automobile development spurred extralocal activity (see Appendix H). The primary conclusion is that, over this period, for whatever combination of technological or other reasons, both in-town and out-of-town (but not distant) events grew in number, with out-of-town events growing somewhat more.

To widen our view of social life yet further, it would be ideal to have some assessment of residents' personal relations or *networks*. Network researchers (including myself) have claimed—with almost no historical evidence—that the geographical span of personal networks has widened over the last few generations, largely because of technologies like the telephone and automobile. In the words of Barry Wellman, "personal communities" have been "liberated" from the locality.[36] We do not have, in this study, evidence on people's friends and relatives, but we can analyze perhaps the ultimate social tie, marriage. To the extent that young people's social activities spilled across town lines, we might expect there to have been an increase in the proportion of marriages formed across those lines.[37]

Norman Moline calculated the percentage of weddings involving couples from different towns in Ogle County, Illinois, around 1900 and again around 1930. To his surprise, he found little change during those 30 years of automobile diffusion.[38] Taking his lead, we counted weddings that took place in our three towns and calculated the percentage of those in which both bride and groom were townsfolk. (See Appendix H.) Figure 18 displays the results.

Looking first at Palo Alto, we find no substantial change in the proportion of marriages that joined two Palo Altonians, whether we count only city weddings or also include those on the Stanford campus. Antioch displays a largely flat trend until the late 1920s, and then Antiochians became less likely to marry one another. They turned, instead, to residents of nearby towns, particularly Pittsburg. None of the Antioch weddings in our sample before 1925 involved a resident of neighboring Pittsburg, but 14 of the 26 marriages with residents of nearby towns in 1925 and 1930 were with people from Pittsburg. These 14 were largely among brides and grooms of Italian origin. For example, Rosa Giovanni of Antioch married Joe Marese of Pittsburg, and Geraldine Faria of Pittsburg wed Fabo Bologni of Antioch. Thus, Antioch's somewhat high rate of in-town marriages appears to have been breached in the late 1920s specifically by ethnic in-marriage that crossed the city line.

San Rafael shows the clearest pattern of declining in-marriage. About 60 percent of marrying San Rafaelians in 1910 wed other San Rafaelians, but fewer than 30 percent did so in 1940. The shift appears to have been from in-town to specifically in-county marriages (in the vicinity, to use our earlier term), rather than to out-of-town marriages generally. Unions between San Rafaelians and San Franciscans, for instance, showed no regular trend up or down.

Thus, marriages in San Rafael suggest an increasing regionalization of social ties, Antioch weddings point to increasing links between Antioch and one other town perhaps peculiar to one ethnic group, and the Palo Alto data show no historical change. Together with Moline's negative results, the evidence shows a mixed bag of trends concerning marriage ties within and without the towns. These particular results also caution us about the variability among towns. Palo Alto's stable marriage patterns may have resulted from its character as an enclave of high status within its region; San Rafael's decline in endogamy may have resulted from its relative stagnation compared to the growth of its region; and Antioch's trend may have resulted from the specific patterns of ethnic residence. We can assume, at the least, that the delocalizing pressures of modernization were, even if real, not strong enough to overwhelm local particularity.

We have looked in this section for evidence that between 1890 and 1940 people increasingly spent their leisure time and formed social

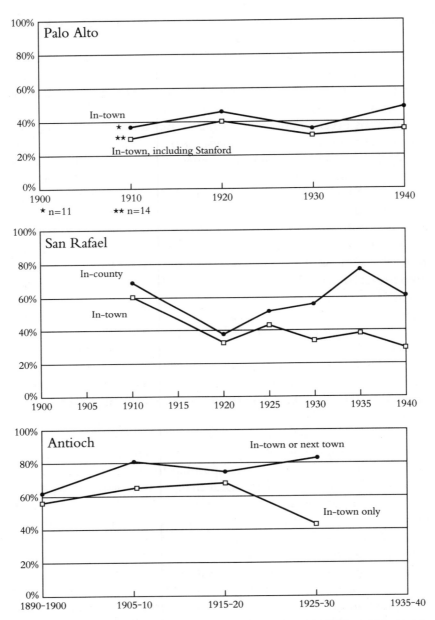

FIGURE 18. PERCENTAGE OF IN-TOWN MARRIAGES, 1900–1940. The chances that a Palo Altonian would marry someone from out-of-town varied little over three decades. Antiochians usually married other Antiochians, but in the late 1920s, they increasingly chose spouses from neighboring towns, especially Pittsburg. San Rafael residents experienced the clearest historical trend toward marrying people from out-of-town, typically from elsewhere in Marin County.

bonds outside their communities. Such a trend was likely experienced most by ethnic minorities (as seen in the expansion of regional ethnic associations and the Italian out-marriages in Antioch), and probably applied to certain leisure activities, notably organized sports and auto-touring.[39] More striking, in the face of rapid technological change, is the modest extent of any social delocalization. Although the trend, if such existed, was toward extra- rather than intra-local social life, the overall pattern seemed to be the expansion of both.

COMMUNITY INTEREST

The next aspect of localism that we will consider is subjective: the degree to which residents attend to their communities rather than to the world outside. This is obviously difficult to estimate for the past.[40] We will analyze three indirect indicators of residents' interest in their towns: what kinds of stories the local newspapers printed, which newspapers the local residents read, and the nature of community celebrations.

Newspaper Coverage and Subscription

The relative weight of local versus outside news in a town newspaper gives some insight into readers' interests — at least, as reflected through the judgement of the editors.[41] The Lynds reported that the proportion of space given to local affairs in the Middletown press declined noticeably between 1890 and 1923, a trend which they interpreted as a sign of increasing intrusion by the outside world, although the *volume* of space devoted to local news almost tripled.[42]

The particulars of the newspaper business explain many of the changes in the news coverage of our three towns. Before 1905, much of the layout of the *Antioch Ledger* was done in San Francisco, so the newspaper contained only fragments of community news. When C. F. McDaniel took over the *Ledger* and its editorship in 1905, he switched its allegiance to the Republican party, moved production to Antioch, and expanded local coverage. Five years later, McDaniel announced that he would triple local news, devoting four pages of the weekly to it.[43] McDaniel was active in local affairs and reported many events in his neighbors' lives.[44] In 1921 a new editor who was less connected to the local elite took over, and the volume of club news and social tidbits seemed to drop. As the *Ledger* became a triweekly in 1929 and a daily in 1937, new editors professionalized it and devoted more space to syndicated and wire-service materials. The shift to daily publication in 1937 may have created a larger "news hole" that editors filled with national features. Also, local news was now mixed in with other news, rather than appearing on separate pages as it had before. It is therefore likely—although we did not estimate column space—that, as in Middletown, coverage of Antioch itself declined proportionately but not in total volume. (The number of items reporting local entertainment, for example, increased over the years; see Figure 16.) Throughout the years, the *Ledger* asked readers to report local social news, as, for example in 1931: "We desire to be a home newspaper in every sense. . . . All items printed without charge. Phone 246–W."[45]

The *Times,* the only survivor of five efforts to start newspapers in Palo Alto, had three different owners between 1895 and 1900. From 1908 to 1918, it had two coeditors, one Republican and one Democrat, both active in local politics. Several reorganizations around World War I expanded out-of-town coverage in the *Times.* New owners appointed an editor who was less familiar with Palo Alto than his predecessors had been. During the war, the newspaper began featuring wire-service reports of national and international events on the front page. In 1923 and 1936, the *Times* Company purchased other peninsula papers, from which it obtained more regional news. Back in 1895, the *Times* had stated that it would cover local news and leave national reports to the metropolitan dailies, but in the 1920s it claimed to provide "two newspapers," one local and one cosmopolitan, for the price of one.[46] As with the *Ledger,* national news increased in the *Times* over the years.

The two San Rafael newspapers had a more complex history. In the early 1890s the weekly *Marin Journal* featured national stories on the front page, but by 1900 its focus had shifted to San Rafael. The *Journal* stressed local and county news into the 1930s, often devoting much space to social events. The *San Rafael Independent,* which became a daily in 1927, did the opposite. It increasingly pushed local news into speciality columns and the back pages in favor of national and international stories. Ultimately, the two newspapers specialized, the *Daily Independent* presenting national reports and the *Journal* functioning as the chronicle of local activities, political ones in particular. The *Independent,* however, did report many social activities, especially on its woman's page, and in 1937 a new co-owner expanded the *Independent*'s local coverage.[47]

The turnover of owners and editors, who typically came from elsewhere, combined with the evolution of wire services largely account for changes in news coverage. A prudent conclusion about general trends would be similar to that of the Lynds: More outside news appeared in the local press as the years passed, occupying proportionately more space and appearing more often on the front page. Yet, the volume of local news also increased, though it was more often segregated in special pages.[48]

To the extent that town residents turn their attention to the wider world, they might decide to read the big-city newspapers rather than their local press. Thus, subscription patterns are another possible indicator of localism. Unfortunately, we could obtain only partial circulation data on the three towns' newspapers, but we did find estimates of how many San Francisco newspapers were sold in the three communities (see Appendix H). Figure 19 shows the per capita circulations of San Francisco's major Sunday, morning, and evening newspapers for each town. In Antioch, per capita circulation dropped in all three categories (perhaps due to the increasing proportion of foreign-born residents and laborers in Antioch). In Palo Alto, there was little clear change. In San Rafael, subscriptions to the Sunday editions increased moderately, whereas subscriptions to the dailies remained flat. For each town, we also have fragmentary estimates of local newspapers' paid circulations. Little weight should be put on the specific trends, but the estimates suggest that the locals at least held their own.

The newspapers provide only rough indications of the subjective localism of residents. But the subscription statistics suggest that the metropolitan press did not in any simple way, if at all, displace the local press. Given the momentous nature of much of the outside news during this era—particularly World War I and the Depression—it is rather surprising that there was not a greater move toward external reports and extralocal newspapers.[49]

Celebrations

Historians have often measured community interest by the activities and symbols surrounding local celebrations. For example, Stuart Blumin found an increasing focus on the local community in the changing texts of a New York town's Fourth of July speeches from the nineteenth century. Roy Rosenzweig concluded that the emergence of class schisms in Worcester, Massachusetts, during the late nineteenth century was reflected in how residents spent Independence Day. Norman Moline suggested that the townsfolk he studied

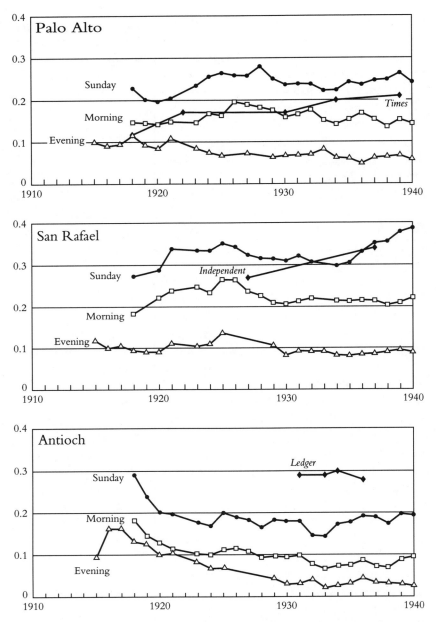

FIGURE 19. PER CAPITA SUBSCRIPTIONS TO SAN FRANCISCO AND LOCAL NEWSPAPERS, 1910–1940. Residents of Palo Alto and San Rafael were about as likely to subscribe to San Francisco newpapers in the 1930s as in the 1910s, but were more likely to subscribe to the local press. Antioch residents became less willing to receive San Francisco newspapers as the years passed. These trends suggest that there was no shift of attention from local to cosmopolitan sources during the period.

increasingly abandoned the town on the Fourth as automobiles became more common.[50] We also looked at ceremonies and celebrations, particularly the Fourth of July, for signs of a changing localism.

One national history of Fourth of July celebrations describes its evolution in the late nineteenth century from a day of patriotic solemnity to a day of outings and amusements. For several reasons, the Fourth reemerged in the 1910s as an occasion for patriotic parades, concerts, and other organized events.[51] The history of the Fourth in our three towns, however, does not fit this chronology.

Around the turn of the century, Antioch alternated with other county towns in hosting Fourth of July activities. Typically, volunteer firemen, women's clubs, and fraternal orders organized the festivities with monies provided by local merchants. (See Photo 24.) But over the years Antioch celebrated the Fourth less and less often. Antiochians held parades and other events in 1903, 1904, 1906, and 1908, for example, but in the 1920s and 1930s the Ledger referred less to local ceremonies and more to "quiet" holidays. One of our Antioch interviewees explicitly blamed the automobile, saying that after the car, "they began moving celebrations and festivals out of town." In 1931 the Ledger printed: "July Fourth will be a quiet one in Antioch as fully two-thirds of the population will be going elsewhere to spend the holiday, since it comes at the end of the week." In 1932 would-be sponsors of a celebration abandoned the effort when they could not raise $1000. In 1936 the Ledger reported that "Antioch will take on the appearance of a ghost town tomorrow" because hundreds of residents planned to spend the holiday at mountain or seaside resorts.[52]

Similarly, Palo Altonians participated in a round of July Fourth celebrations with neighboring towns until about 1910. In 1895, 1901, and 1904, Palo Alto hosted major festivities with as many as 5000 visitors, some coming on specially chartered trains. However, as the years passed, the Times reported fewer Fourth of July activities, more often stating that residents were vacationing elsewhere. In 1909 the paper reported that there was "no attempt at any public recognition of the day" and that 2000 residents took trains to other places. Independence Day 1913 was much the same, except that a letter to the editor declared that Palo Alto lacked spirit and urged that residents stay home on the Fourth instead of going to the mountains or the shore. Celebrations were later revived briefly, as evidenced by a carnival with thousands of people in 1930 and parades and shows in 1931. Yet, in 1932 the Times pronounced the old patriotic spirit to be on the wane. Early, then, the Fourth of July became a private affair in Palo Alto.

Memorial and Armistice Day ceremonies occurred but were largely the concern of veterans and related groups.[53]

Palo Alto's clubs helped run other celebrations that persisted longer than did those for the Fourth of July. In 1915 townsfolk held the first "May Festival." That event largely developed into a children's celebration, the centerpiece being a pet parade, with occasional adult revelry such as dances. The annual Christmas festival had a similar history. Merchants initiated it in 1922 as "the greatest thing in an advertising way that has ever been done here,"[54] but lost much of their enthusiasm by 1924. (They had similarly sponsored an Autumn Carnival in 1907 and then dropped it.) Clubs, schools, and civic groups, under the leadership of the Chamber of Commerce, subsequently became active in the Christmas festivities, and its focus turned to children and Santa Claus—who in 1928 arrived by airplane—although as late as 1935 the *Times* described the event's motivation as "essentially a business one."[55] Unlike the Fourth, these events began and to some extent continued as commercial endeavors. They did, however, seem to rouse the community. Indeed, their increasing focus on children may have been more central to residents' sense of community than was the patriotism of the Fourth.

In San Rafael, excursions to the countryside had become a typical way to celebrate the Fourth dating from the turn of the century. Whereas the neighboring town of Fairfax often hosted countywide ceremonies, San Rafael, according to the press reports, was quiet or abandoned on the Fourth. In 1915 the *Journal* reported that the Chamber of Commerce decided not to have a celebration—because neighboring San Anselmo did—reasoning that doing nothing would be better than "a poor attempt at half a celebration."[56] In 1923, some clubs tried to start a "revival" of the Fourth, but the effort did not take hold.

San Rafael had been notorious in the nineteenth century for "San Rafael Day," a rowdy, Mardi Gras-like festival whirling around bullfighting, drinking, and gambling. It shut down in 1870, but calmer carnivals in or around San Rafael appeared over the years, centered around parades, contests, and sports. After a hiatus of almost 20 years, the Native Daughters of the Golden West sponsored a modest "commemorative" of San Rafael Day in 1934, which continued for a few years afterward. Many town and county celebrations were promotional events. The Grape Festival, for example, was first organized in 1909 as a fund-raiser for the Presbyterian Orphanage. Some events became countywide, such as the Flower Pageant, which originally appeared in San Rafael in 1911 and reemerged in the late 1920s in Fairfax.

Summarized roughly, formal celebrations of the Fourth of July in our three towns waned in favor of private outings. Even before the turn of the century, the towns did not regularly hold local ceremonies, instead alternating with other communities in the region. (To what extent these events were ever celebrations of the locality is unclear.) The change seems to have been from public events on the Fourth to private ones. Other celebrations tended to become more specialized, less like community-wide affairs.

Although some scholars attribute these changes to the automobile, the timing does not fit well. Reports of towns deserted on the Fourth occur before World War I, when only elites had automobiles, and some reports even date from the turn of the century. Also, train excursions for holidays away from town were popular before automobile trips. (The same was true in Moline's Oregon, Illinois.[57]) Perhaps the automobile is responsible for the decline in celebrations—the elite drove off and left no one behind who could organize the Fourth's events—but the timing may simply be coincidental. In any case, the change seems to have involved more of a shift from public to private life than from a local to an extralocal life.

POLITICAL LOCALISM

The final aspect of localism we consider is political: the extent to which residents focused their attention on local or extralocal politics. A redirection of interest from the one to the other may result either from the wider view of the world provided by new technologies or from changing events in community and national politics. First, we will consider the relations between the local governments and outside governments. Second, we will briefly examine voting patterns.

In the early twentieth century there were conflicting trends in local political autonomy. State governments granted municipalities greater freedom of action and financial independence,[58] but also arrogated some decisions that they had previously left to the localities. We saw in Chapter 2, for example, that regulation of utilities became a state rather than a local responsibility. The greatest change occurred during the Depression, when it became apparent that localities, long custodians of the poor, could not deal with the crisis, when the "old ideal of voluntary charity and local responsibility . . . [was] shattered beyond repair. . . . " Between 1932 and 1936, federal contributions to local revenues increased over twentyfold. Moreover, despite resistance, the New Deal government attached strings to the money. The federal authorities gave up much of that control in the mid-1930s, but local independence never returned to previous levels.[59]

Palo Alto's experience in helping the needy is probably archetypal; few American communities were as active. In the nineteenth century, residents contributed to funds for those stricken by misfortune, as well as petitioned the city or county government for ill or destitute individuals. By 1930, a local Red Cross and a Benevolent Society provided more organized assistance to residents. Dealing with transients was more difficult, however. Palo Altonians, like other Americans, viewed them as scroungers at best, criminals at worst. In the winter of 1894–95, for example, the Women's Club and local businessmen supported a "Friendly Woodyard" that employed and helped hoboes for up to three days before sending them on their way. The Great Depression strained this largely voluntary system of clubs and charities to its breaking point. The mayor's 1931 appeal for $10,000 in contributions yielded $4000. The city established a shelter for transients in 1930, but the center lost community support, and in 1934 the city changed it to a community kitchen with a new sign: "Only PA men need apply." Eventually, federal programs largely replaced local ones. Starting in 1934, city and voluntary contributions for relief dropped sharply. The city council abolished its emergency committee in 1935 explicitly because the federal government had taken over the problem of unemployment.[60]

San Rafael's encounter with the Depression was similar. It, too, established a Mayor's Committee on Unemployment Relief, drawing funds from voluntary contributions, and employing men at $4 per day. In addition, associations ran independent projects such as the Lions Club's community kitchen. But the resources soon ran out. In late 1931 the head of the Red Cross blamed transients for the excessive strain and announced that jobs would from now on be reserved for Marin County citizens. However, outside support arrived shortly thereafter. By mid-1933, state-administered federal funds had absorbed 60 percent of the county's caseload.[61]

Antiochians historically had been less able to aid the needy, referring many to the county seat of Martinez for help. In 1919 the *Ledger* expressed a view that was probably representative: "There is work practically all the time for those who desire employment and it is not right that industrious people should feed these leeches." The Depression hit Antioch particularly hard. It came coupled with a drought and a drop in demand for the area's luxury vegetables. Voluntary efforts were not enough. The American Legion closed its shelter after one year of operation when it lost its merchant donors. The local Red Cross required aid from the national organization and in 1934 restricted its own largess to Antiochians. Federal funds were critical to the town's eventual recovery. Even the *Ledger,* which had supported Hoover in 1932, welcomed the New Deal programs in 1933.[62]

These accounts describe a sea change in political localism in the United States, the substantial movement of financial responsibility and authority for the needy from the local to the state and federal levels in the New Deal era. It certainly represented a decline in local autonomy. (Ironically, favorable sentiment toward the locality did not wane, if the preferences given to neighbors over transients by these towns is an indication.) For our purposes, however, the story has further implications. This social change had little to do with technological change. Even the problem of transients predated mass telephony and automobile ownership. The transformation had its roots in economic, governmental, and perhaps cultural, but not technological, modernization.

The politics of liquor have a somewhat different story. The fight over temperance and prohibition was probably the most common, lasting, and contentious battle in local American politics until the mid-twentieth century. In Antioch, beleaguered reformers repeatedly lost campaigns to curtail the saloon culture. Controversy raged for years, but a coalition of saloonkeepers, politicians (often the same), and Catholic "wets" easily frustrated even modest controls. San Rafael, in a region strong with vintners and with immigrants whose culture supported drinking, was also inhospitable to prohibition. Leading citizens—usually wealthy, Protestant, and concerned about residential real estate—nevertheless managed over the years to force some compromises, turning rough saloons into more respectable restaurants. Palo Alto was unique. Founded as a dry town, its citizens repeatedly supported prohibition, and the "wets" largely kept out of sight.[63]

When Prohibition came in 1920, temperance forces asserted national authority over localities. But local communities sometimes subverted this authority through their power to interpret its application. In Palo Alto, which had always been dry, officials and vigilantes alike enforced the laws. In Antioch and San Rafael, however, local officials and citizens scoffed at Prohibition, leaving sporadic enforcement to outsiders. Eventually, with the repeal of the Nineteenth

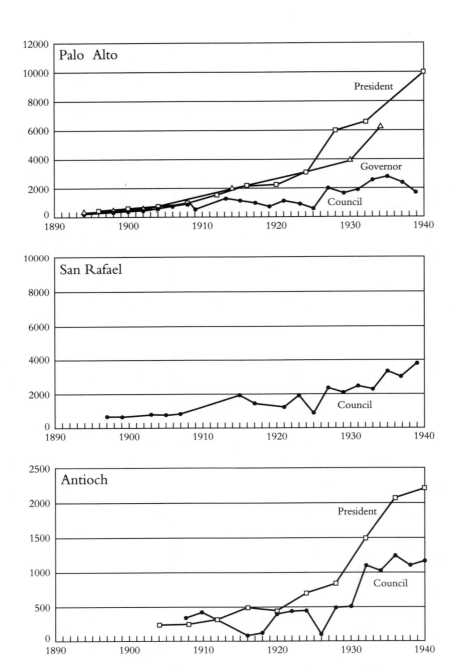

FIGURE 20. VOTER TURNOUTS FOR LOCAL AND GENERAL ELEC-
TIONS, 1900–1940. In Palo Alto and Antioch, voter turnouts for presidential
and gubernatorial elections increased much more rapidly than did voting for local
elections. (We have no similar data for San Rafael.) These findings imply that,
from the turn of the century to World War II, interest in wider-scale politics
expanded faster than did interest in local issues.

Amendment and new state legislation, much authority over liquor—though not all—reverted to the locality.

The cases of both welfare and liquor point to ratchet-like increases in the power of the federal government, but they also indicate the nonlinear pattern of such changes. Particular events led to dramatic expansions in the range of activities controlled by the national government. These were followed by a return of authority to the local level, but in neither case did communities regain their initial degree of autonomy. Both occurences resulted in serious changes, yet neither was attributable to technological change.

The changes in welfare and liquor policies indicate the nature of political localism in whole communities. What of individual citizens' interest in the local polity? It is well known that Americans today turn out to vote in far fewer numbers for local elections than they do for statewide or presidential races.[64] Has that differential remained constant over time, or did it change between 1900 and 1940?

We can compare the turnouts in our towns for local elections to those for federal or state races. Estimates are presented in Figure 20. (See Appendix H for details. We have relatively complete data on the turnouts for council races; the data on the other election turnouts are incomplete—largely missing for San Rafael.) The results for Antioch and Palo Alto are similar and clear: Over the years, the number of voters in general elections grew much more rapidly than did the numbers turning out for city elections. Looked at another way, general election turnout increased in proportion to population, but local election turnout per capita stayed the same. (See Appendix H. The pattern for San Rafael city elections, however, shows a proportional as well as an absolute increase in local election voting.) These findings support the contention that as time passed, voters' interest in higher levels of government grew faster than did their interest in local politics.

At both the individual and community level, it is therefore fair to say that politically, localism gave way to some form of statewide or national orientation as the twentieth century passed. We were unable, however, to demonstrate a link between these changes and the diffusion of the space-transcending technologies.

CONCLUSION

Let us bring together the different strands of evidence regarding localism in our three towns between 1890 and 1940. Commercial activity seems not to have become noticeably more or less localized in this period (excepting, perhaps, the failure of a few Antioch merchants in the 1920s.) Out-of-town shopping and advertising were common even before mass telephony and automobility. In the realm of social life, no clear geographical trend appeared in the activities of the voluntary associations. The number of recreational events held both in and outside of the towns increased, the latter at a somewhat greater pace. Organized sports were the only public events that showed a significant shift toward out-of-town locations. There are hints that minorities widened their geographical span of activity over time. It is also likely, based on anecdotal evidence, that townsfolk, in general, expanded their private leisure activities beyond the town limits. Although people had taken trains for such trips, the automobile probably made outings to the country and other places of entertainment considerably more common. Whether car travel, as well as telephoning, widened personal social networks is more difficult to tell. Our strongest evidence is that marriages spanned San Rafael town limits more often as the years passed. In Antioch, however, change was limited to more marriages between Antiochians and residents of neighboring Pittsburg. In Palo Alto, there was no tendency for residents to wed across town lines more often than before the widespread diffusion of the telephone or the automobile.

Turning to more subjective indications of localism, we found that the local press expanded its coverage of national and international news over time, but also covered more local news. The balance probably shifted toward the out-of-town reports, but not dramatically so. There was no sign that residents began preferring metropolitan newspapers over their local press. Local celebrations apparently waned, but not necessarily in favor of extralocal ones. Rather, private activities seemed to replace public ones. That development apparently preceded mass telephoning and driving, but may have been accelerated by both technologies. State and national politics increasingly subsumed town politics, largely due to significant economic and political events. (We looked directly at the Depression and Prohibition but could also have included World War I as a precipitant of political extralocalism.) Similarly, citizens' interest in national politics, as indicated by voter turnout, expanded more than did their interest in local campaigns.

In sum, we found but a few modest changes in localism. The net trend was in the direction of greater attention to the outside world. Yet, rather than indicating a *displacement* of local interest, these changes

suggest a simultaneous *augmentation* of local and extralocal activities.*
Given how radically people's access to the wider world expanded between 1890 and 1940, the modesty of increased involvement in extralocal events is remarkable. Moreover, the source of some changes
we noted—news coverage and political autonomy, for example—lies
more directly in political and economic events than, if at all, in new
technological devices. It seems that the residents of Antioch, Palo Alto,
and San Rafael made use of the new technologies to supplement their
activities—with car trips to Yosemite, for example—but did not abandon local ones (excepting, perhaps, town celebrations). In this sense,
they used the devices for varied ends in complex fashions; they were
not uprooted by the new technologies. People located their activities
and interests somewhat more often outside the towns, but mostly they
expanded their total activities. Perhaps, the characterization of a move
toward privatism rather than extralocalism best describes the bulk of
the changes.

The Cooley argument cited at the top of the chapter, that wider
contacts were expanding, seems valid, but the contention that these
wider networks diminished people's "spiritual community" within
the locality receives much less support, at least from northern California in the early twentieth century. The Willey and Rice speculation
regarding the "intensification" of local cultures also has some validity, although outside contacts may have intensified to a greater degree.
One is struck with the multitudinous ways in which people used these
devices, sometimes to innovate new lifestyles, more often to sustain
older ones, with a net result of only subtle movements toward modernity, at least in the domain of localism.

If the complex picture drawn in this chapter is faithful to reality,
then neither the telephone nor automobile can be substantially credited or blamed for undermining localism in the early twentieth century.
Certainly, both enabled Americans to participate in activities more
frequently and more easily outside their localities. And so they did.
People called kin long-distance, took more trips to tourist spots, followed their sports teams to more games, and so on. Yet, Americans
also seem to have augmented, if not by quite as much, their local activity using both technologies. Although the net balance of change was
in the direction of the wider world, it was not a weighty shift—not as
substantial as the increase in total social activity.

*For example, local voter turnout remained, per capita, about the same, although
national turnout increased. Total leisure events increased, too.

CHAPTER EIGHT

Personal Calls, Personal Meanings

An elderly Antioch man recalled to us that, as a youngster before World War I, he sometimes rode a horse to a wealthy neighbor's home. "One day at lunch the phone rang. It was a hand magneto phone that they had just installed. Mr. Henry answered it and he reveled at it. 'I was talking to Concord! Just like I'm talking to the guy in my own living room!' he said. He had to scream into the thing, though. . . . It didn't impress *me* too much. I always took innovations and advancements for granted. [But] it was a heck of a machine." From Mr. Henry's excitement to our interviewee's nonchalance, personal reactions to the telephone varied, as did its personal consequences.

This chapter assesses how Americans in the first half of the twentieth century used the telephone and what personal meanings it had for them. Speculations about those meanings have ranged widely and wildly. Industry flacks claimed that telephoning empowered people and preserved family life; critics charged that the telephone's ring jangled nerves and shred privacy; media guru Marshall McLuhan drew images of electronic transcendence; and a recent deconstructionist analysis found Freudian significance in the fact that ears, the receivers of calls, are orifices.[1]

How can one appraise even modest claims about the historical psychology of the telephone? How can we overhear the conversations of, and weigh the implications for, people long gone—implications even they themselves may not have fully appreciated? We have a few tactics at our disposal. One is to draw inferences from our findings in the previous chapters about early telephone users. Another is to consult studies of telephone users today. This second strategy must be employed with great caution, however, for the social psychology of the telephone has probably changed over time, as subscription has become a requirement instead of a luxury. Yet, lacking comparable evidence from the past, current research adds to our understanding. A third tactic is to draw on the recollections of now-elderly people. Although it draws on two published oral histories, this study depends largely on 35 interviews we conducted in the mid-1980s with residents of our three towns. We talked with men and women who ranged in birthdates from 1888 to 1917. Oral histories are not magic windows to the past, but they do provide otherwise unattainable information.[2]

For simplicity, I have divided discussion of the telephone's personal implications into two broad categories, social and psychological. Did having a telephone in some way alter social relations? Did it alter what the French historians call *mentalité?* These questions must be qualified by another: altered in comparison to what? Some writers imply that telephoning is properly compared to letter writing. Yet, by far, most calls were and still are immediately local. The more relevant comparison, in most instances, is between telephoning and face-to-face conversations. Another comparison will be, as through most of this study, with the automobile. A second qualifier to the social and psychological questions is to stipulate that we are looking at implications in America. There is a small but suggestive literature claiming national differences in how people use telephones—for example, that Greeks call more often for sociable chats than do the British, the French more often than Americans.[3]

THE TELEPHONE AND SOCIAL LIFE

AT&T advertisements claimed that telephones nurtured a "close-knit, personalized society" and "simultaneously provided a means of overcoming distance and complexity by reestablishing simple, immediate, person-to-person contacts." Marshall McLuhan implied much the same: "By electricity, we everywhere resume person-to-person relations as if on the smallest village scale." (The vocabulary of "reestablishing" and "resume" underlines a point made by Roland Marchand that some observers saw modern technology as a way to recapture an ideal past.)[4] Other enthusiasts linked the telephone to family life in particular. In an essay for an AT&T magazine, Margaret Mead extolled the capabilities of the telephone for binding together kin. Telephone songs in the late nineteenth century sentimentalized the telephone connection with ballads like "Kissing Papa Thro' the Telephone"; "Love by Telephone"; and "Hello, Is This Heaven? Is Grandpa There?" Many commentators—government commissioners, industry spokesmen, and authors of popular magazine articles (some no doubt ghosted by industry publicists)—praised the telephone as a way to alleviate rural isolation (see Chapter 4).[5]

Others, however, present less sanguine views. One concern has been that the telephone, by allowing people to substitute electronic communication for face-to-face encounters, sustains only a semblance of "real" relations. A story published in 1893 forecast America in 1993: Families would live on scattered homesteads, neighbored only by people of like "sentiment and quality," would conduct their work electronically, and would meet one another only on ceremonial occasions. (Today's futurists who predict a country of scattered "electronic cottages" are not so inventive, after all.) The problem many find with such telephonic neighborhoods is that they are "larger but shallower kinds of community." A sociologist of technology, Ron Westrum, recently claimed that the "coming of the telephone began the unraveling of social processes. . . . [P]eople became willing to accept physical separation as along as contact could still be maintained by telephone. But telephone contact is not the same as being there, and it creates a different kind of society. . . . " A related worry is that telephonic relations are inherently inauthentic and will, if they become customary, impair other interactions, too. Sociologist Peter Berger, for example, has claimed that

> [t]o use the phone habitually also means to learn a specific style of dealing with others—a style marked by impersonality, precision, and a certain superficial civility. The key question is this: Do these internal habits *carry over* into other areas of life, such as nontelephonic relations with other persons? The answer is almost certainly yes. The problem is: just how, and to what extent?[6]

A different concern is that the telephone may enable too much or the wrong kinds of social interaction. In 1899 an Englishman noted that the day when every household could call every other was to be feared "by the sane and sensible citizen." An American professor fulminated in 1929:

> We are largely at the mercy of our neighbors, who have facilities for getting at us unknown to the ancient Greeks or even our grandfathers. Thanks to the telephone, motor-car and such-like inventions, our neighbors have it in their power to turn our leisure into a series of interruptions, and the more leisure they have the more active do they become in destroying ours.

Malcolm Willey and Stuart Rice also concluded that "[p]ersonal isolation — inaccessibility to the demands of others for access to one's attention — is increasingly rare, and, when desired, increasingly difficult to achieve." The most notorious offenders were telephone salesmen. A contributor to *Readers' Digest* in 1937 complained that "there is not a room in the house so private that he cannot crash it by telephone."[7]

The wrong kinds of telephone sociability, for some, included chitchat or gossip, the "exchange of twaddle between foolish women." (More on women later in this chapter.) As discussed in Chapter 3, many people, both inside and outside the industry, considered "idle" conversations an inappropriate invasion of the household. Similarly, some worried that the telephone would permit indiscretions, especially between unsupervised women and strange men; would lead to inappropriate contact from people of the lower classes; or would simply allow any outsider free access to the family. People have also worried about the loss of privacy to listeners-in, be they others in the same room, party-line neighbors, nosy operators, or government officials.[8]

Was the new technology of the telephone, then, a means of building a wider and richer community? Or was it a seductive device that ultimately impoverished social life? Or both? The historical record cannot resolve all the nuances of these arguments. Still, we can assess whether, how, and with what apparent import people used the telephone for sociability. First, we ask, to what degree did people use telephones for conversation and for sustaining social relations?

Surveys done in the last three decades suggest that people today most often telephone from home for social or vaguely personal reasons rather than for practical matters. AT&T research shows that half of the calls from any given residence go to only five numbers, indicating that repeated conversations are held with a small circle of friends and family. In 1975 a fire in New York City knocked out thousands of household telephones for three weeks. In a subsequent survey, most respondents said that they largely missed calls to and from friends and relatives. In a 1985 poll, Californians estimated that about three-fourths of the local calls they made from home were for such purposes (against only one-eighth for household affairs). Of Americans interviewed in 1982 about their leisure activities, almost half talked on the telephone with friends or relatives virtually every day—fewer than the number that watched television or read a newspaper daily, but more than those that exercised, read books, shopped, drank alcohol, or had sex daily. In other nations today, even less-developed ones, most calls are made to friends or family.[9]

For our period, we have the eavesdropping research done in Seattle in 1909, first reported in Chapter 3. Of the total calls surveyed, 30 percent were "idle gossip," fifteen percent were invitations, and twenty percent were from home to office—presumably some of those were from wives to husbands. Roughly half, then, had some social content, at a time when only about one-third of Seattle households had telephones. Also, those calls averaged over seven minutes in duration (compared to about four minutes for calls today), again implying social conversation.[10]

Lacking any better statistical estimates of social calling for the early years, we must rely on the comments of contemporaries and the recollections of the elderly. The most dramatic and consistent testimony in the first few decades of the twentieth century indicated that rural people, especially farm women, depended heavily on the telephone for sociability, at least until they owned automobiles (see Chapter 4). These women used the telephone to break their isolation, organize community activities, keep up on news, help their children maintain friendships, and so on. Observers repeatedly claimed that telephoning sustained the social relations—and even the sanity—of women on scattered homesteads.[11] Industry men were among such observers. For example, the North Electric Company stated in 1905, "The evil and oppression of solitude on woman is eliminated." An officer of an Ohio telephone company wrote the same year:

> When we started...the farmers thought that they could get along without telephones.... Now you couldn't take them out. The women wouldn't let you even if the men would. Socially, they have been a god-

send. The women of the county keep in touch with each other, and with their social duties, which are largely in the nature of church work.[12]

Government inquiries bemoaned the isolation and boredom of rural existence for women but pointed to the telephone—and to the automobile—as a way to provide community life.[13]

How many calls and how much telephone time people devoted to social purposes is unclear, but the social function in rural areas—even if it was not the reason for initially subscribing—was readily apparent.[14] Similar testimony for urban residents is, however, sparse.

Most of the 35 northern Californians whom we interviewed recalled considerable use of the telephone for sociability. Two of them had managed switchboards as youngsters. One, a San Rafael man born in 1902, said that most of the calls he put through were personal rather than business. A San Rafael woman, born in 1903, who had run a country switchboard reported that people often "visited" back and forth. A Palo Alto woman, born in 1892, was the daughter of a doctor in San Francisco. Although she had been taught "not to hold the line" in case a medical emergency arose, she frequently telephoned her friends. She reminisced about walking home from school with a chum, separating at the street corner, and then calling the friend when she got home. Others, particularly from Palo Alto, reported using the telephone often as children. One man born in 1893 remembered that few of his friends had telephones in the first years of the new century, but that he regularly called those who did. Another, born in 1908, told us, "I used the phone for talking to my father in his office, calling the boys I played with, calling a girl to make a date." A woman born the next year said that it "seems we always had a telephone. Both my mother and I used it, mostly to talk to friends." A San Rafael woman born in 1907 recalled that her mother and father rarely used the telephone but that the children regularly talked to their friends. Others remembered using the telephone socially when they came of courting age or first set up their own households. These included one Palo Alto woman, born in 1900, whose father had forbidden a telephone in her childhood home. A San Rafael man born in 1908 recalled that as a bachelor he called to arrange dates and that after marrying, his wife called to shop and to talk with friends and family.

Others who remembered telephone "visiting" claimed it was someone else in the family who did it, usually a mother, sister, or wife. An Antioch woman born in 1906 said that her mother phoned regularly to talk to friends. "One lady, every morning at 9 a.m., they'd get in touch with what was going on." A San Rafael man born in 1909 said that his family got a telephone around 1915 and that his sister held conversations, although he used it only for business or making dates. An Antioch man born in 1911 recalled his family having a telephone by 1920. He didn't use it much, since he lived near his friends, and his father used it largely to call the office. His sister and mother, though, were active:

> Mother talked for hours sitting on the piano stool. . . . My father said, "Get off the phone and go visit them." I guess it was a great way to kill time. I didn't believe in long conversations on the phone, but everyone has their own ideas.

A minority of our interviewees described social calling as limited. A very few did not have telephones until well into adulthood. Others described being constrained from or uninterested in telephone chats. The daughter of a Palo Alto doctor, born in 1907, told us that nobody used the telephone much in her childhood home since they were supposed to keep it free for medical calls. A woman who grew up in the Antioch countryside recalled:

> I wasn't to touch it. "Don't you talk on that and don't you touch that!" I guess I didn't use it very much. Kids didn't talk to each other on the phone because there might be an important message coming in and you shouldn't tie up the line. We had a phone for my mother because she would call into the grocer's or the butcher's and tell them what she wanted and they would deliver it.

A few others described themselves, in the words of an Antioch woman born in 1911, as not being "a telephone person." Although her dairy-owning family and their friends had telephones, she did not call much. She visited friends "over the back fence" and didn't go many places, so she had little use for it.

Many, perhaps most, of those who reported not using the telephone much still associated it with sociability, as did the woman just quoted. Because she could visit friends she did not need to phone. A San Rafael woman born in 1902 recalled that her parents rarely used their telephone because they did not speak English. When she lived elsewhere, first as a college student and then as a teacher, she seldom called because she hardly knew anyone in her new towns. After marriage, she and her husband rarely called because they socialized little. A Palo Alto woman born in 1895 lived part of her childhood in San Jose. Although her father telephoned often for business purposes, she said, "I didn't have any good friends from San Jose, so I wasn't especially interested in the phone." In these sorts of recollections, telephone use correlates with friendship.

Many respondents contrasted telephone "visiting" in their youth to the flood of talk today. An Antioch woman born in 1903 said, "We certainly didn't depend on the phone the way we do now-a-days. . . . No, phones weren't used that much. You would go to see people and visit them in person." Social conversations for some only developed over the years. A San Rafael woman born in 1910 said that unlike people today, her family did not use the telephone much for socializing, although she did arrange dates by phone. She recounted that conversation was limited when few people had telephones. As more subscribed, social calling increased. It also increased as her siblings left home and her mother tried to stay in touch with them. A San Rafael man born in 1913 remembered that telephone conversations with friends were rare in the early 1930s, but as social life picked up, calling increased. After he got married it increased further, although not, he said, to the level of today's frequent and long conversations.

The several older women Lana Rakow interviewed in rural Wisconsin had similar recollections. A few said people did not telephone as much as nowadays, especially for chatting:

> We use our phone more now. It was a great convenience in the first years when I was too busy raising the family, gardening, canning, whatever there was. It was used for the purposes it was meant to be, to aid you in the things that came along in the day. I really don't care to go to the phone and just stand and visit. But you get some of that from the other end of the connection so I guess that's why I don't do it.

The early telephone, because of party lines and poor sound quality, was an instrument for practical purposes, not sociability. And so, Rakow's rural interviewees said, people did not miss it much when they lost service in the Depression. Also, the "women those days did not have time to spend on the telephone the way we do now," said one. "I'm sure [my mother] felt lonely at times, but the poor thing had so much work to do." Still, even those who downplayed use of the telephone admitted some sociability: "I didn't use the phone very much then," said a farm woman, "just to talk to the neighbors." Another recalled that "the women in those days did not have the time to telephone the way we do now. . . . But [mother] and Mrs. B— did keep track of each other and talk. . . . " A woman who had been a rural operator recalled the need to enforce a 10-minute limit on subscribers occupying party lines.[15]

The women Rakow interviewed commonly complained of notorious "talkers" who misused the telephone for long conversations "just to go over things that were going on in the country. They didn't have radio. They didn't have anything but the telephone or an outdated newspaper that came a day or two too late. . . . For those who liked to use the phone to keep up their relationships with people, it was very much an improvement in our way of living."[16] Rakow's informants looked upon the talkers of yesteryear and the gabby women of today as morally suspect and made it a point to assert that they themselves were different. Recall that telephone companies, at least until the late 1920s, taught that the instrument was only for the business of the household, not for "frivolous" chatter.

It is unclear why Rakow's interviewees seemingly recalled less sociability than did ours in Antioch, Palo Alto, and San Rafael. Perhaps differences in the region of the country, access to telephones, or interview method account for these results.[17] Still, the Wisconsin women and their families were using the telephone regularly for purposes beyond practical needs, even if not to the extent of the Californians.

Most of several elderly, rural women interviewed in an Indiana oral history project reported telephone sociability. Only a minority claimed to have restricted calls to those of practical necessity. Two referred to "enjoying" the telephone. One recounted (originally quoted in Chapter 3), "When the men came in and needed to use the phone, and somebody was talking, they would just ring the ring and say, 'I need the phone,' and the women would stop until they used the phone, and they would ring back again. They would have their chance at it again."[18]

In sum, people in the past obviously varied in their use of the telephone, as they do today, and perhaps more so. Some disdained it or saw it only as a device for "serious" matters. Others "enjoyed" the telephone and freely chatted on it, though these were probably fewer in number than at present. Most striking, however, people called often for social purposes, and frequently even for simply "visiting," as far back as the 1910s. "People" may not, however, be the most precise noun. The quotations cited here and in the earlier discussion of the industry's marketing (see Chapter 3) suggest that "women" ought to be the operative subject.

Gender and the Telephone*

A prefatory comment: This topic has a controversial side. From Mark Twain to current *New Yorker* cartoons, women on the telephone have been the butt of jokes. For many, and surely for leaders of the early telephone industry, chatting on the telephone represented "one more female foolishness." Social scientists know better. Conversation, even gossip, is an important social process, serving to sustain networks and build communities.[19] Those who would dismiss the topic as either trivial or sexist would be repeating the dismissive attitude of industry leaders, journalists, and other male critics who ignored the seriousness with which women themselves took and take conversation.

North Americans believe that women talk on the telephone more than men do. This stereotype happens to be correct. People in the industry associated women with the telephone. For turn-of-the-century experts, "[t]alkative women and their frivolous electrical conversations about inconsequential personal subjects were contrasted with the efficient, task-oriented, worldly talk of business and professional men."[20] Recent studies more reliably associate women with extensive use of the telephone. Research, largely by AT&T, shows that American women today are likelier to have telephones at home than are men, that the number of women or teenage girls in a household better predicts how often calls are made than does the number of males, and that women dial most residential long-distance calls. An Australian study showed that women made longer telephone calls than did men. A large French study found that women spent far more time on the telephone than did men, employment status notwithstanding. An English survey found that women called their kin and friends much more often than men called theirs. An Ontario survey of people 40 and older showed that women were two to three times more likely to telephone their friends than were men. A survey of the elderly in New York found that telephone conversations between an older person and his or her "helper" was more common if either was a woman. A survey of social networks in Toronto found a correlation between the frequency with which people called one another and the proportion of women in the network.[21] These diverse studies all confirm that women

---

*This section abbreviates but also updates C. Fischer, "Gender and the Residential Telephone." See the article for greater theoretical and empirical detail.

today are much heavier users of the residential telephone than are men.

Returning to the period of this study, time-budgets filled out by suburban New Yorkers before World War II revealed that women reported spending an amount of time on the telephone fourfold that of men.[22] The 1900 to 1936 household surveys we examined in Chapter 5 and the 1918 national living standards survey and the 1924 Dubuque farm family listings analyzed in Chapter 4 point to a similar conclusion: The higher the proportion of adult women in the household, the higher the chances were that it had a telephone.[23]

Lana Rakow concluded from her interviews in rural Wisconsin that "[b]oth women and men generally perceive the telephone to be part of women's domain."[24] Our interviewees, too, most often mentioned women together with the telephone. True, some men recalled talking to friends and making dates by telephone. One San Rafael man born in 1914 reported that his gregarious father often called church friends. Yet, recollections generally linked women to the telephone. Other bits of evidence even suggest that men were shy of the telephone, that wives commonly made calls for their husbands.[25]

What does all this calling by women mean? Ann Moyal recently interviewed over 200 Australian women about that subject. She concluded that women in all areas of the country "attach high importance to telephone conversation and its essential place in their personal affairs" and that the telephone was "a key site for kin-keeping, caring, friendship, support, voluntary activity and contact with the wider world."[26] To what extent does this description apply to the telephone's early years in North America?

When telephone salesmen began marketing the device for residential use (see Chapter 3), they promoted a particular vision of how women should use it. Into the 1920s, their dominant suggestion was that women, acting as "chief executive officers" of the household, should telephone to order goods and services. This approach was consistent with the image of the housewife-administrator that was emerging in both advertising and home economics.[27] The companies also stimulated such use from the other end by encouraging merchants to organize, invite, and advertise ordering by telephone.[28] Industry men considered the clichéd and lampooned woman's sociable telephone call—a conversation with friends or family—as a problem, and they initially tried to suppress it. By the late 1920s and the 1930s, however, telephone advertising increasingly depicted women using the telephone for sustaining social contacts and even for conversation.

For what purposes, in fact, did American women use the telephone between 1900 and 1940? No doubt, many used it as the industry men originally intended, for emergencies and for shopping. But some evidence suggests that most women only occasionally used it for shopping. Bell's own surveys in the 1930s imply that fewer than half of women even liked shopping by telephone.[29] None of the several contemporary advice books on household management that were consulted for this study favored using the telephone for shopping. One, by the popular Christine Fredrick, recommended against it: "The telephone habit encourages the lack of knowledge of conditions and prices. . . . "[30] Although several of our interviewees recalled that they or their mothers ordered groceries by telephone, most did not mention it.[*]

If women a few generations ago were developing an "affinity" for the telephone but not employing it much in household management, then for what purpose were they using it? Conversation with family and friends is the answer Canadian scholar Michèle Martin concluded from an examination of contemporary advertisements, newspaper columns, and industry reports. Despite the companies' efforts to direct their use of the telephone, women cultivated their own purposes — "delinquent activities" — primarily social visiting.[31]

More concrete evidence in favor of this assertion comes from an unusual time-use study of homemakers conducted in 1930. As part of a widespread, if unsystematic, survey of how women spent their time, government home economists asked "Seven Sisters" alumnae to fill out time-budgets. The forms encouraged women to record all their activities over a whole week.[32] I randomly sampled and examined the forms that 62 respondents had completed for a total of 250 days. The 250 forms listed only 83 telephone calls. Considering the great likelihood that almost all these high-status women had telephones, the low number implies that for most, telephoning was an unremarkable event, not worth recording. (Only one woman noted using a neighbor's telephone.) Of all the reported outgoing calls, 30 percent to 50 percent apparently involved orders for goods and services, and

---

[*]Ann Moyal ("Feminine Culture," 10) found that teleshopping was not popular among her 1988 sample of Australian women.

another 30 percent to 50 percent concerned personal or social matters. Of all noted calls, both incoming and outgoing, 25 percent to 40 percent were commercial and 30 percent to 50 percent were social. This is probably a conservative estimate of the frequency of social calling.[33]

We have already seen the testimonies that farm women used the telephone to sustain social activities and help create community bonds in rural areas. In urban areas, middle- and upper-class women likewise used the telephone for organizational activities, as did those in the Palo Alto Women's Business and Professional Club (noted in Chapter 6).[34] Young urban women used it for courting, as some of our interviewees noted. In the Indiana interviews, one woman remembered: "We were the only ones in the neighborhood with a telephone, and [the next-door neighbors] had several girls and would get a good many calls. I would just raise the window and holler, and they would. . . come on over." An etiquette column in 1930 warned "Patty" that to ensure that her boyfriend would "respect and admire her, she does not call him up during business or working hours. . . . [and at home] she should not hold him up to the ridicule of his family by holding an absurdly long telephone conversation." In 1934 the telephone company in Palo Alto had to add a switchboard to the Stanford Union because "with eighty women residing this year in the Union, the telephone congestion has been so great during the 'dating' hours around lunch and dinner that service has been slow."[35]

The evidence indicates that women called more often for sociability—making social arrangements and making conversation—than for other ends, especially after the telephone's earliest years and especially on private lines (i.e., not party lines). It confirms the old stereotype that women have a greater affinity for the telephone than men do, particularly with respect to social conversation.* Why the gender difference?

Three answers seem plausible. First, modern women have been more isolated from adult contact during the day than men, so they have grasped the telephone as a device for breaking that isolation.[36] Second, married women's duties have usually included the role of social manager—making appointments, preparing events, staying informed about kin and friends and keeping them informed about the family, and the like; men neglect those tasks. Indeed, by many ac-

---

*The French study cited earlier found that women's calls were far more often for "relational" purposes and men's for "functional" ends (Claisse and Rowe, "The Telephone in Question").

counts, a wife typically maintains the family's communications with the husband's kin as well as with her own. As Rakow puts it, "telephone talk is work women do to hold together the fabric of the community. . . . "[37] Third, North American women are more comfortable on the telephone than are North American men because they are generally more sociable than men. Research shows that, discounting their fewer opportunities for social contact, women are more socially adept and intimate than men, for whatever reasons—psychological constitution, social structure, childhood experiences, or cultural norms. The telephone therefore fits the typical female style of personal interaction more closely than it does the typical male style.[38] To underline the point, some evidence suggests that women's advantage over men in sociability is greater for telephone contacts than it is for face-to-face interaction.[39]

Some argue that women, by using the telephone to carry out their duties as families' social secretaries, have entrenched themselves all the more deeply in that time-consuming role.[40] There is little evidence, however, that such tasks occupied more—or less—of women's time because of the telephone. A few women told Lana Rakow that having a telephone made them vulnerable to requests for help—to advise, comfort, organize, and so on. The workload of the confidant may indeed have expanded. For the calling women, however, the enhancement of the ability to request aid may have been a boon. Using the telephone probably facilitated the social work that men and women alike, be it fair or not, expected women to do. We cannot calculate the net balance of comfort and burdens without more and comparative evidence in the form of data, for example, showing that in comparable places without telephones, women's social duties were more or less burdensome.

The most reasonable conclusion is that from the first decades of the twentieth century women used the telephone, and used it often, to pursue what *they,* rather than men, wanted: conversation. The testimonials and other evidence suggest that people experienced use of the telephone more often as a pleasure than as an ordeal, a feeling expressed by women more often than men.*Ann Moyal refers to a

*Lana Rakow, who is skeptical of the technology, writes: "Despite social practices which enact gender on a daily basis, we cannot neglect to account for the pleasure, solace, and companionship that many women derive from the telephone. Because they are generally less mobile and less independent financially and more likely to be isolated from other adults than men, many women have found the telephone to be a lifeline to mothers, sisters, and friends. . . . We should not dismiss the telephone,

"pervasive feminine culture of the telephone in which kinkeeping, nurturing, community support, and the caring culture of women forms a key dynamic of our society. . . . " This, therefore, is an instance of a machine, so often identified as masculine by its very nature, that men often shied away from but women aggressively turned to their own ends. Michèle Martin draws the further conclusion, that "women subscribers were largely responsible for the development of a culture of the telephone, as they instigated its use for purposes of sociability."[41]

The Nature of Telephone Sociability

If North Americans prior to World War II — especially women — largely used the home telephone for social calls, how did that use affect the nature of their social ties? Did it, for example, replace face-to-face conversations? Urban planners today frame this question, usually with respect to business, as the "communication-transportation trade-off": Does telephone service reduce in-person travel by allowing business people to accomplish the same tasks by calling instead? Or do such calls actually increase travel by generating more business?[42] Our interest is closer to that of the Knights of Columbus quoted in Chapter 1: "Does the telephone make men . . . lazy? Does the telephone break up . . . the old practice of visiting friends?" To be more precise, did the adoption of the telephone between 1890 and 1940 lead Americans to one or more of three customs: First, did people replace visiting with telephoning, so that their total social intercourse remained the same but more of it was by wire? Or, second, did people have conversations by telephone that they would otherwise not have had at all, so that they added more conversations to a constant number of in-person visits? Or, third, did people, stimulated and aided by their telephone calls, make even *more* in-person visits than they would otherwise have made?*

The first contention, that telephone contacts replaced face-to-face meetings, has both a weaker and a stronger form. The weaker form postulates that people became lazy and began calling neighbors and friends instead of dropping by. The stronger form posits that having the telephone encouraged people to live farther apart from one an-

---

then, as another source of women's oppression, but recognize the complicated role it has had in the shifting plane of ideology and gendered experience" ("Gender, Communication, and Technology," 81).

*There is another logical but unlikely possibility: that telephone calling disrupted social relations so much that total contacts (face-to-face plus telephone) diminished.

other. For example, a grown daughter could move to town and leave her elderly parents on the farm once she had the ability to reach them by telephone. Our fragments of evidence best address the first, weaker form, that people just called instead of visiting.

The Lynds argued that "neighborliness" diminished in Middletown during the years up to 1924 and implied that the telephone was partly responsible. Several women whom they interviewed described a decline in visiting, either in their adulthoods or in comparison to their mothers' day. Two of the women attributed the decrease to having children, others to people being more independent, and one to social clubs, but a few mentioned the telephone. One remarked, "Instead of going to see a person as folks used to, you just telephone nowadays." Another noted, "When the 'phone came, it took up a lot of time, since you were within reach of so many more people, but it saved all the time formerly spent with women who 'ran in' on you while you were trying to do your morning's work."[43] In a different context, some observers thought the telephone had reduced the frequency with which farmers visited town and thus cut down on their community involvement.[44]

Our elderly informants from northern California drew a more complex picture of the connection between calling and visiting. Several implied that people in the early twentieth century substituted telephoning for dropping in. An Antioch woman, born in 1903 and quoted earlier, claimed: "We certainly didn't depend on the phone the way we do nowadays. . . . No, phones weren't used that much, you would go to see people and visit them in person." An Antioch man recalled of the 1920s, "Neighbors and friends always visited over the back fence in those days, so there was no need for a phone." And another Antioch woman, born in 1915, recalled telephone visits between her family on the ranch and grandparents in town: "They would call back and forth. It was easier to do that than to go the five miles just for a short visit." Such comments suggest that the telephone call constituted an alternative to the face-to-face conversation and, perhaps, that having the telephone inhibited personal visits.*

Other interviewees implied that by calling, people increased their total conversations. Some remembered that they called their friends

---

*It could also be argued that extraneous reasons (say, more work commitments) led people to reduce their in-person visiting and telephone instead. Hence, telephone calls were the response to, not the cause of, less visiting. This contention would be hard to document.

frequently as youngsters. (Today, one common call is between teenagers who have just seen one another at school[45] — just like the calls of over 80 years ago recounted by the San Francisco doctor's daughter.) Similarly, several remembered their mothers regularly chatting with neighbors on the telephone. Although some of these chats may have been substitutes for dropping in, they were probably often conversations that would otherwise not have happened. A few of our interviewees also noted that their mothers' use of the telephone increased as children moved out on their own. That is, mothers spoke to grown children who were infrequently seen. The common perception that telephones helped break the isolation of rural women also implies that calling added to total social interaction. In addition, many long-distance calls, certainly, were conversations that would not have otherwise occurred.

Finally, recollections of using the telephone to arrange dates, trips, and meetings suggest that calling assisted, even if it did not generate, many in-person encounters. Although these people might have found other means of making arrangements, many meetings would probably never have taken place without the calls. In the same vein, a San Rafael woman born in 1907 commented that with everybody driving, one needed to call ahead to make sure someone would be home for a visit.

All three types of relationships between calling and visits seem to have developed during the period of our concern. People substituted telephone visits for some in-person ones; they made or received calls when they would or could not have otherwise met in person; and they used the telephone to arrange face-to-face encounters. We cannot measure from such accounts the relative volume of each of these three changes. The balance would, however, appear to be that the number of total contacts stayed constant or increased.

Researchers nowadays have estimated the communication-transportation trade-off for personal interaction. A few studies show some intersubstitution of telephoning and visiting. For example, in the survey of New York City subscribers who temporarily lost telephone service in 1975, 34 percent reported more frequent visiting during the disruption. After a bridge collapsed between two parts of Hobart, Tasmania, telephone calls increased.[46] But in other studies done in the United States, England, and Chile, people with telephones reported more total social contacts than did those without telephones; subscribers visited and wrote letters more often than did nonsubscribers. These correlations support the claim that telephone use multiplies all forms of contact.[47]

Did the telephone, then, "break up . . . the old practice of visiting"? That is too strong a conclusion. Telephoning probably changed visiting practices moderately during the first half of this century. People forewent some visits that they otherwise would have made, particularly unannounced drop-ins. (The elite practice of "calling" during designated at-home hours, printed card in hand, was probably in decline anyway.) Telephone users altered the character of other visits by telephoning ahead to arrange and confirm them. Finally, people probably made rendezvous they otherwise would not have, particularly appointments at public places, because they could telephone. Thus, telephoning perhaps had a limited effect on visits to another's home but made a greater difference in the ability to arrange an appointment outside the home. Furthermore, different people responded in various ways. Whether the sum total of face-to-face conversations with people outside the household declined because of the telephone we will never know, but it is much likelier that the total volume of social conversation increased notably. The telephone probably meant more talk of all kinds.

Little of the evidence, however, addresses the stronger version of the substitution argument, that over the years the availability of the telephone encouraged people to live farther apart, necessarily transforming face-to-face relations into telephonic, presumably weaker, ones. This is, for instance, sociologist Ron Westrum's charge: Communications technologies "allow the destruction of community because they encourage . . . far-flung relationships."[48] We have no evidence that the telephone encouraged separation. (Residential mobility in the United States has actually decreased over the last century and longer.[49]) None of our interviewees indicated any such tendency, although it may not have been on their minds. Nor does our evidence address the converse to the substitution argument, that people moved apart for other reasons, such as job searches, and the telephone allowed relations that would otherwise have withered to continue.

Nevertheless, the assumption that much face-to-face interaction has become largely telephonic stirs worries that such relations are emotionally thin. Not even a telephone company publicist could assert that telephone calls capture the intimacies conveyed by eye contact and physical touches, or that telephone friendships can plumb the same depths as sharing meals, taking walks, or just being together. But the question is not whether a telephone conversation is as rich as a face-to-face one. It probably is not.[50] The question is, can telephoning sustain a relationship, or does it provide only an "inauthentic" intimacy? Malcolm Willey and Stuart Rice, otherwise cold-blooded analysts, worried that more social contacts had become brief and impersonal (due to the telephone and automobile) and believed that this entailed a loss "of those values that inhere in more intimate, leisurely, protracted personal discussion."[51]

As keen an issue as this is, we cannot directly resolve it with the evidence at hand. Recent studies show that people report feeling "closer" to those friends and relatives who live farther away than to nearby associates, even associates whom they see regularly. Although people depend on those nearby for certain kinds of sociability and practical help, they turn as often to distant kith and kin for critical emotional and practical support. As a final note, Americans say they would just as soon have some distance between themselves and their friends.[52]

The critique contends, however, that even though people may believe that their relationships are intimate, honest, and committed, they do not realize that the relationships they handle by telephone actually lack those qualities and are "inauthentic." Personal testimony cannot, therefore, adjudicate the issue. Yet personal testimony is almost all we have. Lana Rakow concluded from her interviews: "Even for those who do use the telephone for companionship and talk it is not always viewed as an adequate replacement for face-to-face talk." The "always" implies that her interviewees usually viewed it as adequate. In a Canadian survey of middle-aged and elderly telephone users, about two-thirds of those over 55 agreed "a lot" that "I feel I only have to lift the telephone and I can be right there with my family." Among those 54 and younger, 55 percent of the women but only 37 percent of the men agreed. Ann Moyal's Australian interviewees believed that "the telephone played a key and continuing role in building kin and friend relationships."[53]

Assessing the quality of social relations over a half-century ago requires relying even more than these recent studies do on personal testimony. Our own interviewees did not say that they had found telephone relations unfulfilling in the 1920s and 1930s. Some were critical of too much idle talk on the telephone, but none critiqued the authenticity of telephone relationships. (We did not, however, expressly pursue this question with them.) This issue awaits more subtle research.

Finally, what about too much sociability—having one's home invaded by calls or one's conversations overheard? Busy people, news editors, and, in his old age, Alexander Graham Bell himself, complained of telephone interruptions. Household efficiency expert Lillian Gilbreth encouraged women to organize their lives to avoid or to work around telephone calls.[54] Yet, ordinary people seemed to have few such complaints for the years covered in this study. A few women told Lana Rakow that they had to listen to other women's problems because they were reachable by telephone. A Middletown woman quoted earlier said that more people reached her because she was available (but she also said that this interruption was better than being dropped in on).[55] But complaints about such intrusions were not raised in the general reminiscences about the early telephone previously cited or in our own interviews. Nor were unwanted calls much of an issue for the etiquette writers reviewed in Chapter 6; they were more concerned about excessive "chatter" on wanted calls. (The level of concern about unwanted calls might be higher today, of course.)

Eavesdropping, on the other hand, posed a frequent worry, especially on rural lines. From the beginning of telephony, people expressed concern that they were being overheard, at first simply by others in the same room—one had to speak loudly—and then by operators or fellow subscribers on a party line.[56] Several rural Indiana women cheerfully recalled listening in on the farm lines: "I was just as bad to eavesdrop as anybody else was, but I heard a lot that way," said one. Another commented, "We sure did visit and eavesdrop. That eavesdropping was funny. Everybody knew what was happening, and what all the neighbors were doing." One of our interviewees recalled that her aunt who ran the country switchboard listened in on "everything." Another half-joked that although her family didn't call much, they did use the telephone to eavesdrop. As noted in Chapter 4, listening-in on rural lines caused some uproar. Eavesdropping may have contributed to the most dramatic case of telephone controversy, the decision of the Pennsylvania Amish to ban the device. According to one account:

> ...then a couple women got to talking about another woman over the phone and this woman also had the phone in and had the receiver down and heard what they said, this made quite a stink and at last came into the gma (church) to get it straightened out, then the Bishops and ministers made out if that is the way they are going to be used we would better not have them.[57]

Summary

Americans in the early twentieth century did not use the telephone to recreate the personal relations of arcadian villagers, notwithstanding the beliefs of Marshall McLuhan, country life reformers, and AT&T

copywriters to the contrary. Still, the adoption of the telephone probably led people to hold more frequent personal conversations with friends and kin than had previously been customary, even if it also led them to curtail some visits. Sustaining personal relations by telephone was probably rare before the turn of the century but became common in the middle class and on farms by the 1910s or early 1920s. Such talk turned torrential after mid-century, when telephones became nearly universal. To this day, some Americans, particularly men, have never become "telephone people" and rarely chat on the telephone. A small proportion probably abhorred the device in its early days; some probably still do.

Middle-class and farm women more often carried on personal relations by telephone; therefore the arrival of the technology probably altered their lives most. (Urban working-class women were not among the heavy callers in these years, and we know much less about their telephone usage.) Although conversations over the telephone do not duplicate face-to-face talks, many people seem to have found it a satisfactory way—sometimes the only way—to keep "in touch." The authenticity of relations that rely on the telephone is much harder to assess, but the claim that it is lacking is, so far, mere assertion— an assertion most callers themselves did not make. On a yet broader point, we have inadequate evidence to judge whether people, sensing the telephone's utility for staying socially close, chose to physically distance themselves from family and friends.

THE PSYCHOLOGY OF THE TELEPHONE

What are the psychological consequences of the telephone? Some analysts argue that it produces feelings of empowerment—as claimed by AT&T*—or alienation, or even infantile sexuality. Psychological characterizations of the telephone have changed historically. According to John Brooks, theatrical and literary authors used the telephone as a symbol of sophistication and wonder before World War II and afterward as a symbol of threat, violence, and powerlessness. (For example, a telephone on stage, like a gun displayed in the first act,

---

*For example, a 1909 advertisement entitled "The Sixth Sense—the Power of Personal Projection" told businessmen that the telephone "extends his personality to its fullest limitations." Another ad that year read: "If *any man* in the Union rings the bell of his Bell Telephone at his desk, any other man *at the most distant point* is at his instant command" (in the Ayer Collection, National Museum of American History).

goes off before the final curtain.)[58] We will look at two psychological themes common before 1940: briefly, that the telephone was an emblem of modernity and, more fully, that the telephone was a source of tension.

Being Modern

Advertisers argued that the telephone both exhibited and produced a psychological quality of modernity. A 1905 ad told women that "The Modern Way is to save one's time and temper by telephoning." A 1909 ad called the Bell System sign that hung near pay telephones, "The Sign Board of Civilization." Scenes in advertisements for other goods often included a telephone in order to associate the products with modernity and power. Other people also linked the telephone to modernity. The Amish and Mennonite communities divided over allowing the telephone, in part because it threatened to bring them too much contact with the modern world.[59]

A few of our interviewees described the telephone in tones suggesting awe. The Antioch man quoted at the beginning of this chapter indicated such a feeling in his story. A Palo Alto man born in 1892 recalled: "I still remember when the first phone was installed at a neighbor's. Everybody was so anxious to talk on the phone." A Palo Alto woman born in 1895 also said: "I remember distinctly getting a phone. That was really something!" However, the younger among our interviewees, especially those raised in towns, were more nonchalant. They were likelier to say, as did a Palo Alto woman born in 1909, "Seems we always had a telephone." An Antioch woman born in 1903 noted: "Telephones were no big deal. It wasn't like you never saw a telephone. They had one at the store or at the neighbors."

According to our interviewees, the telephone was, except perhaps for rural households, commonplace in northern California by the early 1910s. It neither frightened nor amazed people, and it symbolized nothing special about subscribers except perhaps that they were well-to-do. Recall, too, from Chapter 5, how quickly the local newspapers lost interest in the telephone. At some level, the connection between the telephone and modernity probably remained, subtly and below people's consciousness, available for exploitation by advertisers. A kitchen with a telephone would appear more up-to-the-times than one without it. The evidence suggests, however, that few Americans found the telephone dramatic beyond about 1910.

Pace, Tension, and Anxiety

The telephone, some observers assert, sped up the pace of life, forced people to be alert, and thus created a lasting feeling of tension. In 1899 an English newspaperman argued, "The use of the telephone gives little room for reflection, it does not improve the temper, and it engenders a feverishness which does not make for domestic happiness and comfort." An AT&T-sponsored author asserted in 1910 that the telephone made life "more tense, alert, vivid." In his 1976 history of AT&T, John Brooks claimed that the early telephone "was creating a new habit of mind—a habit of tenseness and alertness, of demanding and expecting immediate results, whether in business, love, or other forms of social intercourse."[60] These judgements make sense intuitively (and will get empathetic nods from busy individuals who feel harassed by telephone calls). If people become anxious when they are eager for someone to answer their call or are constantly on edge because their telephone can ring at any time, then they are probably more tense than they would be without the telephone. We are familiar with people today who seek vacations from their telephones. But what evidence is there to support the proposition that Americans in the first half of the century generally were made tense by telephony?

We can search the comments of old-timers for expressions of anxiety or irritability associated with the telephone. One elderly Indiana woman recalled that the telephone "used to scare me," and another that her mother was afraid of the telephone because she had been knocked down by lightning. Two more Hoosier women simply said that they did not enjoy the telephone and used it only of necessity. A few of our own interviewees expressed some disdain for the telephone—"I am not a telephone person"—but whether that reflected tension is unclear. A San Rafael woman born in 1902 reported that her husband disliked the telephone and hated to find her or their children on it when he came home from work, but again, whether his displeasure stemmed from nervousness or some other source is not evident. Of our three dozen interviewees, only this handful made comments at all implying that the telephone provoked anxiety. For the most part, tension was not a vivid part of people's recollections of the telephone.

One cause of such tension could be the fear of ill tidings. A woman told Lana Rakow that during World War I people who hadn't recently heard from their sons in Europe hated to hear the telephone ring. An AT&T study found that elderly people felt the sound of the telephone bell to be unpleasant because they feared it brought bad news. Yet, middle-aged and young people in that same AT&T study thought of it as stimulating because it "promised relief from boredom."[61] There seem to be more comments of the latter sort in the oral histories of the telephone. An Indiana woman said, "We appreciated [the telephones] so much; we just enjoyed them." Another echoed, "We enjoyed the telephone."[62] A few of our interviewees raised on farms appreciated telephones in the sense that having the devices increased their sense of security. One Antioch man born in 1911 rhapsodized about the machine itself:

> You could [call] from ranch house to ranch house and could get the main phone service in town. I was awestruck with that. My father would open up the phone and I would look at it. He'd open up the box and remember the colors—beautiful colors—that identified the circuits. And he would try to identify this thing for me.

Such positive comments outnumbered the few expressing anxiety.

A rare, systematic survey done in the 1930s examined people's feelings toward three forms of communication. A market research firm, sponsor unrecorded, interviewed 200 men and women from four American cities. Interviewers first asked respondents how they thought people in general felt and then how they personally felt toward telegrams, telephones, and letters. (Answers to the general and personal questions were similar.) The key probe was: "Without considering expense or time, do you think there are people who feel slightly uncomfortable or mind in any way _____ ?" The following list gives the percentages of affirmative responses:

Making telephone calls	33%
Receiving telephone calls	33%
Sending telegrams	42%
Receiving telegrams	62%
Writing letters	70%
Receiving letters	8%

Of the 66 people who expressed discomfort about making telephone calls, almost half complained that it was difficult to hear or to be heard, or that there were other service problems, such as inconvenience in reaching people; 28 percent felt self-conscious or did not know what to say; and 15 percent thought calling was a waste of time because people talked too much. Of the 66 people who expressed discomfort at receiving calls, several complained about hearing poorly, several felt self-conscious about speaking on the telephone, and 38 said it was "time-wasting, bothersome, disturbing." (The report lumped the last three comments together.) The best estimate is that less than 20 percent of the original 200 respondents expressed anxiety about using the telephone. Add those who felt shy about talking and then perhaps 25 percent, at most, expressed some tension about using the telephone.* By contrast, most people disliked receiving telegrams because they feared bad news; many disliked sending telegrams because they might scare the recipients; and most hated writing letters because of the time and effort required or because they felt inarticulate. Few people objected to receiving letters. Asked their preferred modes of communicating, respondents ranked telephone calling first and writing letters last. (Note, however, that this survey was conducted near the end of our period and may reflect a late-developing comfort with the telephone.)[63]

We can also consider a few recent surveys, but only with some reservations due to the probable change in reactions to the telephone as the technology became more common. The AT&T survey in which young and middle-aged respondents found the telephone ring stimulating was previously mentioned. Benjamin Singer asked 138 residents of London, Ontario, how they reacted when the telephone rang at dinnertime. Forty-four percent said they did not mind, 15 percent were mildly annoyed, and 9 percent said they got angry. For calls that interrupted television watching, fewer than 10 percent reported irritation. Recalling the point made earlier that the proper comparison for the telephone is with face-to-face meetings, considerably more respondents, 30 percent, objected to people dropping in unannounced. Calls in the middle of the night, however, upset or angered 63 percent. When asked what the disadvantages of the telephone were, about half pointed to interruptions or unwanted calls. Thus, in general, even though people labeled nuisance and ill-timed calls as a problem, they expressed only modest amounts of irritation about such interruptions.[64]

We can again consult the interviews from the Manhattan study of 190 people who lost telephone service for three weeks in 1975. Large majorities reported feeling out of control, uneasy, isolated, and frustrated without service. Yet, almost half also reported that "life felt

---

*I assume that complaints about sound quality, operators, or other people's gabbiness, although they express irritation, do not show the kind of emotional anxiety that is the subject of this discussion.

less hectic." A substantial minority, therefore, felt that having a telephone simultaneously made life hectic and enhanced their control over their lives. (Ironically, after reporting these findings the authors concluded that "although it may reduce loneliness and uneasiness, [the telephone's] likely contribution to the malaise of urban depersonalization should not be underestimated"—with no evidence at all demonstrating "the malaise of urban depersonalization.")[65]

We can fit these fragments together into a few tentative conclusions. Most people saw telephoning as accelerating social life, which is another way of saying that telephoning broke isolation and augmented social contacts. A minority felt that telephones served this function *too* well. These people complained about too much gossip, about unwanted calls, or, as did some family patriarchs, about wives and children chatting too much. Most probably sensed that the telephone bell, besides disrupting their activities (as visitors might), could also bring bad news or bothersome requests. Yet only a few seemed to live in a heightened state of alertness, ears cocked for the telephone's ring—no more, perhaps, than sat anxiously alert for a knock on the door. Some Americans not only disliked talking on the telephone but also found having it around disturbing, but they were apparently a small minority.* Perhaps a few of the oldest felt anxious around the telephone, but most people—according to our interviews, perhaps almost all of those born after the turn of the century—seemed to feel comfortable or even joyful around it. (An Australian researcher has suggested that a telephone call also raises recipients' feelings of self-esteem; it shows that someone cares.[66]) Sociologist Sidney Aronson may have captured the feelings of most Americans when he suggested that having the telephone led, in net, to a "reduction of loneliness and anxiety, an increased feeling of psychological and even physical security."[67]

THE AUTOMOBILE

Commentators have placed far more social and psychological effects at the wheels of the automobile than at the base of the telephone, but also with little evidence. Observers of farm life have credited the

*Perhaps today, and perhaps earlier, too, the people who most complain about telephone calls harassing them are those with already packed schedules. While many essayists probably fit in this group—and the author and many readers of this book, too—we do not necessarily represent Americans in general.

automobile with expanding community sociability, yet they have presented largely anecdotal testimony. Most writers concerned with the automobile, however, have focused on the family when discussing its social consequences during the first half of the twentieth century. Although some have claimed that car trips—touring, camping, Sunday "spins"—reinforced togetherness, a majority have argued that the automobile, by enabling young people to escape home, go on joy rides, and visit movies and roadhouses, weakened family bonds and spurred generational conflict. (Similar complaints about bicycling arose in the 1890s.) In *Middletown*, the Lynds reported that the automobile most often united families, but warned that this tendency seemed to be a "passing phase," that youth were chafing to go their own ways. The most consistent and specific complaint of the time was that youngsters used cars to avoid parental and community supervision and to indulge their sexual desires. This allegation was based partly on the simple observations that teenagers could use cars to get away from home and that the closed car allowed them privacy. One essayist in 1950 admitted, however, that it was impossible to know for sure how many such opportunities adolescents really exploited. The sexuality claim was also apparent in reproaches made by contemporary officials and journalists in the 1920s, such as the Muncie, Indiana, judge who claimed that the "automobile has become a house of prostitution on wheels," and the editors of the *Tennesseean* who wrote that the automobile was the principle "medium used to ruin girls of tender ages."[68]

Despite their constant repetition of this charge, scholars have presented no substantial evidence that the automobile stimulated more pre- or extramarital trysts. In her history of American courtship, Ellen Rothman argues that the sexual license of the 1920s emerged from cultural changes set in motion twenty years earlier, before mass automobile ownership. She also notes that "going out" began in the late nineteenth century, facilitated by trolleys, trains, and bicycles. Perhaps mass automobility accelerated dating, but her account denies that it initiated a substantial change.[69]

The automobile activities most commonly recalled by our interviewees were family trips—excursions to vacation spots such as the Russian River, Lake Tahoe, or Yosemite National Park, weekend trips to see relatives, and outings to picnics and the like. (Recall that for a generation the automobile was primarily a recreational device.) Our oldest interviewee, a Palo Alto man born in 1888, told of buying a $500 Model T at age 27 to drive to coastal Carmel on his honeymoon.

Two others recalled adventuresome trips into the Sierra Mountains. Another Palo Alto man, born in 1903, declared that "having a car changed life because it made it possible to be away from home more. It broadened the potential of what you might see, like driving into Yosemite Valley." A Palo Alto woman born in 1905 concluded that the automobile didn't change "our life at all . . . except being able to go on camping trips, which we hadn't done before." Warren Belascoe has documented the touring and camping boom that mass automobile ownership first stimulated in the 1910s. Campers found Yosemite Park especially attractive.[70] Our oral histories confirm the importance of auto-touring. They also confirm a portrait of family togetherness.

Yet, when our interviewees recalled getting their own automobiles as young adults in the late 1920s and early 1930s, either after they moved out or even while still living with their parents, they often mentioned youth-only activities. One San Rafael man born in 1914 suggested that having a single car encouraged families to do things together, but when families had more than one car, parents and children went their own ways (an insightful observation supporting the theory that the consequences of a technology may be nonlinear). Although teenagers and young adults also drove with friends on country outings, more often they drove to dances or movies. Most recollections were of groups of same-sex friends traveling together. A San Rafael woman born in 1913, for example, regularly drove with girlfriends to dances around Marin County and eventually met her husband at one. These stories suggest that the automobile functioned to accelerate sociability. The same Palo Alto man who mentioned Yosemite also remarked that "the development of the auto for a pleasure vehicle had an enormous impact. . . . It made it practical to go and see friends in the evening, to go out to theater. . . . " Others were not so sure it made such a difference. Another Palo Alto man, born in 1908, observed that "cars widened the area you could visit people—not that we didn't manage before that." An Antioch man born in 1911 said it was just a "nice convenience." And there was the Palo Alto woman quoted earlier who claimed that few changes resulted from the automobile, with the exception of camping.

The issue of gender differences resurfaces in the discussion of the automobile. As with the telephone, women seemingly took to the automobile as a device for sociability. Scholars have reported rural women's testimonials that having a car was crucial to breaking their sense of isolation. Similar evidence, although in less quantity,

suggests that middle-class urban women also found such social pleasures in the automobile.[71] Among our interviewees, a few recalled
that their mothers did not drive. One said her mother even feared rid-
ing in a car; two said that their fathers did not believe women should
drive; and one remembered that "women didn't drive; mother didn't
have to." More of the northern Californians we spoke to, however,
recalled mothers who did drive, sometimes enthusiastically. These interviewees described how their mothers shopped by car and frequently
drove to social events. For example, a woman born in 1915, who
lived outside Antioch, recounted: "Mother preferred a car. She was
very lodge-minded, the Rebecca Lodge, and she was always going
into town to do lodge activities. . . . When I got into grammar school
she drove the 'bus.' It wasn't the bus, it was our car." In more than
one family, it was the mother, not the father, who regularly drove
the family on outings. In addition, almost all the women interviewees
themselves drove, sometimes from an early age. An Antioch woman
born in 1903 reported that she was the first in the family to drive. A
Palo Alto woman born in 1895, said that neither of her parents learned
to drive, so she became the family chauffeur. An Antioch woman remembered getting a car in 1928 as a *grammar*-school graduation gift. An
Antioch woman born in 1906 recalled a girlfriend who at age 14 gave
her friends rides in her father's grocery truck. For younger women
in the period, driving seemed to be part of the adolescent culture.
The links between sociability, gender, and the telephone previously
discussed may apply to the automobile as well.

Finally, what did our interviewees indicate about the automobile
and sexuality? None made any explicit connection between the two.
(We did not ask them bluntly about sex.) A few did allude to dating.
An Antioch man born in 1905 bought a "cutdown" Model T at age 20,
which was "hot stuff," but "we could never get a girl to ride in them."
A Palo Alto man born in 1908 noted that "cars make it a helluva lot
easier to take girls out." And an Antioch woman born in 1911 remembered getting her license in 1929 despite her father's resistance, and said
that cars were "a good way to court." Several others remarked that
cars helped them get to places where they could meet young people of
the other sex. Yet, no one mentioned sex itself, despite its association
with the automobile in popular culture. Perhaps our elderly respondents were simply being discreet, or perhaps licentiousness was not as
widespread as commentators of the day believed.

Culture critics and advertisers alike have vested the automobile with
myriad psychological overtones — virility, femininity, seductiveness,

power, status, dynamism, materialism, and so on. A common theme regarding the years covered here is that driving a car reflected and perhaps taught a "modern" sensibility. Driving proved that one was up-to-date, a master of the new technological world. The automobile, with the industry's vast advertising efforts, instilled in people the sense that the new and the mechanical were sources of status and admiration.[72]

In reality, was this so? Several of our interviewees could remember, some with nostalgic affection, their parents' first cars. But only the oldest reported awe inspired by automobiles. Among our three oldest, all Palo Altoans, one man said, "I remember seeing the first car coming down Alma Street—gosh, it was an excitement!" The woman who drove by age 12 remembered, "We'd drive carefree all over, just for the drive . . . and for the envy and looks!" A man who started driving in 1910 said the "auto was the biggest change there was. I don't know what life would be like without [it]." A slightly younger Antioch woman reported that "the car opened up a whole new life for us." Finally, a younger San Rafael man, the one who bought the "hot stuff" Model T, remarked that he became an auto mechanic because it was the most prestigious job after doctor and lawyer. A few others noted that cars often represented wealth. For instance, an Antioch man born in 1911 stated that in "those days those that had the ability to buy one were showing off anyway." Other interviewees, however, were dismissive of the automobile's portentousness, even when we asked whether the car made a difference in people's lifestyles. A Palo Alto woman born in 1905 commented: "It all came along easy. Instead of a horse, you had a car."

Essentially none except perhaps the oldest interviewees expressed any of the veneration or titillation that commentators would have expected. They often remembered their first automobiles in some detail, even lovingly, but few embellished these vehicles with greater power than their functional uses.* Moreover, the most common functional uses were outings to distant nature spots or to nearby recreation centers, usually with family or friends, rather than activities connected with sexuality or power.

The theme of modernity, even with its widest penumbra, is but one of many psychological themes that culture critics have associated

---

*This conclusion does not deny that some people now—and, no doubt, then as well—invest themselves deeply in their cars, as perhaps most vividly described by Tom Wolfe in the *The Kandy-Kolored Tangerine-Flaked Streamlined Baby*. Yet another debatable issue is whether these people typify most Americans.

with the automobile. This study cannot hope to treat even a small set of them. Yet, perhaps it can show that such themes ought to be empirically assessed not through aesthetic readings of the technology or its representations in art or literature (which are valid as studies of cultural products but not necessarily as studies of people's behavior), but through the testimony of contemporaries. Our evidence suggests that as the initial years of novelty passed, Americans incorporated the automobile into mundane life; and in the early years of the century they linked it primarily to recreation and sociability. This assimilation into commonness was not nearly as full as that of the telephone. Automobiles were and are big, brassy, dangerous, expensive, and controversial. Still, the psychological aura of the automobile has probably been exaggerated.

Today, the entire technological system surrounding the automobile—interstate highways, billion-dollar service industries, suburban sprawl, exhaust-choked traffic jams, the mechanics of the cars themselves—is truly awesome. In the years before World War II, this system was only developing, as governments surfaced roads, entrepreneurs built motel chains, and working-class people bought used cars. Personal life, however, was not yet entangled by all the later, second-order developments such as shopping malls and rush-hour radio programming. For some in those days—salesmen, doctors, repairmen—the automobile represented a work-related machine. At the same time, for them and for many more who could afford the cost, it was a device for recreation and sociability, like the home telephone. Given that the automobile loomed psychologically larger than the telephone, people assimilated it with striking ease into the taken-for-granted character of daily life.

CONCLUSION

In her book *When Old Technologies Were New,* Carolyn Marvin recounts the wonder with which some nineteenth-century people regarded new electric devices, including the telephone.[73] Perhaps for the first twenty years, when few Americans had them in their homes, the telephone may have carried such an aura. When automobiles were playthings of the wealthy during the first decade of the new century they, too, probably carried a powerful symbolism. Yet, by the time the telephone was common in the homes of the middle-class—the 1910s and early 1920s, at least outside the South—it had become mundane.

(The lack of interest social scientists have had in studying the telephone is further testimony to its shortage of charisma.) The automobile, although as common by the mid-1920s as the telephone, was not so easily rendered ordinary. It was an expensive investment, required costly care and feeding, came in various brands, sizes, shapes, and colors—a differentiation advertisers constantly reinforced—and was used for a variety of tasks. Still, it, too, was soon taken for granted by most middle-class people. These technologies, more so the telephone, lost their mystique and became two of the commonplace tools people used to pursue their prosaic private lives.

Americans by the 1910s and 1920s were using the home telephone largely for sociability. This does not include all people. It is true for women more than men, the younger more than the older, and the gregarious more than the shy. What mode of communication did Americans displace with the telephone? Besides curtailing telegrams and hand-delivered notes, telephoning probably cut into casual, drop-in visits. Yet, telephoning at the same time helped arrange other kinds of meetings. In total, calling probably led to more social conversations with more people than before. Perhaps these calls substituted for longer visits or chats with family members, or perhaps they simply took up time that would have been spent alone. A few people seemed to regret a loss of face-to-face contact, attributing it to the telephone, but they were a clear minority.

The automobile eventually cut a wider swath through social life. Well-off Americans used the automobile in place of horse and buggy rides in the first decade or two of the new century. They, middle-class Americans and even farmers, used the automobile in place of train trips for touring. By the 1920s, many in the urban middle class had abandoned streetcars for automobiles when commuting or shopping in town. Some observers felt that driving displaced other activities as well, such as attendance at church, family evenings around the hearth, and courting on the veranda.* The evidence suggests that typical car-owning families before World War II bought cars and drove them largely for recreation and sociability. These automobile activities no doubt displaced other forms of recreation—with people, say, traveling to see distant friends rather than visiting neighbors—but driving seems to have added to the total sum of social activity.

---

*In a review of time-budget studies, researchers failed to find evidence that the diffusion of the automobile—in sharp contrast to the spread of television—significantly altered how people spent their time (Robinson and Converse, "Social Change Reflected in Use of Time").

In this sense, both the telephone and automobile before World War II were, in their domestic use, "technologies of sociability" (and thus perhaps especially "feminine"). The net result of their use was to expand the volume of social activity and, in that way, add to the pace of social life. Most people seemed to, at least consciously, welcome that.

Our theme would be more dramatic if we could implicate the telephone in the emergence of some aspect of psychological modernity—rationality, angst, anxiety, dehumanization, whatever. The available facts, which indicate that Americans absorbed the telephone into mundane life, seem deflating. But there is something yet more profound in seeing people as active participants, assimilating a major material transformation into their lives. Those lives were not left unaltered, to be sure, but the alterations were largely the conscious product of people employing things, not of things controlling people.

CHAPTER NINE

Conclusion

In 1909, when about one-fourth of American households had telephones, an AT&T ad described the Bell System as "A Highway of Communication," passing by "every home, every office, every factory, and every farm in the land." Near the end of the century, in a world of virtually universal and constant electronic connectivity, of cellular and satellite telephones, how primitive that 1909 highway seems. The technological changes have been startling and sweeping; the accompanying social changes, however, have been more modest and circumscribed.

On the eve of Pearl Harbor, about 40 percent of American households had telephone service. Telephone diffusion had almost reattained its pre-Depression peak of 42 percent. Some Americans—typically non-Southerners, city-dwellers, the well-to-do, and white-collar workers—were more likely than others to have telephones. During World War II, telephone subscription expanded—unlike automobile ownership. Subscription continued to grow rapidly after the war, such that it surpassed automobile ownership around 1950. By 1990, all but several percent of American families had telephones. The device was effectively universal, and controversy surrounding it was largely passé.* Debate focused instead on the newer electronic technologies—satellite up- and downlinks, cable television, computers, and the like—and what they portended for life in the twenty-first century. This chapter briefly reviews the development of the telephone system after 1940 and some projections for its future, summarizes the accounts from prior chapters regarding the social history of the telephone's first two or three generations, and concludes by considering the implications of those findings.

AFTER WORLD WAR II

Enhancements of telephone technology and service came rapidly after the war: long-distance direct-calling and other automation of operators' work, touch-tone dialing, satellite transmission, answering machines, simplified wiring, the multiplication of the number of conversations that could pass simultaneously over one connection, fiber optics, call-waiting, cordless and cellular telephones, faxes, "800" collect-call service, and so on. These developments made calling faster, simpler, more convenient, and cheaper—especially long-distance. Yet, like the automobile, the technology as users actually experienced it did not change in the latter half of the century. For a modest price, one could pick up a simple device in one's home, dial or punch a few numbers, and speak almost instantly to people near and far.

For several decades the industry too, continued in the pattern set in the 1920s. AT&T held a monopoly on long-distance service and, through its regional operating companies, on local service in most communities as well. Independent companies exercised their local monopolies typically in small towns and rural areas, serving but a small portion of American subscribers (17 percent in 1970). These independents, in fact, depended on Bell in many ways, for long-distance connections, for technical advances, for political lobbying, and so on. State agencies regulated local and intrastate service and rates, whereas the FCC monitored Bell's interstate system. AT&T was an industrial colossus, easily doubling the financial size of the next largest American corporation and employing almost one million people.[1]

This industrial structure crumbled. The federal government repeatedly pressed AT&T about its monopoly, particularly about its

---

*A notable exception to the taken-for-granted status of telephony was the dispute over the financial and regulatory structure of the industry, brought to a head by the divestiture of AT&T in 1984, which I will discuss in the following section.

sole-source purchasing of telephone equipment from its own subsidiary, Western Electric. Allegedly, this practice allowed AT&T to shut out other manufacturers and to inflate the costs of equipment, charges which it then passed on to consumers via regulated rates. In a 1956 consent decree, AT&T agreed to license its patents to others, to confine Western Electric production to telephone equipment, and to stay out of telecommunications enterprises other than telephony. On another front, this one more fateful, technical and regulatory developments opened the door for competition in long-distance service.

Technical advances sharply reduced the real costs of long-distance calling. At the same time, regulatory agencies kept long-distance rates up while they held down those for local service, in effect increasing the so-called cross-subsidy from long-distance revenues to local service.* This left an opening, aided by FCC and court decisions to expand competition, for others to offer long-distance service at rates below Bell's. These competitors, sometimes using microwave transmissions, served mostly large businesses with inter-office connections but later solicited general long-distance users on heavily trafficked lines. Bell regarded this strategy as "cream-skimming." New operators, such as MCI could undersell AT&T to attract lucrative clients. AT&T, as a regulated utility, had to serve everyone, including Aunt Millie in Muskogee, at "nondiscriminatory" long-distance rates, rates inflated to subsidize basic local service. Bell, however, resisted this burgeoning competition legally and politically.[2]

AT&T's resistance and public opinion favoring more competition in telecommunications stirred the embers of antitrust action in the Department of Justice, a smoldering, decades-long challenge to the Bell monopoly. In 1974 President Ford's Attorney General filed suit against AT&T to break up the Bell System. Legal skirmishing continued for

---

*The cross-subsidy issue is this: Given that local and long-distance calls are charged and regulated separately, what proportion of the common total telephone plant that links telephone to telephone—wire, poles, switchboards, trucks, tools, buildings, paper clips, and so forth—should be attributed to local calls and what proportion to toll calls? Those estimates, in turn, form the bases regulators use to set rates for primary local service on the one hand and for long-distance tolls on the other. There is no simple answer. Critics have long asserted that Bell used excessive profits from the regulated, basic service—from local callers—to finance high-profit operations like long-distance and other telecommunications activities. Bell's defenders have countered that efficiencies in long-distance service have cross-subsidized local rates through the caps regulators have put on charges for basic telephone service. The latter appears to be the conventional understanding of economists and the one I assume for this discussion. See the endnote to this paragraph for futher discussion.

several years* before resulting in the divestiture agreement that became effective on 1 January 1984.

Divestiture and the related changes that accompanied it had many complex consequences. At the industry level, AT&T's operating companies became separate corporate entities, even competitors with one another and with AT&T in some activities. The manufacture, sale, and maintenance of telephone *equipment* was transformed into an open market that almost anyone could enter. Long-distance service also became, in effect, an open, albeit regulated, market. The new "Baby Bells," along with the independents, maintained their regulated monopolies on local telephone *networks* — on the service of interconnecting local telephones. At this writing, however, some regulators are moving to open local service up to competition as well.[3]

At the consumer level, before the 1970s, typical residential customers leased their telephones and received service from, made both local and long-distance calls through, and paid their bills to one company — effectively, "Ma Bell." Starting after divestiture in 1984, most customers assumed responsibility for their own telephones and the wiring in their own homes (although leasing a telephone remained an expensive option). In addition, they made local calls and paid local bills to one company, usually a "Baby Bell," but chose among several separate long-distance companies. Consumers likewise witnessed several shifts in costs. Cheaper telephones, and a variety of them, became available, and long-distance rates dropped. However, basic local service, including charges for installation, got comparably more expensive. Consequently residential customers were upset, and subscription rates dropped for a few years.[4]

Despite the dislocations of the 1980s, over the long stretch, residential subscribers enjoyed substantial reductions in the real cost of basic telephone service, a cut of almost half between 1955 and 1980. Also, Bell in 1963 began discounting long-distance calls during off-peak hours. Later, regulators required "lifeline" rates for poor customers. Starting in the late 1940s, the availability of financial subsidies such as below-cost federal loans for telephone cooperatives, along with technological improvements, made rural telephony more feasible and even cheaper than urban service.[5]

---

*The debate was, in part, about the history of the industry, leading AT&T to commission historical research for its lawyers. Historians have recently presented some of that material in monographs. Thus are we scholarly sparrows fed.

These structural changes, combined with social changes occuring over the same period, such as increasing affluence in the 1950s and 1960s, resulted in a vast expansion in the number of American households with telephone service: 62 percent in 1950, 78 percent in 1960, 90 percent in 1970, and 93 percent in both 1980 and 1990. This is fewer than the number of households that had televisions (98 percent in 1990) but more than had motor vehicles (88 percent in 1987).[6] In addition, the number of conversations multiplied over sevenfold between 1950 and 1985 and continued to accelerate in the 1980s.[7] Telephone shopping attained, at least in some realms, the level Bell salesmen had dreamed of generations earlier.* The mail-order process gave way to one of phone orders and mail delivery.[8] Some towns and some people remained outside of the American telephone system, but they were newsworthy precisely because they were so rare.[9]

For almost all Americans today, the telephone has become an "anonymous object," part of the everyday environment, like houses, streets, and grocery stores. Intellectual interest, which paid little attention to the telephone in the past anyway, now focuses on more recent innovations and on futuristic scenarios. One recent innovation, for example, is the answering machine. By the end of the 1980s, perhaps one-third of American households had one. Some commentators are ready to invest that device with heavy psychological overtones. More typical Americans probably wonder about the etiquette of using answering machines and advice-givers like "Miss Manners" are ready to help.[10] Realistically, it is too soon to speculate about the ultimate social meaning of the answering machine, but its very development serves as a cautionary tale (reminiscent of points made in Chapter 1): The social implications of a technology cannot be simply deduced from its operational features. One claim about the telephone, discussed in Chapter 8, has been that its insistent ring demands instant response, leading in turn to constant tension. The answering machine—a minor modification of the technology—allows people to ignore the ring without losing the message. Has the social meaning of the telephone been so easily reversed?[11]

Futuristic scenarios project radically new ways of life inspired by recent developments in electronic technologies. Some, for instance, have described a twenty-first-century America organized around

---

*Yet, at the same time that telephone shopping grew, so did in-person shopping, specifically through the great expansion of suburban malls.

"electronic cottages"—families living in homes scattered about the countryside, working via linked computers, and learning, shopping, relaxing, and socializing by video telephones.* The more prosaic reality—at least in the early 1990s—is that home computerization has not really occurred, despite great marketing efforts, and that "telecommuting" remains exceptional.[12] (To be sure, telecommuting is common among those who write about it, like this author; that should make us even more suspicious about the hoopla.)

History, too, suggests caution regarding extrapolations into the future. Commentators in the past made many broad predictions about the effects of the telephone and related technologies. The telegraph, as well as the telephone, was supposed to make "one neighborhood of the whole country." First, the telephone and then the mass media were supposed to erase local dialects. (Even in England, where the BBC has consciously promoted its brand of the Queen's tongue and largely controlled the broadcast media, speech standardization has not occurred.) A few predicted that there would be an airplane in every garage (Alexander Graham Bell among them). Similarly, a long dissertation could be written about all the technologies that were heralded as the solution to farm crises. In 1891 the Postmaster General predicted that Rural Free Delivery would keep youngsters on the farm by bringing them the entertainment of newspapers and magazines. Later, others argued that the telephone and the automobile would stem the exodus of the young by enabling them to have a social life while residing on the farm, and that radio would save the family farm by reducing boredom there. Throughout, save briefly during the Depression, the farm population continued to dwindle.[13]

One implication of the failure of such predictions is that basic social patterns are not easily altered by new technologies, that they are resilient even to widespread innovations. Whatever the structural and cultural factors that encouraged youth migration from farms—agricultural overproduction, the expansion of education, the "bright lights" of the cities, and so on—that pattern has persisted from the early years of the Republic through the whole twentieth century, continuing through the diffusion of steam power, railroads, telephones, automobiles, movies, television, and much more.

---

*The reader may recall from Chapter 8 that this prediction was made a century ago, too (p. 224).

THE ROLE OF THE TELEPHONE

The Knights of Columbus Adult Education Committee, quoted in Chapter 1, asked in 1926 whether the telephone weakened character, made people lazy, broke up home life, and reduced visiting among friends. Our concerns in this study were phrased in a similar vein: How did Americans come to use the residential telephone, with what social and personal consequences? (Like the Knights, this study bracketed the issue of the telephone in manufacturing or commerce.) To help answer these questions, it was important, as a first step, to understand who adopted the telephone, why, and for what ends. In this section, I draw some broad conclusions from the evidence presented in the previous seven chapters.

The reasons for which Americans in the decades before World War II subscribed to residential telephone service are not obvious. The telephone companies themselves encountered difficulties in understanding how to sell this product. The evidence about early subscribers suggests that some people wanted a home telephone for job-related reasons. These people typically included doctors, certain types of businessmen, farm owners, and specific white-collar workers. Others wanted a home telephone for primarily social reasons. This group contained disproportionately many rural and suburban residents, young people, and especially women. Most probably also wanted the home telephone for use in emergencies, even if such occasions were rare. These patterns show us the important role of need and taste—more abstractly, of agency—in explaining the spread of telephony. The patterns of demand sometimes differed from the preconceptions of industry men, as the stories of rural telephony and womens' use of the telephone demonstrate. The industry did not determine the demand for home telephones. Nor did the technology's logic simply impel its use. People decided whether or not to subscribe based on their tastes and needs. Furthermore, people changed their tastes over time. Into the 1920s, millions became more willing to subscribe. Yet, as the decline of rural telephony after 1920 and the unusual weakness of telephony during the Depression implies, others decided to drop the "telephone habit."

Wanting telephone service, however, was not the same as obtaining it. Subscription also depended on affordability and availability. Clearly, the more affluent households were the earliest subscribers to telephone service; working-class people subscribed quite late. Choice was also constrained by availability. Where and to whom the companies, particularly Bell, chose to market telephone service, the type of service they provided, and the rates they charged—all these factors helped determine who would have telephone service and when. Changes in the extent of competition and the industry structure— monopoly, followed by fierce rivalry and then regulated oligopoly— affected the spread of the telephone.

Similarly, decisions of government influenced the ability of people to subscribe. One source of this influence stems from the partial control regulators had on pricing, technical requirements, and the like. Another can be seen in comparing the spread of the telephone, where subsidies came late and were small, to that of other technologies, where subsidies came early and were sizeable. We noted in particular the strong governmental support automobiles received. The role of government is evident on an even larger scale in cross-cultural comparisons. Other nations developed telephone systems later and smaller than that of the United States, but some of those systems were more evenly spread than American telephony. Industry and government, in addition to the technology of telephony, helped form a structure of telephone availability that permitted some who wanted telephones to obtain them while preventing others from doing so. (The most dramatic stories in this regard are those of the rural towns that pled for service in the earliest years.) Finally, this changing structure of access may have later compelled some people who would otherwise not have wanted telephone service to get it. At least, it is our speculation that telephone subscription has in some ways become required precisely because it has become so common.

Knowing who subscribed during the first decades, and why they did, does not directly reveal how Americans used the telephone. Also, uses can change over time (as suggested by the apparent increase in telephone "visiting" over the century). Still, the evidence yields some conclusions about how Americans used home telephones during the first half of the twentieth century. Residential subscribers called for help in emergencies. More often, they called to conduct those commercial activities recommended by the telephone companies (and some they did not recommend, like ordering liquor). Most striking, however, given the telephone companies' hesitancy about, even resistance to, such use, was the early and expanding use of the telephone for sociability—for invitations and greetings, but especially for conversations.

In more general terms, Americans apparently used home telephones to widen and deepen existing social patterns rather than to alter them.

Despite the romance of speaking over long distances, Americans did not forge new links with strange and faraway people. Despite experiments with novel applications of telephony, they did not attend concerts, get healed, hold town meetings, or change lifestyles via the telephone. As well as using it to make practical life easier, Americans — notably women — used the telephone to chat more often with neighbors, friends, and relatives; to save a walk when a call might do; to stay in touch more easily with people who lived an inconvenient distance away.* The telephone resulted in a reinforcement, a deepening, a widening, of existing lifestyles more than in any new departure.

What, then, of the relationship between the telephone and "modernity?" Although the "modern" way of life obviously incorporates the telephone, whether contemporary lifestyles have social, cultural, or psychic distinctions that can be tied to the telephone — or to other contemporary technologies — remains unproven. A common contention is that modern telecommunications have created a "placelessness" or "rootlessness," or, put more positively, a liberation from geographical constraints. Yet Americans, always unsettled compared to most peoples, today seem no more so, surprisingly enough.** Their social circles may now be more spatially expansive than before, but this has not necessarily made them rootless, nor cosmopolitan.

To clarify the links between the telephone and modernity, we will consider some specific propositions about the telephone and social change. Again, I set aside the telephone in manufacturing, commerce, and the like. Many commentators have addressed those spheres, suggesting ways that the telephone expedited the rise of the American corporation, mass production, nationwide product distribution, urban decentralization, and so forth. Development experts make similar points when they advocate improving Third World telephone systems to stimulate economic activity and curb urban growth there. These arguments seem plausible, although they still lack empirical proof.[14] If they are valid, then the diffusion of the telephone seriously altered the economic circumstances of Americans' lives. My focus here, however, has been on how Americans before World War II used the telephone in their personal lives and what may have directly resulted therein.

---

*But not necessarily with people who lived far away. Long-distance calling from home was relatively rare in the first several decades.

**I noted in Chapter 7 that residential mobility has declined. "Surprisingly" refers to the fact that this stability exists in spite of other changes that have encouraged footlooseness, such as the expansion of higher education, cheaper costs of travel, and retirement pensions.

Isolating the changes brought on by popular use of one technology from those brought on by others during the same period—the automobile, electrification, radio, and so on—and from other secular changes—increasing affluence, industrial restructuring, or even autonomous cultural movements—is difficult. But the effort is necessary. As argued in Chapter 1, these technologies are not interchangeable expressions of some common transformation, but rather, constitute distinct developments. Although, for example, it is likely that use of the automobile greatly amplified the process of suburbanization originally stimulated by horsecar and streetcar travel, the role of the residential telephone in suburbanization appears to have been marginal, at best. Conversely, although it is probable that wide diffusion of the telephone accelerated the pace of commercial buying and selling, the personal automobile probably had a much smaller influence. Such separation of the historical parts played by each distinct technology and by other secular changes is a complex task. Nevertheless, we venture some conclusions about the telephone.

One argument is that use of the telephone had direct psychological consequences. Most commentators in this vein claim it increased alertness, tension, and anxiety, although a few argue that it resulted in a greater sense of security and, thus, calmness. The evidence shows that Americans in the first generation of telephony may have had some emotional reactions to the new device, ranging from awe to trepidation. For the most part, though, Americans in the first decades of the twentieth century seemed to quickly take the technology for granted. There was little serious controversy about the telephone, unlike the automobile, toward which many people were intially fundamentally hostile. Moreover, the telephone left little lasting impression on the consciousness of Americans compared to, say, the automobile. If there has been some general change in the American psyche during the twentieth century, the telephone is not responsible.

(Some writers, such as Kern, Meyerowitz, and Ronnell, have speculated about the implication of the telephone at deeper levels of the American psyche and of American culture than I have treated here. Ronnell has suggested, for example, that the "ringing [of a telephone] corresponds to a deeper, more primal voice within us, perhaps a parental voice. . . . We cannot resist the command."[15] They may be right, but it is a challenge to find reliable, relevant evidence. Thus, these sorts of arguments are difficult, if not impossible, to evaluate empirically.)

Did use of the residential telephone undermine local community? Did it foster detachment from place? Or, as a few observers have suggested, did telephone contacts intensify local interaction? Our general conclusion regarding the history of localism from 1900 to 1940 found that it did not change much. Northern Californians increased their interest and involvement in the world outside their immediate towns, but they also expanded their local activity at the same time.

Localism is a sphere in which disentangling concurrent influences is particularly difficult. Some part of the spatial expansion of Americans' lives must be attributed to nontechnological changes, such as political and economic events—World War I, the Depression, the New Deal, and so on. The automobile is also responsible to some degree, as exemplified in the fashion of auto touring during the 1910s and 1920s. Telephony, however, seems to have played either a small role or, perhaps more accurately, a complex one. It enhanced participation of all kinds, local and extralocal, thus expediting the expansion of all social activity.

The telephone appears to be implicated more in another trend, that of increasing privatism. By privatism, I do not mean a retreat into the nuclear family, which many writers see as a result of suburbanization in particular. I refer instead to the participation in and valuation of private social worlds as opposed to the larger, public community. Suburbanization probably contributed to privatism by promoting more geographically isolated and more culturally homogenous communities. The automobile, likewise, probably played a role, not only indirectly through suburbanization but also directly by facilitating more private travel, more private recreation (recall the increase in holiday outings over public celebrations), and more frequent visits with kin and friends. In several ways, the telephone may also have assisted this trend. In practical matters, the home telephone allowed subscribers to conduct from the privacy of their homes some otherwise public activities, such as calling the doctor, ordering liquor, or courting a lover. It permitted people, especially women, to more intensely pursue organizational involvements. Although strictly speaking club and church work are public activities, they are usually restricted to the immediate social world.* The home telephone allowed subscribers to maintain more frequent contact with kin and friends by chatting

---

*That is, there is little evidence that the telephone enabled people to become involved in distinctively new organizational commitments. The implication in most accounts is that it allowed people to be more active in their usual churches and clubs.

briefly perhaps a few times a week instead of at greater length once a week. There is little sign that telephone calling opened up *new* social contacts. Finally, in many circumstances, a telephone conversation was more private than a face-to-face one.* In these ways, telephone use may have contributed to the development of a more compartmentalized society during the first half of the twentieth century, one in which people were more active in their parochial worlds than in the wider, public realm.

Finally, did the telephone calling end isolation and increase "gaiety" as telephone publicists claimed, or did it undermine human relations by substituting impersonal media for face-to-face contacts? A major theme throughout this study has been that Americans used the telephone for sociability and used it for that purpose early on. The relationship of women to the telephone emphasizes the connection with personal relations. The telephone was, as argued in the previous chapter, a "technology of sociability."

The telephone allowed Americans to maintain relations with people in other neighborhoods or nearby towns more easily than would otherwise have been possible. Even today calling remains overwhelmingly local, although the area that is termed "local" may have expanded several square miles over the course of this century. With the telephone, Americans did not have to so easily give up contact with kin or friends from whom they had separated.** Yet, did having the telephone stimulate such separation? In a few cases, probably so. It is unlikely, however, that the availability of the telephone provided the major impetus for population dispersal and decentralization. Are relations sustained by telephone contact diluted or distorted by their medium of communication? Some have argued that telephone calling undermines the "authenticity" of personal relations, but what evidence we have suggests that calls more often serve to reinforce relations. (Most telephone relations, even today, are combined with frequent personal meetings. The archetypal telephone tie, one carried on solely through long-distance calls, is unusual.) The best estimate is that, on the whole, telephone calling solidified and deepened social relations.

---

*This was obviously more so for those with individual than with party lines. Still, despite all the stories about listening in on party lines and listening in by operators, even for those on party lines a call could avoid the exposure of openly walking to a person's home or meeting in public.

**The distances involved in such separations are usually rather modest, several rather than several hundred miles. Although many essays on telephony have portrayed an image of coast-to-coast relations, telephones have always been overwhelmingly used to call people in the local exchange area.

In assessing the role of the residential telephone in American society, we can add the foregoing uses to those more mundane ones about which there is little debate, those that make aspects of daily life more convenient—such as calling to tell someone you will be late, reserving a table for dinner, or arranging a babysitter—and the use of the telephone in emergencies. People accomplished these purposes without the telephone, but it greatly enhanced their efficiency. Add the use of the telephone in commerce, and one can make the case that the technology has played a wide and significant role in people's lives during this century. More interesting than the dimensions, however, are the subtleties of the telephone's role, particularly its relationship to modernity.

If "modernity" is defined in contrast to "community," as entailing atomism and hollowed-out social ties, then the present findings imply that the telephone was anti-modern. (Whether that general characterization of modern society is at all correct is another matter.) The evidence suggests that by using the telephone, people—especially women—more easily maintained and reinforced social relations; that the telephone was a device supporting parochialism.* Meanwhile, other changes—say, declining family size, increasing affluence, mass media, and expansion of higher education—may have reinforced individualism. There is no need to assume that social change was consistent. As for the telephone, it may have been *not,* in the philosophical use of the word, "modern."

In sum, the residential telephone is implicated in the development of what we call modern life and has even become necessary to sustain it. The size of its role may not approach that of the family automobile. Automobile production and supply form a major sector of the economy; important life decisions, such as choice of home and job, depend on having a car; communities are built around traffic; and society struggles with the social and environmemental by-products of driving. Yet, a society in which telephone communication were unfeasible would be substantially different than the one in which we live. Focusing on personal matters, the home telephone seems to have added considerable convenience and security. It also seems to have

*This point is similar to Willey and Rice's (*Communication Agencies*) suggestion in the 1930s that the telephone and automobile may have had localizing influences, but with a major difference: The parochial circles of the late twentieth century may be less spatially determined than those of earlier eras. People's social worlds may be less tied to particular places—rather, they may be organized around age peers, occupation, leisure tastes, and so on—but be nonetheless parochial.

expanded a dimension of social life, the realm of frequent checking-in, rapid updates, easy scheduling of appointments, and quick exchanges of casual confidences, as well as the sphere of long-distance conversation. No doubt, such social calls have displaced some face-to-face contacts (as well as letter-writing), but overall, total talk has most surely expanded. Americans—and here we are speaking far more of women than of men—enrich their existing social relations by frequent telephonic "touch."

Telephony, of course, has its serious frustrations. Aside from annoyances, such as sales people and abusive callers; aside from problems of service, pricing, and equity; and aside from the harassment some people feel from receiving too many calls—a key drawback of the home telephone is that very same expanded sociability. To have access to others means that they have access to you, like it or not. Increased sociability can be a mixed blessing.

IMPLICATIONS FOR THE STUDY OF TECHNOLOGY

In Chapter 1 (pp. 6-21), I questioned some of the literature on technology and society. I criticized perspectives that treat technology as determinative of social life, a specific version of which is the "billiard-ball" model that depicts the cascading "impacts" of an innovation. This type of model makes no contribution toward the understanding of the history of the telephone. This study could find no clear sequence of dramatic social changes attributable to its introduction. Also, the simple economic rationality that is assumed by the determinist model does not explain the diffusion of the telephone. The struggle of marketers to identify uses, the fumbling steps toward a sales strategy, the cultural blinders industry men wore, the active role of consumers—all of these elements are too complex to fit into the old schema crediting devices with remaking society. Another determinist approach, which I labeled the "impact-imprint" model, tacitly assumes that properties of a technology directly transfer to mental states and social action. This model also does not correspond to the evidence. This study was unable identify psychological changes that simply mirror the characteristics of the telephone (for example, that the suddenness of calls creates tense personalities). Nor did it find the sorts of social changes that some analysts deduce from the operation of the telephone. People did not, for example, lose their regional accents or their local prejudices as the result of a "placelessness" inherent in the device. Nor did human relations become more impersonal because the telephone is a machine.

Another major approach to technology and society, which I designated as "symptomatic," treats all modern technologies merely as parallel expressions of underlying cultural tendencies. The story of the telephone, however, is different from that of the automobile and other contemporaneous technologies. At various points, the telephone's adoption (and abandonment), use, and social consequences were distinct and perhaps even in contradiction to those of contemporaneous technologies. Although, for example, Americans widely used the automobile to explore new worlds by auto touring, they used the telephone largely to reinforce their existing worlds.

Rather than follow these approaches, I have argued for a "user" focus, a perspective that stresses the agency of consumers in adapting new technologies to their own ends. This study has shown, in the cases of farmers, women, and sociability, the role of taste in shaping demand for the telephone. It has also illustrated, in the patterns of adoption and use, the autonomy of consumers from the pressure of vendors and from any supposed technological imperative. Simultaneously, the users' autonomy was limited by structures beyond their control. Who subscribed depended in part on income, prices, where companies marketed the service, the role of government, and other external factors. One should not minimize the constraints on choice. At the same time, however, consumers pressed the limits, as the story of women and telephone sociability suggests, and as does the story of rural telephony's rise and fall.

Users try to put a new technology to their own ends, which can lead to paradoxical outcomes not easily deducible from the straightforward logic of the technology. Subscribers found many seemingly impractical uses for the telephone industry's practical device; automobile owners put that intended plaything to practical work. AT&T's copywriters touted the telephone as a "highway" to the wide world and the telephone system as a "web" that joined the people of our nation. Yet, residential subscribers were about as likely to use the system to remain provincial as to become cosmopolitan, talking to those they already knew rather than forging new ties. This history can best be understood if technologies are viewed from the vantage point of the user rather than from the perspective of the inventor or marketer.

The interplay between technology and society can also be better understood, I argued, with the cumulation of detailed social-historical studies of how average users employed new devices. Although many rich accounts exist telling of society's role in technology—that is, studies of social influences on technological development—there are few such detailed analyses of a technology's role in society. One reason for this imbalance is the burden of the determinist and symptomatic approaches I critiqued earlier, which presume facile answers to complex empirical questions. Another reason is that such studies are difficult to accomplish. Evidence about ordinary use is hard to obtain and is often open to varying interpretations, as this book probably well illustrates. Still, these studies must be undertaken to achieve a full understanding of the difference these devices make in society.

If one implication of this study is that Americans used the telephone to enhance the ways of life to which they were already committed, that conclusion should not be simply extrapolated to all technologies in the same class.* We have seen here that the social history of the automobile, although similar to that of the telephone in some respects (for example, its popularity in rural America), was different in others (for example, its popularity among the working class). The television provides another illustration. Some persuasive arguments have been made that watching television represented a serious alteration in how Americans spent their time, to take but one aspect of television's possible implications.[16] Perhaps, then, telephone calling reinforced social patterns and television watching revolutionized them. But these sorts of conclusions require comparably complex studies of the uses of the various technologies.

Similarly, the reader ought not to extrapolate these relatively benign conclusions about residential telephony to a pollyannish view of all technologies. Many specific technologies can have regretful consequences. This is so, not because Technology—with a capital T—is "in the saddle," but because of the ways in which users quite reasonably employ new devices. The personal automobile is, again, a good example. The jury is still out on whether, summing up across various domains, the family car is more of a boon or a bane for America.[17] Most of the negative aspects arise from the collective irrationalities inherent in the mass use of this machine that Americans have found to be so beneficial to them as individuals. These irrationalities include pollution, huge numbers of injuries and deaths, traffic jams stretching

---

*What such a class would be is still open: Is the telephone best categorized as an electronic technology, a space-transcending technology, a networking technology, a nineteenth-century technology?

for miles, inefficient town planning, and so on. Too much of a good thing can add up to a bad thing. Many technologies, such as smoking and television addiction, may also be, ultimately, "social traps." But the particulars of each technology's role in society need to be studied closely. In the case of the home telephone, contentions about irrationalities and traps do not seem to be valid for the first half of the twentieth century.

The American way of life as the twentieth century ends is, of course, very different from that at the end of the nineteenth, when home telephones first appeared in meaningful numbers. It is obviously different materially—a world of suburban sprawl, shopping malls, television, automobiles, and certainly telephones. The social structure of the population has changed. Americans are richer, divorce is common, families are smaller, mothers typically are employed, a college education is routine, and so forth. Cultural changes are less obvious, but one can point to sexual liberalization, the expansion of rights, and other expressions of individualism. These are among the many characteristics found on various observers' lists of what makes *fin-de-siècle* America "modern" (or even "postmodern"). Still, one questions whether all these changes form a common, unitary phenomenon. There is no necessary reason they should cohere or operate in parallel, no necessary reason to think that there exists one social and cultural syndrome called "modernity." It is plausible, for instance, that expanded individual freedom brought with it, not atomism, but sociability.

Even if there is a core lifestyle that merits the appellation "modern"—say, individual autonomy undergirded by financial security and guarantees of the modern state—it does not logically follow that every social change necessarily supports that culture. In the case of technology, we may find that some particular manifestations promote such modernity and some undercut it.

This study opens only a small window on the historical progression labeled "modernization." Yet, even though debate flares about the cultural symptoms of modernity, all must acknowledge that Western history for the last several generations includes radical changes in material culture. Americans living between 1880 and 1940 witnessed a vast array of material changes, many of which happened abruptly. Technological innovations transformed everyday life. The life of the present—with its cars, telephones, televisions, radios, packaged foods, vitamin pills, air-conditioning, stay-pressed clothing, disposable diapers, and so on and so on—is very different from that of a century ago. Whether or which cultural and psychological changes accompanied these material transformations is a matter of great debate. Scholars of the subject seem to be moving away from the old, simple, global dichotomies—that the world of the past was community, the world of today impersonal society, or that the people of the past were unthinking expressors of tradition, the people of today rationalists. Instead, contemporary culture probably includes combinations of, even contradictions among, elements that have been labeled traditional and modern. In this more complex view, we might consider a technology, such as the telephone, not as a force impelling "modernity," but as a tool modern people have used to various ends, including perhaps the maintenance, even enhancement, of past practices.

APPENDIX A

Bibliographic Essay

Despite the ubiquity of the telephone in American homes, the literature on the subject is quite thin in contrast to the volumes available on other technologies. Attention has grown in recent years with the Johns Hopkins/AT&T Series in Telephone History edited by Louis Galambos (cited below), a few scattered dissertations, and more recently, a handful of social historical works. In this essay I mention mostly books, monographs, and archival sources, leaving specific articles and book chapters largely to the notes. There are, no doubt, more fugitive sources to be found, especially in the archives of individual telephone companies.

GENERAL HISTORIES

A few books have appeared during the twentieth century that recount the history of the telephone in America, but most must be approached with caution since they have been in some fashion subsidized by AT&T and extol the industry. John Brooks's *Telephone* (1976), although aided by special access to AT&T archives, is the most solid and useful narrative of the industry's first century. The books by Herbert Casson, *The History of the Telephone* (1910), Arthur Pound, *The Telephone Idea* (1926), and M. Dilts, *The Telephone in a Changing World* (1941), are suspect but informative. H. B. MacMeal's *The Story of Independent Telephony* (1934) presents telephone history from the viewpoint of AT&T's competitors. Robert Chapius, *100 Years of Telephone Switching* (1982), provides a useful technical history. (A few volumes have appeared recently dealing with the divestiture of AT&T, but those go far beyond the focus of this study. There is also a literature on telephones and economic development that I do not review here.)

CORPORATE, FINANCIAL, AND ECONOMIC HISTORIES

Several studies have been done on the history of AT&T, often with polemical intent. Hostile New Deal investigations led to several reports by the Federal Communications Commission, a *Proposed Report* in 1938, a version cleaned up in response to Bell protests, *Investigation of the Telephone Industry* (1939), and a free-lance book using much the same materials, N. R. Danielian, *A T.&T.: The Story of Industrial Conquest* (1939). AT&T defenses include H. Coon, *American Tel & Tel: The Story of a Great Monopoly* (1939) and A. W. Page, *The Bell Telephone System* (1941).

There are many histories of specific U. S. telephone companies, virtually all classifiable as vanity publications, but which sometimes include useful details. In this category are: F. L. Howe, *This Great Contrivance* (1979); R. S. Masters et al., *An Historical Review of the San Francisco Exchange* (1927); J. Nagel and M. Nagel, *Talking Wires* (n.d.); James Crockett Rippey, *Goodbye, Central; Hello, World* (1975); W. A. Simonds, *The Hawaiian Telephone Story* (1958); and J. L. Walsh, *Connecticut Pioneers in Telephony* (1950). Many more can be found in pamphlet or manuscript form in various company archives, such as in Ralph L. Mahon, *The Telephone in Chicago, 1877–1940* (c.1955).

Scholarly studies of the telephone business include those in the Johns Hopkins series: Robert W. Garnet, *The Telephone Enterprise* (1985); Kenneth Lipartito, *The Bell System and Regional Business* (1989); George David Smith, *The Anatomy of a Business Strategy* (1985); and Neil H. Wasserman, *From Invention to Innovation* (1985). Other important studies include: William P. Barnett, "The Organization Ecology of the Early Telephone Industry" (1988); G. W. Brock, *Telecommunications Industry* (1981); I. M. Spasoff and H. S. Beardsley, *Farmers' Telephone Companies* (1922, 1930); and J. W. Stehman, *The Financial History of the American Telephone and Telegraph Company* (1925). Alan Stone's *Public Service Liberalism* (1991) appeared too late to be used in this study.

SOCIAL STUDIES

Sociological and social historical studies are rare. The most important volume is Ithiel de Sola Pool's edited collection, *The Social Impact of the Telephone* (1977), which is notable more for its theoretical articles than for its empirical findings. Pool's *Forecasting the Telephone* (1983) is a compilation of speculations made early in the century about what the telephone's social consequences would be. A recent German symposium on the telephone includes some English-language articles: Forschungsgruppe Telefonkommunikation (Hrsg), *Telefon und Gesellschaft* (1989). Carolyn Marvin, *When Old Technologies Were New* (1988), reviews some of the nineteenth-century debates about the telephone's place in society. Roy Alden Atwood, "Telephony and Its Cultural Meanings in Southeastern Iowa, 1900–1917" (1984), traces the introduction of the telephone in that region. Lana Rakow, *Gender on the Line* (1991), uses oral histories to explore the meaning of telephony for women. Michèle Martin's *"Hello, Central?" Technology, Culture and Gender in the Formation of Telephone Systems, 1876–1920* (1990) does the same with archival materials. The fullest exploration of the telephone's social role is the 1933 book by Malcolm Willey and Stuart Rice, *Communication Agencies and Social Life.*

Any study of telephone development in the United States relies on the Bureau of the Census' reports covering every fifth year from 1902 to 1937 (starting with U. S. Bureau of the Census, *Special Reports: Telephones and Telegraphs 1902,* 1906). All the reports provide detailed statistics and the first few also contain extended descriptive material and analyses.

Important sociological and historical articles about the telephone that can be found in American periodical literature include: Aronson, "The Sociology of the Telephone" (1971); D. W. Ball, "Toward a Sociology of Telephones and Telephoners" (1968); R. H. Glauber, "The Necessary Toy: The Telephone Comes to Chicago" (1978); George A. Griswold, "How AT&T Public Relations Policies Developed" (1967); J. V. Langdale, "The Growth of Long-Distance Telephony in the Bell System: 1875–1907" (1978); Milton Mueller, "The Switchboard Problem" (1989); Michael L. Olsen, "But It Won't Milk the Cows: Farmers in Colfax County Debate the Merits of the Telephone" (1986); and Joel A. Tarr, with Thomas Finholt and David Goodman, "The City and the Telegraph: Urban Telecommunications in the Pre-Telephone Era" (1987).

FOREIGN TELEPHONY

Christopher Armstrong and H. V. Nelles, *Monopoly's Moment* (1986), includes a solid history of Canadian telephony. Michèle Martin, *"Hello, Central?"* (1990) gives a more critical history, especially of Bell in Quebec. Other Canadian monographs are internal or subsidized publications: Bell Canada, *The First Century of Service* (1980); Tony Cashman, *Singing Wires: The Telephone in Alberta* (1972); Robert Collins, *A Voice from Afar* (1977); E. B. Ogle, *Long Distance Please* (1979); and William Patten, *Pioneering the Telephone in Canada* (1926). Canada's Department of Trade and Commerce issued "Telephone Statistics" for 1922 on. See also Robert Pike, "Kingston Adopts the Telephone: The Social Diffusion and Use of the Telephone in Urban Central Canada, 1876 to 1914" (1989).

Other accounts of foreign telephony include: Attman et al., *L M Ericsson 100 Years* (Sweden, 1977); F. G. C. Baldwin, *The History of the Telephone in the United Kingdom* (1938); Catherine Bertho, *Télégraphes et Téléphones* (France, 1981); A. N. Holcombe, *Public Ownership of Telephones on the Continent of Europe* (1911); J. E. Kingsbury, *The Telephone and Telephone Exchanges* (United Kingdom, 1915); articles in Renata Mayntz and Thomas Hughes (eds.), *The Development of Large Technical Systems* (France and Germany, 1988); Ann M. Moyal, *Clear Across Australia* (1984); J. H. Robertson, *The Story of the Telephone* (United Kingdom, 1947); and J. F. Ruges, *Le Téléphone pour Tous* (France, 1970). See also, articles in Joel A. Tarr and Gabriel Dupuy, eds., *Technology and the Rise of the Networked City in Europe and America* (1988).

ARCHIVES

There is a wealth of original materials to be mined from archives of telephone companies, both large and small, around the continent and also from branches of the employees' volunteer association, the Telephone Pioneers of America. Some of the recent scholarship cited in this essay has made use of these sources. In this study, I used five major depositories. Only one or two could be said to have systematized their records, with Bell Canada being the most sophisticated. But all contain much more richness than I could exploit. The archives are: AT&T's Corporate Archives, later renamed Historical Archives and Publications, located at the time of my research at AT&T corporate headquarters in New York City and now in Warren, New Jersey; Bell Canada's Historical Services in Montreal; the Museum of Independent Telephony in Abilene, Kansas, which contains papers donated by independent companies, including in-house histories; Illinois Bell's Information Center in Chicago, which contains many original documents on microfilm; and the Telephone Museum of the Pacific Telephone Pioneers of America (since reorganized as the Telephone Pioneer Communications Museum of San Francisco, Archives and Historical Records Center), which houses the files of former Pacific Telephone executives. I found the curators of these collections to be open and generous. In addition, the Business Americana collections of the National Museum of History in Washington had supplementary materials on the telephone industry.

Many telephone companies published useful in-house magazines. I drew, for example, on the *Pacific Telephone Magazine*. Two most useful national publications are the *Bell Telephone Quarterly* (although Bell has other publication series as well) and *Telephony,* the journal of the independent companies for about a century. Other, short-lived, industry publications can be found, as well, in the larger libraries or archives.

Statistical Analyses of Telephone and Automobile Distribution

ACROSS THE STATES, 1902 AND 1917

The analysis reported here draws on the data-set used in Fischer and Carroll, "Telephone and Automobile Diffusion," U.S. Bureau of the Census surveys of telephone companies conducted every five years from 1902 through 1937, and other census data describing state-level characteristics.

For each state, I regressed the absolute number of telephones reported in 1902 on various demographic and economic indicators interpolated for 1902. (All measures were logged to base 10.) Given the multicollinearity of variables at the state level, it is not surprising that alternative combinations of indicators yield similar R^2 estimates. This result required some informed choices to be made in selecting one equation over another. The most difficult such decision concerned whether to use a measure of *urbanization*, the number of people living in urban places, or a measure of *commercialization*, the number of people over ten years old employed in trade or transportation. Commercialization correlates with urbanization at .98 and with other measures of economic activity and development at similar levels. All these measures can be thought of as one complex. Commercialization, as the number of trade and transport workers in a state, seems to "work" better (indeed, population counts generally drop out), but more states are missing data on that variable. Consequently, I explored two parallel models, one using commericalization and one using urbanization. Their implications turned out to be similar. In addition, I examined residuals for outliers and "influential" cases, adjusting the equations accordingly. Table B–1 gives the "best" equations in terms of variance explained, robust estimates, and substantive meaning. (Significance tests, which have no real meaning in such a sample, are provided as rough guidance.)

The most influential factor in equations A–1 and A–2 is the (log) number of workers in trade and transportation, a predictor largely of Bell and of business telephones. Measures of industrialization, such as the number of manufacturing workers or total wages, have no independent correlation of significance. Beyond trade and transport, the state-level distribution of telephones in 1902 largely reflected conditions of agriculture: the value of the average farm property; the extent of farm ownership; to a lesser degree, farm tenancy; and negatively, the number of nonwhite farmers. The latter is highly correlated with total black population but is a better correlate of telephony; it may reflect transiency and

TABLE B–I

REGRESSION (LOG-LOG) OF TELEPHONES
ACROSS THE STATES, 1902
(unstandardized regression coefficients)

Equations:	A–1	A–2	B–1	B–2	B–3
Constant	-2.18^{**}	-1.74^{**}	-3.70^{**}	-3.27^{**}	-2.04^{**}
Number of Workers, Trade and Transportation	$.81^{**}$	$.77^{**}$			
Number of Farms Operated by Owners	$.30^{**}$	$.26^{**}$			
Number of Farms Operated by Tenants	$.09^{*}$	$.16^{**}$			
Number of Farms Operated by Nonwhites	$-.05^{*}$	$-.06^{**}$			
Average Value of Farmland and Buildings (log $1000)	$.31^{**}$	$.22^{**}$	$.54^{**}$	$.40^{**}$	$.30^{**}$
State Located in Deep South		$-.11^{*}$		$-.20^{**}$	$-.17^{**}$
State Located on Pacific Coast		$.23^{**}$		$.16^{**}$	$.24^{**}$
Urban Population			$.53^{**}$	$.46^{**}$	$.50^{**}$
Nonurban Population			$.57^{**}$	$.68^{**}$	$.46^{**}$
Massachusetts (an "influential" point)				$.23^{*}$	$.26^{*}$
Number of Telephone Systems					$.15^{**}$
R^2	.97	.98	.96	.98	.98

$^{*}p < .10$
$^{**}p < .01$

poverty, or perhaps a shunning of predominantly black regions by telephone entrepreneurs. Beyond these controls, states in the Deep South were still disadvantaged and the Pacific Coast states advantaged. These may indicate unmeasured
socioeconomic or industry factors—aggressive marketing by Pacific Telephone,
passive marketing by Southern Bell companies.

Equations B–1 through B–3 use population counts instead of industry and
farming data. Equation B–2 (which controls for statistically "influential" Massachusetts) suggests that telephony was more responsive to the nonurban than
the urban population, and responded to farm affluence. (Breaking down urban
population into two components, population in cities over 100,000 and in cities
below 100,000 made no appreciable difference.) The inclusion of the number of
telephone systems in a state, Equation B–3, indicates that nonurban and farm
population influenced telephony via the creation of numerous independent lines.

(An estimate of *profit*, telephone companies' gross revenues minus their gross
expenditures, is significant if added to these equations—less profit per telephone
coincided with more telephones per capita. But it is not included here because
of potential circularity in the measures.)

I applied equations A–2 and B–3 to predict number of telephones in 1907,
adding to the list of predictors the number of telephones for 1902. (Results not
shown.) With model A–2, the results indicate that states having many farm
owners and affluent farms in 1902 experienced especially large increases in telephones during the subsequent five years. Using equation B–3 shows that those
states destined to add many telephones had relatively large nonurban populations
(and, other analysis shows, smaller rather than larger cities) and valuable farms,
and that the South continued to fall behind. The thrust of the analysis is that
in these early years, the spread of telephony largely responded to rural conditions. Subsequent data analysis suggests, furthermore, that it was independent
telephones that increased according to these rural conditions. The number of
Bell telephones grew in response to commercialization, and it expanded in states
where independents had been previously successful.[*] On patterns of diffusion
after 1907, see Fischer and Carroll, "Telephone and Automobile Diffusion."

I conducted a similar analysis of state-level automobile diffusion for 1917,
when the distribution of automobiles was comparable to that of telephones in
1902. The correlation between telephones per capita and automobiles per capita
in 1917 is .79.

The regression analysis runs into the same multicollinearity problems discussed in the case of the telephone. In particular, several indicators (logged to
base 10)—urban population, total manufacturing wages, total value of manufacturing production, number of workers in trade and transport, total property
wealth, and total personal income in a state—are so highly intercorrelated that
in a factor analysis, all the measures load at .96 to .99 on the first principle
component.[1] In exploratory regression analyses, total personal income had the
highest coefficients, and it is used in Table B–2. It should, however, be thought
of as an indicator of a complex of urbanization, commercialization, and wealth.

---

[*]This last trend may have been due either to Bell aggressively competing with
independents or to Bell buying them out.

TABLE B—2

REGRESSION (LOG-LOG) OF AUTOMOBILES
ACROSS THE STATES, 1917
(unstandardized regression coefficients)

Constant	−.89*
Total Farm Population	.23**
Average Value of Farmland and Buildings	.52**
Total Personal Income (log $1 million)	.71**
New England States	.19**
Midwestern States	.12**
Illinois (an "influential" point)	−.26**
R^2	.98

*$p < .10$
**$p < .01$

The other variables used include interpolated farm statistics and dummies for unexplained region-specific effects.

Automobiles were most numerous where urban wealth, population, commerce, and industry were greatest and where farm wealth (the number of farmers multiplied by the average farm value) was greatest. The advantage of the Midwest, excepting Illinois, may have something to do with lower shipping costs from Detroit.[2] The source of the advantage of New England remains mysterious. Among the factors that were *not* important were land area, location in the Deep South (beyond what can be explained by its poverty), the average wage of workers, and measures of road development.

This model is simpler than that for telephones in 1902. In particular, automobiles increased along with rises in the number of farmers and their wealth, rather than differentially by farmers' race and tenure. This distinction may point to the greater "portability" of automobiles—that migrant tenants were as likely to buy them as farm owners, income held constant—and also, given the lack of any effect for location in the South, to a lesser importance for the diffusion of the automobile of the industry marketing structure.

I applied the same model, together with the number of automobiles, 1917, to predict automobiles in 1922. Relatively large increases in automobiles over that five-year period occurred disproportionately in wealthy urban states and the well-to-do farming states; the regional effects drop out. For 1922, I could differentiate farm from nonfarm automobiles. The number of farms with automobiles increased as the number of farm owners, farm tenants, and farm wealth increased. The number of automobiles off the farms increased as the complex of urban population, commerce, and industrialization increased. For more on state-level automobile diffusion, again, see Fischer and Carroll, "Telephone and Automobile Diffusion."

ACROSS CALIFORNIA COUNTIES, 1930–1940

Since the following analysis of rural telephony was conducted after the publication of Fischer, "Technology's Retreat," which dealt only with states, I briefly report it here.

County-level data on farms with telephones were available only for 1930 and 1940, so I could look only at the process of change during the 1930s. Even then, not all the state-level data were available for counties (for example, I could not distinguish number of telephones by type of company). I used the California

TABLE B–3

REGRESSION (LOG-LOG) OF TELEPHONES AND AUTOMOBILES ON FARMS, ACROSS CALIFORNIA COUNTIES, 1930 AND 1940
(unstandardized regression coefficients)

		Number of Farms with Telephones, 1930	Number of Farms with Autos, 1930	Number of Farms with Telephones, 1940	Number of Farms with Autos, 1940
Constant		−1.73**	−.93**	−.79	.29
Number of Farms with Telephones	1930	—	—	.86***	
Number of Farms with Automobiles	1930	—	—		.57***
Number of Owner-Operated Farms	1930	.88***	.75***	−.35*	−.33***
	1940			.52***	.65***
Number of Tenant Farms	1930		.27***		−.09**
	1940				.21***
Average Value of Farms	1930	.17**	.12**	−.35***	−.16***
	1940			.38***	.14***
Percentage of Population that is Urban	1930	.20		−.50	
	1940			.88*	
San Francisco County			−.85***		−.41***
R^2		.960	.995	.989	.999

*$p < .10$
**$p < .05$
***$p < .01$

Notes: N = 58 counties; "Owner" includes managers; Urban percentage is percentage of whole county's population.

data because it offered a large and complex state and because of the California focus in the susbsequent research for this book. All the data come from the censuses. I examined several models, in various forms, and tested for robustness by weighting cases, assessing autocorrelation (none), exploring "influential" cases, and making jackknife estimates. The models in Table B–3 seem robust. All values are logged.

In 1930 the counties that had many farms with telephones were those with more owner-operated and more valuable farms. The same was true for automobiles, except—as we found in the state-level analysis—more tenant farms also meant more farm automobiles, a condition which which was not so for telephones. San Francisco—a virtually all-urban county—trailed in farm automobiles. During the 1930s, the counties that added more farm telephones were those with the greatest growth in owner-operated farms, farm value, and urbanization. Farms with automobiles grew in number along with farm owners, farm value, and farm tenants (and San Francisco fell further behind). There was no effect of automobiles on telephones or vice-versa.

APPENDIX C

Telephone Subscription
Among Iowa Farmers, 1924

Jon Gjerde, a historian at the University of California, Berkeley, came across an unusual directory of Iowa farms for 1924, the *Prairie Farmers' Home and County Directory of Dubuque County*. It lists farmers, their spouses, their children both at home and away, others in the household, length of residence in the county, addresses, acreage owned or tenanted, the telephone company to which they subscribed (if any), and, in separate listings, the number of automobiles, tractors, and radios they owned (if any).[1]

We drew a random one-third sample of the listings (dropping any that were of nonfarmers, such as local mill operators) and coded the available data.[2] Then, I explored various logit regression models for predicting whether or not a farmer had a telephone listing.[3] The best-fitting models are presented in Table C–1. The entries are derived estimates that represent the change in the probability of having a telephone for each unit change in the predictor variable, holding all others constant at their means.

Table C–1 gives results for the sample as a whole and for only the male-headed households; the latter subsample highlights the role of women in the household. For each, there are two equations, differing only in how the farm's location is indicated. The first measurement divides the counties' townships roughly by their distance from Dubuque, the major town. The results (Equation 1) indicate that, all else equal, the odds of subscribing were least in Dubuque and in the first ring and highest in the farthest ring. (The estimate for Dubuque itself must

TABLE C–I

LOGIT REGRESSIONS OF PROBABILITY OF FARM
HAVING A TELEPHONE
(derived estimates)

	Total Sample		Male-Headed Households	
N	745		704	
Overall proportion	.698		.703	
	Eq. 1	Eq. 2	Eq. 1	Eq. 2
Constant	−.10	−.16	−.06	−.13
Location				
Dubuque	.18	.72**	.17	.74**
First Ring	−.40**		−.42**	
Second Ring	−.16*		−.18**	
Third Ring	0		0	
Northern Part of County		−.15**		−.15**
Proportion of Bell Telephones		−.34**		−.36**
Farm				
Acres (in hundreds)	.15**	.16**	.14**	.14**
Owner of Farm	.08*	.09*	.08*	.09*
Household				
Wife Present	.21**	.19**	.25**	.23**
Other Adult Males	−.12*	−.14**	−.12*	−.15**
Other Adult Females	.08	.09	.13*	.13*
≤ 30 Years, No Children	.24**	.27**	.20*	.23**
Children at Home, None Away	.08	.11	.04	.08
Children Away	.01	.03	−.04	−.02
> 30 Years, No Children	0	0	0	0
Interaction				
Wife × Dubuque	−.52*	−.52*	−.54*	−.54*
L	−358	−358	−337	−337
−2 × L	197	196	183	183
pseudo-R^2	.209	.208	.207	.207

*$p < .05$
**$p < .01$

be combined with the interaction term at the bottom), Wife × Dubuque. For households without a wife present, location in Dubuque—and there were only seven such farms—was associated with a .18 increase over third ring townships in probability of a telephone; whereas for households with a wife, Dubuque location reduced the odds of a telephone by .52, for an overall probability of

−.34 for the town. These results can be interpreted in two different ways. The rings are roughly associated with farmland quality, so that the findings may show that farmers on richer lands — with farm size held constant — were more likely to have telephones. The rings are also associated with more dispersed farms, so that the findings may reflect a greater need, and therefore demand, of sparsely-settled farmers for the telephone.

The second method of locational coding allocates townships according to three criteria: Dubuque township versus others; northern part of the county versus southern; and Bell company dominance. The latter indicator is the proportion of the farm telephones in a township that were Bell's, estimated from the sample of telephone subscribers themselves. The results (Equation 2) indicate that holding northern location and Bell dominance constant, location in Dubuque was positively associated with having a telephone (especially for bachelor or widower households[4]). This result differs so much from that in Equation 1 because Dubuque is in the north and was Bell country; thus, Dubuque farmers were relativley likely to subscribe compared to others in northern, Bell neighborhoods. The results show that residents of northern townships were less likely to subscribe. That, again, may reflect the poor quality of north county land. Finally, the parameter for Bell shows that in townships totally dominated by Bell, the chances of a farmer subscribing were much lower, .34 less, than in those townships with no Bell lines, all else equal. This speaks to the argument in the text that Bell was particularly unwilling to serve farmers or priced its farm service significantly higher than the other systems.

The results for farm variables show that, for each 100 acres of land tilled, the odds of telephone subscribing increased .15 or .16. Owners were .08 or .09 more likely to subscribe than tenants.

The results for the household variables show, first, the effect of wives. Outside of Dubuque, a wife increased the odds of having a telephone by about .20.[5] More precisely, this means that the minority of farms in the county without telephones were disproportionately those without wives. The results for other adult males and females show a curious result — curious, but consistent with other findings reported in Chapters 4 and 5. The more adult males in the home — kin or unrelated — the less likely there was a telephone. The more adult females, the greater the chances of a telephone, significantly so if we look only at male-headed households. Although arguments are made in a few places in this book for explaining the effect of females (especially Chapter 8), the reason why having more men in a household should reduce subscription is not clear.

In the absence of solid life cycle measures, I created a rough equivalent, using years residing in the present location as a proxy for age. Under 30 years, without any children listed, indicates a new, young family. Slightly further in the life cycle are families with children at home but none listed as away from home. Third are households old enough to have children out of the nest. The final, residual category is made up of older households without any children listed. The results indicate that the youngest families were especially likely to subscribe, the oldest and childless households least likely. (The gender composition or number of children seemed to make little difference.)

In sum, telephone subscription tended to be more common among farm families that, all else equal, were located far from the urban center, were served

TABLE C–2

LOGIT REGRESSIONS OF
PROBABILITY OF FARM HAVING
AN AUTOMOBILE
(derived estimates)

N	745
Overall proportion	.827
Constant	−.16*
Location	
North County	.06*
Farm	
Acres (in Hundreds)	.09**
Household	
Wife Present	.16**
Husband Present	.11*
Other Adult Females	.06
Number of Sons at Home	.01
Number of Sons Away	−.03**
Years in County	−.0018*
L	−298
−2 × L	88
pseudo-R^2	.106

*$p < .05$
**$p < .01$

by telephone companies other than Bell, worked large farms, owned their farms, were young, and had relatively more adult women than men.

For a brief comparison, Table C–2 displays the best-fitting models to predict automobile ownership in Dubuque County. Nearly everyone—83 percent of the farmers—had automobiles. The exceptions were the smaller farms, those missing either a male or female head of household, and the older households. These are straightforward findings. Less easily interpretable are the results showing that households in the southern part of the county and those who had many sons away from home were also likely to be among the few without cars. In contrast to telephone subscription, owning or tenanting did not matter (as we found in the state-level analyses; Appendix B), gender mattered less, and of course, telephone companies did not matter.

Summary of Expenditure Studies by Household Income or Occupation

Date and Population	With Phone (%)	With Auto (%)	Avg. Annual Expenditure, Phone ($)	Avg. Annual Expenditure, Auto ($)	Source of Study
1895, Chicago					Rosher, 1968
General population	< 1%	—	—	—	
"Elite"	24	—	—	—	
1903, San Francisco (dwelling units, by "class of residence")					Davis, 1903
Bottom quartile	10%	—	—	—	
Second quartile	13	—	—	—	
Third quartile	37	—	—	—	
Top quartile	56	—	—	—	
1905, Chicago					Rosher, 1968
General population	3%	—	—	—	
"Elite"	66	—	—	—	

Date and Population	With Phone (%)	With Auto (%)	Avg. Annual Expenditure, Phone ($)	Avg. Annual Expenditure, Auto ($)	Source of Study
1911, Kingston, Ont. (ratio of group's share of subscribers to its share of total population)					Pike, 1989
Laborer, unskilled	1 : 10	—	—	—	
Skilled, trades	3 : 10	—	—	—	
Propr., manager, agent, owner	6 : 1	—	—	—	
Semi-professional, white-collar	6 : 10	—	—	—	
Professional	10 : 1	—	—	—	
1915, New England (families in "typical" southern NE residential exchange, by income)					Harrell, 1931
Under $1300	3%	—	—	—	
$1300–$2700	22	—	—	—	
$2700–$4000	65	—	—	—	
Over $4000	95	—	—	—	
1918–19, 92 cities (12,096 white families of "wage-earners" and "small salary" men, by income)					Reanalysis of U.S. Bureau of Labor Statistics, 1924
Under $1200	10%	1%	$17	$105	
$1200–$1500	17	6	20	169	
$1500–$1800	25	9	21	222	
Over $1800	37	10	23	281	
1921, San Francisco Bay Area (82 families of typographical union members)	60%	21%	$27	$400	Peixotto, 1929
1922, Berkeley, CA (96 families of faculty at UC Berkeley)	98%	58%	$40	$673	Peixotto, 1927
1924, Chicago (884 families of unskilled and semi-skilled workers)	20%	3%	—	—	Houghtelling, 1927
1927, National sample (331 high-status families)	98%	88%	—	—	Morgan, 1939
1927, Michigan (4521 families in unspecified area, by "rental class")					Wilson, 1928
Under $35	15%	41%	—	—	
$35–$45	30	66	—	—	
$45–$60	48	69	—	—	
Over $60	65	88	—	—	

Date and Population	With Phone (%)	With Auto (%)	Avg. Annual Expenditure, Phone ($)	Avg. Annual Expenditure, Auto ($)	Source of Study
1927, 36 cities (families, excluding "working class," by income)					R.O. Eastman, 1927
Under $2000	55%	—	—	—	
$2500–$3000	75	—	—	—	
$3500–$5000	81	—	—	—	
$5500–$9000	89	—	—	—	
Over $9000	100	—	—	—	
1930, New England (families in a "typical" southern NE residential exchange, by income)					Harrell, 1931
Under $1300	22%	—	—	—	
$1300–$2700	58	—	—	—	
$2700–$4000	90	—	—	—	
Over $4000	97	—	—	—	
1931, Pittsburgh (ca. 11,000 families)					Lynd, 1933
Lower half	38%	17%	—	—	
Next 45%	76	48	—	—	
Top 5%	96	83	—	—	
1933, National sample (331 high-status families)	92%	93%	—	—	Morgan, 1939
1934–36, 42 large cities (12,903 families of wage-earners and clerical workers not on relief, by per capita income)					Williams and Hanson, 1941
Under $300	14%	26%	$23	$ 96	
$300–$500	29	40	25	142	
$500–$700	45	50	28	193	
Over $700	57	59	31	340	
1935–36, Small cities (native-white families of wage-earners or white-collar workers not on relief, by region and income)					Kyrk et al., 1941
North Central					
Under $1000	24%	47%	$23	$ 86	
$1000–$1500	46	65	23	133	
$1500–$2250	75	79	25	206	
Over $2250	93	89	29	336	

Date and Population	With Phone (%)	With Auto (%)	Avg. Annual Expenditure, Phone ($)	Avg. Annual Expenditure, Auto ($)	Source of Study
Southeast					
Under $1000	8	32	20	94	
$1000–$1500	35	49	20	188	
$1500–$2250	57	76	23	234	
Over $2250	91	92	26	310	
Pacific Northwest					
Under $1000	29	63	26	127	
$1000–$1500	35	73	26	176	
$1500–$2250	59	81	28	275	
Over $2250	85	88	27	401	
1941, National sample (ca. 1300 units—urban families and singles, by household income)					U.S. Bureau of Labor Statistics, 1945
Under $500	8%	12%	$28	$ 72	
$500–$1000	15	21	24	75	
$1000–$1500	36	42	22	122	
$1500–$2000	41	52	26	133	
$2000–$2500	54	73	29	159	
$2500–$3000	64	70	36	188	
$3000–$5000	82	81	37	237	
$5000–$10,000	86	88	47	333	
Over $10,000	100	94	81	589	

Sources and Annotations: Davis, 1903: Letter from Joseph P. Davis to F. P. Fish, 23 October 1903. In "Telephone Statistics, 1902–1904," Box 1312, AT&THA. The data are based on a procedure linking dwelling units to insurance maps. Note that San Francisco had highest telephone development in the nation at the time.

Harrell, 1931, "Residential Exchange Sales in the New England Southern Area."

Houghtelling, 1927, *The Income and Standard of Living of Unskilled Laborers in Chicago:* Families had to include at least one child, and heads had to be fully employed. Names were drawn from employee lists at major firms. The author was surprised at the low levels of telephone subscription and automobile ownership.

Kyrk et al., 1941, *Family Expenditures For Housing and Household Operation, Five Regions* (Urban and Village Series), and Monroe et al., 1941, *Family Expenditures for Automobile and Other Transportation:* The samples were drawn from seven North Central small cities; two Southeastern small cities; and four Pacific Northwest small cities. Families had to include husband and wife, both native-born whites, married at least one year, resident in the community at least nine months, and with nonfarm earnings but no relief during the year. As a consequence, the sample represents *less than half* the residents of the communities, disproportionately excluding lower-income families. Automobile expenditures represent the average of purchase and operation costs for *family* use only.

Lynd, 1933, "The People as Consumers," 896: Table reproduced from R. L. Polk survey.

Morgan, 1939, *The Family Meets the Depression:* The 1933 sample is based on a re-questioning of the original sample from 1927. Both were drawn from homemaker graduates of home economics programs. Heads of household were overwhelmingly professional and managerial. Both were mail-in surveys.

Peixotto, 1927, *Getting and Spending at the Professional Standard of Living*.

Peixotto, 1929, "Cost of Living Studies. II. How Workers Spend a Living Wage": The sample was drawn from a list of members of the San Francisco International Typographical Union. The sample included only families, had a low response rate and many "unusable" interviews.

Pike, 1989, "Kingston Adopts the Telephone": Pike compared the distribution of residential subscribers to the distribution of wage-earners in the community.

R. O. Eastman, 1927, *Zanesville and 36 Other American Communities*, 92. The original source reports income categories rounded off as shown in the table.

Rosher, 1968, "Residential Telephone Usage among the Chicago Civic-Minded": "Elite" were defined as members of the "Blue Book."

United States Bureau of Labor Statistics, 1924, "The Cost of Living in the United States": Based on a reanalysis of the data, as reported in Chapter 4 and Appendix E. To be surveyed, families had to include husband, wife, and at least one child; be English-speaking or U.S. residents at least five years; and have no resident boarders (and no more than three lodgers). I defined having a telephone as spending at least $5 per year on telephones, and having an automobile as spending at least $25 per year on automobiles. (The tables in the original report included *any* expenditure on telephone service and also lumped automobiles together with bicycles and motorcycles.) The study also showed that respondents in the West exceeded national levels of telephone subscription by a few percentage points and automobile ownership by several.

United States Bureau of Labor Statistics, 1945, "Family Spending and Saving in Wartime": Units are "consumer units," families or singles. The study covers a sample of urban places.

Williams and Hanson, 1941, "Money Disbursements of Wage Earners and Clerical Workers. 1934–36": Families had to have both husband and wife present, receive no relief help, exceed $500 per year in income, and for clerical workers not exceed $2400. Having a telephone is defined here as *any* expenditures on telephones, but is probably close to telephone subscription. The statistics are calculated from table on p. 258. Other results show that in New York City, 37 percent of the sample had telephones and 15 percent had automobiles. This data-set is currently being prepared for distribution by the Inter-University Consortium for Political and Social Research, Ann Arbor, Michigan.

Wilson, 1928, "Sales Activities": Both the area in Michigan surveyed and the method used were unspecified.

APPENDIX E

The 1918–1919
Cost of Living Study

In 1918 and 1919, the U.S. Bureau of Labor Statistics (BLS)[1] interviewed 12,817 wives about their families' incomes and spending. We randomly drew one-fifth of these interviews to analyze, a subsample of 2588 households (66 of which had missing data on key variables).

The BLS sampling procedure is opaque; it seemingly relied on the procurement by field-workers of lists of employees from large firms:

> The group of families selected for these interviews was intended to represent the urban working class of the United States. Interviewers were instructed to choose families of wage and salary earners (with the latter making no more than $2,000 per year[2]). Only husband-wife families with at least one child were to be selected and families chosen had to have resided in the community of residence for a year prior to the interview. Families with more than three boarders or lodgers were to be excluded . . . as were "slum" or charity families or non-English speaking families who had been less than five years in the United States. . . . Sample families were selected from employer records.[3]

Other restrictions also applied:

> The object in so doing was to secure families dependent for support, as largely as possible, upon the earnings of the husband. . . . [A] large number of families [were] therefore excluded where much income came from other than the husband's earnings.[4]

Our ability to generalize from the data is limited, given the narrow range of occupational types and income, with no truly poor and no wealthy families, and given only conventional, largely native, nuclear-family households with a dominant male earner. Also, the clustering of the sample makes generalization to cities, states, and even regions problematic.[5] Regional differences in telephone subscriptions depicted in these data are at odds with telephone census data.[6]

Interviewers asked wives how much their households had spent on telephone service in the prior year. Twenty-seven percent had spent something on telephones, ranging from 10 cents to $39. It is unlikely that in the remaining 73 percent of the households no one had used a pay telephone for one year; probably, many housewives did not bother to estimate such episodic expenses.[7] We defined subscribers as those households that had spent more than $5 on telephones in the year; 20 percent qualified. Since even cheap urban rates were $12 a year, the $5 definition may err by including a few active users of pay telephones, but it also ensures that many who had subscribed for only part of the year would be included. (Using a $10 cut-off makes little difference in the results. Using any telephone spending at all as the criterion makes a few subtle differences, noted in the next section.)

We systematically explored the attributes of households that might presumably affect having a telephone, using both regression and logit analyses (which yielded similar results). Table E–1 presents a summary equation that provided the best and most robust fit for estimating telephone subscription; it also shows that model's fit for automobile use and for having electricity in the home.[8]

The measures used in the model were defined as follows:

Telephone includes spending more than $5 in the past year on telephones.

Automobile includes spending more than $25 in the past year on automobiles.

Electricity includes reporting the use of electricity in the home for any purposes; lighting alone was overwhelmingly the main use.

Region: Mountain and Pacific states form the residual category. The distribution is, following the list in the table, 10%, 17%, 31%, 24%, and 16% (adding to 98% due to rounding error).

Urbanism: the population size in 1920 of the city, or the center-city if the town was an immediate suburb; the residual category is over one million. The distribution is 22%, 54%, 16%, and 8%, from smallest to largest category.

Occupation comprises four groupings: category one includes professionals, managers, builders, etc. (13%); category two includes white-collar other than category one (13%); category three includes craft workers and manual unskilled workers (56%); and the residual category contains service workers, including workers in transportation (18%). Coding of occupation was done by Michele Dillon, adapting the scheme reported by Hershberg.[9]

Head's earnings is the annual earnings of the male head, in hundreds of dollars.

Other income is the difference between head's earnings and total annual household income, in hundreds (note that the sampling procedure eliminated households where this figure exceeded head's earnings).

Number of adult males and **females** counts only members of the nuclear and extended family age 18 or older who had resided in the home at least 27 weeks. (Because of the sampling procedure, relatively few households had more than one additional adult kin.)

Male lodgers is a simple count, up to three-plus, of male lodgers who had resided in the home at least 27 weeks (female lodgers were rare).

Detached single-family house is a simple dummy variable (55% of households), as is **apartment or flat**, which includes only flats in row or duplex structures (15%). The residual category includes semi-detached and row houses and flats in detached buildings (30%).

Annual housing value is the annual rent reported by renters or the rent imputed by homeowner respondents, in hundreds of dollers.

ANALYSIS OF TELEPHONE SUBSCRIPTION

The regional effects are puzzling, although robust. I attribute them to the sampling procedure.[10] (See Table E–1.)

Respondents in small communities were *more* likely than those in larger places to have telephones: 30 percent in towns under 50,000, versus 12 percent in cities over 500,000. Table E–1 also indicates that under statistical controls, the relation appears largely negative, although residents of the largest cities have higher residualized rates. The latter is solely due to Chicago, where almost half reported telephones—as compared with New York and Philadelphia, where almost no respondents did. The higher rates in smaller towns may reflect factors similar to those in rural America: lower charges, more competition, and perhaps greater need. If one counts *any* expenditures for telephones as indicating a telephone-using family, the advantage of smaller communities weakens considerably, perhaps implying that city-dwellers more often used pay telephones.[11]

White-collar families were likelier than blue-collar ones to have telephones (30 percent versus 17 percent), but the details are complex. Professionals and managers were not especially likely to have service (the logit estimates are unstable), but these are unusual professionals and managers given the ceiling on earnings required for being part of the white-collar sample. Lower white-collar families were the major subscribers in this study: 50 percent of commercial agents, 43 percent of bookkeepers, and 34 percent of clerks subscribed. Least likely to subscribe were those in production or craft jobs—for example, 6 percent of fitters—and unskilled workers—for example, 7 percent of porters and of "laborers." (Craft workers here are employees rather than independent operators.) Service, government, and transportation workers formed a residual and mixed category. The collar difference is robust and not a reflection of income. It may indicate work needs, such as keeping in touch with the office, status concerns, or other considerations discussed in Chapter 4.

LOGIT REGRESSION OF TELEPHONE, AUTOMOBILE,
AND ELECTRICITY USE
(derived estimates)

	Telephone	Automobile	Electricity
Overall proportion	.203	.057	.395
Constant	−.29**	−.17**	−.58**
Region			
New England	−.02	−.05*	−.72**
Mid-Atlantic	−.20**	−.05**	−.75**
Midwest and Plains	.01	−.02*	−.40**
South	−.06**	−.03**	−.42**
Mountain and Pacific	0	0	0
Urbanism			
Population < 50,000	−.01	.03	.37**
50,000–500,000	−.06	.02	.27**
500,000–1,000,000	−.16**	.03	.17*
Over 1,000,000	0	0	0
Occupation			
Professional, Managerial	−.04	.04*	−.03
Commercial, White-Collar	.03	.03*	.05
Blue-Collar Manual	−.08**	.02	−.04
Service	0	0	0
Income			
Head's Earnings ($100 increments)	.02**	.004**	.03**
Other Income ($100 increments)	.02**	.002	.005
Household			
Number of Children	−.04**	−.004	−.04**
Number of Adult Males	−.05	−.02	−.02
Number of Adult Females	−.02	−.006	−.10*
Number of Male Lodgers	−.11	−.03	.04
Housing			
Detached Single-Family	.07**	.02*	.12**
Apartment or Flat	−.08*	−.01	.02
Row Housing	0	0	0
Annual Housing Value ($100 increments)	.06**	.004	.24**
L	−987	−475	−1160
$-2 \times L$	569	154	1063
pseudo-R^2	.184	.058	.296

* $p < .01$
** $p < .001$
$N = 2522$

For each additional $100 earned by the head of household, the chances of having a telephone increased by about 2 percent. (A linear function of income fit the data best.) These chances also increased by about 2 percent for each $100 brought into the household some other way, largely by employed, adult children living at home. Note, however, that the sampling procedures excluded families with substantial income from other earners besides the father.

The more children, the less the chances of a telephone. I interpret this as an indication of financial strain. But having *adult* kin or male lodgers (fewer than 1 percent had female lodgers) led to a more complex pattern. Overall, households with more than one adult male were a bit less likely, whereas those with more than one adult female were significantly more likely, to have a telephone than those with only one.* In multivariate analyses, coefficients for these variables are unstable due to the small number of extended households, statistically significant in some models and not in others. The estimates are also confounded with father's and household income. When these complications are untangled, it appears that males, both kin and lodgers, indirectly contributed to having a household telephone through their additions to household income but otherwise directly *depressed* subscription; adult females, aside from the wife, contributed to subscription through their income, but had no direct effects. (If the criterion for household telephone use is spending *any* money on telephones, these effects wane—which suggests that number of adults influenced only subscription, not pay-phone use, females positively, males contradictorily.) Had this sample better represented the wide range of household types—as our three town samples do (see Chapter 5 and Appendix F)—the parameters would have been more stable and probably the contrast between males and females sharper.

All else constant, families in detached single-family homes were more likely and families in apartments less likely than those in row houses to have telephones. And for each additional $100 in housing value, the likelihood of a telephone jumped 6 percent. One interpretation of these findings is to consider the housing measures as indicators of wealth, rather than as independent factors, but this seems to be a minor explanation.[12] Another interpretation is to consider the housing measures as reflective of neighborhood conditions: the availability, cost, and marketing of telephone service. Where more expensive housing existed, companies made more efforts to sell. Still another explanation is local need: People in lower-density neighborhoods (neighborhoods of single-family houses) may have had less access to services and other people and therefore may have been more willing to subscribe. Finally, some apartment buildings may have provided a common telephone.**

---

*Males: two or more, 17 percent, versus only one, 20 percent (not significant); females: 46 percent versus 20 percent ($p < .05$); and male lodgers: any, 9 percent, versus none, 20 percent ($p < .10$). Here, the issue of pay telephones is relevant. If having *any* expenditures for telephone service codes a household as a telephone user, then adult males and male lodgers had no effect and adult females a modest one (only one woman in the household, 24 percent subscribed, versus more than one, 32 percent; $p < .05$). Households with lodgers were especially likely to spend between 10 cents and $5 on telephones per year.

**Households in apartments were likelier than others, by 9 percent versus 4 percent, to spend between 10 cents and $5 on telephones—an indication of pay telephone

THE TELEPHONE VERSUS OTHER TECHNOLOGIES

In the BLS sample, unlike the town samples discussed in Chapter 5 and Appendix F, we can compare people who spent money on telephones to people who spent money on other technologies (see Table E–1). Some aspects of households affected all three purchases similarly. But telephone subscription was distinctive in a few ways. (This summary also draws from parallel analyses using ordinary least squares.) Compared to their levels of automobile ownership and electric use, New Englanders and Midwesterners were likely to have telephone subscriptions. Residents of large cities (500,000 to 1 million) were comparatively unlikely to have telephones. Manual workers, all else equal, were comparatively less likely to have telephones than automobiles or electricity. Contributions to household income from other than the household head subsidized only telephone subscription and additional adult men reduced only the likelihood of having a telephone. Finally, living in an apartment or flat discouraged only telephone service (although this last result is confounded with the effect of living in the largest cities).

Automobile owners and electricity users were, all else equal, more likely to subscribe to telephone service than were nonowners and nonusers.[13] This suggests either correlated error, some untested causal factor common to all three, or that there may be a family "type" drawn to spending its income on newer technologies.

SUMMARY

1. In the BLS survey of blue-collar and lower-white-collar households, telephone subscription was unusually high, relative to other technologies, in the Midwest and in New England.

2. Households in smaller cities were relatively likely to adopt the telephone; those in larger cities, excepting especially Chicago, were relatively unlikely. Small-town families may have subscribed more often because the charges were lower there, because there was more often competition there, or because lower housing densities made for greater need.

3. White-collar families were likelier to subscribe than blue-collar families, even when wealth was similar. Several explanations are possible, ranging from job-related needs to class cultures (see pp. 115–16).

4. Wealth determined telephone subscription, as well as other purchases. Distinctive to telephones, however, was the income contributed by others than the household head. This suggests that these additional earners—mostly adult sons and daughters—needed or wanted part of their money spent for telephone ser-

---

use. But this effect disappeared under statistical controls. Overall, residents of apartments or flats were less likely than others to have telephones. Yet, apartment and flat residence was strongly confounded with the specific city: 93 percent of the New York City respondents and 54 percent of the Boston ones lived in apartments or flats; elsewhere, no more than 43 percent (St. Louis) and usually far fewer did.

vice. They may have been motivated by practical reasons, job needs, or by social reasons. There are some weak indications that women encouraged subcription more than did men.

5. Residents of expensive housing and single-family homes were more likely, all else equal, to have telephones. Perhaps this simply indicates wealth, or alternatively, advantages of the neighborhood.

APPENDIX F

Who Had the Telephone When?

This appendix reports a statistical analysis of which households in Palo Alto, San Rafael, and Antioch had telephones in five different years spanning 1900 to 1936. The analysis is based on a procedure linking households sampled from census manuscripts and other lists to telephone directory listings. We aimed for a random sample of about 100 households living in each town in each of 1900, 1910, 1920, 1930, and 1936. (We chose 1936 instead of 1940 to give a glimpse of the Depression.) To make sure that we included sufficient numbers of telephone subscribers, we also drew a supplementary sample of households of town "notables." We coded demographic and occupational information about each drawn household and then checked in the PT&T telephone directory for that year to see whether or not the household was listed as having a telephone.

SOURCES AND SAMPLING: THE 1900 AND 1910 CENSUSES

We randomly sampled every xth household, as listed in the census manuscripts, by setting x to the value needed to yield about 100 households in each town each year and selecting a starting point at random. (For Palo Alto, we sampled both the city and adjoining Mayfield, which was eventually annexed in 1926. Palo Alto is best read as Palo Alto *and* Mayfield.)[1]

We drew supplementary samples of notables by selecting residents whose names were mentioned in the town newspaper for each year.[2] Especially in the early years, some notable households had already been drawn in the random selection. One bias in this procedure was to disproportionately exclude working-class and minority notables, whose activities were less often reported. Analysis of these supplementary samples is largely restricted to 1900 and 1910.

The information we recorded from the census materials are not without error: Aside from households that census-takers missed, spelling variations of names created problems, as did occasional illegibility; and some information, notably addresses for many households in 1900, was absent. But the census manuscripts describe the great majority of the towns' households as of 1 April 1900 and 1 April 1910. We also traced the sampled households four years hence, in city directories and voter registration lists for 1904 and 1914. Many from the original sample were lost because people had moved or died or because the secondary sources were much less complete.[3]

Most of the information in the census is relatively straightforward. Occupation, however, required coding. Where the head of household was not currently employed, we substituted the occupation of another employed adult in the home (using the oldest employed male's occupation first). I coded occupation initially into about 35 categories and then collapsed them.[4] Some census-takers' occupational descriptions were incomplete or ambiguous—for example, "own account," "hotel", "employee at X"—and so some error inevitably crept in. For the 1904 and 1914 follow-ups, we used city directories as sources in Palo Alto, the 1905 directory and the 1914 "Great Register of Voters" for San Rafael, and the voting registers for Antioch.

SOURCES AND SAMPLING: 1920–1936

Because original census records are sealed for 70 years to maintain confidentiality, we could not use the censuses for 1920 and beyond. We turned therefore to city directories where possible, voter registration lists otherwise. For Palo Alto, we could draw on city directories for each pair of years (1920 and 1924; 1930 and 1934; 1936 and 1940), supplemented by the voter lists. For San Rafael, we depended on the voter lists, except for 1924 (we used the 1925 city directory) and 1940 (the 1939 directory). For Antioch, we could use city directories for the base years—albeit off a bit, with the 1919, 1931, and 1937 city directories—and voter lists for the four-year follow-ups.[5]

Directories are more problematic than the census—the extent depends on how professionally the directories were done.[6] Listing procedures varied—by household or individual, by street or alphabetically—and we had to adjust our sampling accordingly. Spelling and alphabetization errors add to possible confusion, as do the uncertain dates of the directories' listings. The information is severely limited: we can only discern family members if they are listed at the same address and of the same surname; occupation is occasionally missing, especially in the 1919 Antioch directory; and ethnicity is not provided. Although I coded last names for ethnicity, using a reference on name origins, I could not classify many of them and garnered little from the effort. Voter registration lists have similar limitations, in addition to biases due to voting qualifications and self-selection. (By 1920, women had voted in California for six years, so systematic exclusion of women was not a problem.) Despite these difficulties, it is striking that the results for the 1920 to 1936 samples are consistent with the 1900 and 1910 samples and are consistent with one another across different sources. (The greatest difficulty arises from using the incomplete 1919 directory of Antioch.)

CODING TELEPHONE SUBSCRIPTION

We used telephone directories to determine who had a telephone.[7] We looked up each sampled household in the directory closest in date to the sample source. We coded whether we found a certain, probable, or possible match of names. In the data analysis, I treated only "certain" and "probable" as matches. We coded whether the telephone was in the name of the head of household or of another household member. And we coded whether the telephone listing was for a residential or a business instrument (occasionally, households had both). During this era, residential telephones were indicated by an *r* after the name and address. In Palo Alto, around 1910 this notation became uncertain, and I had to adjust the coding.[8]

ANALYSIS OF DATA: 1900

The 1900 census provides a range of data on the households and their members. I explored the relationship between the variables we coded and the probability of having a telephone in the home. Given the highly skewed nature of the dependent variable—only 12 of the *randomly* sampled households had telephones—some quirky effects appeared. I applied logit regressions to deal with the skew.[9] The best-fitting models appear in Table F–1. The table presents "derived estimates," which can be read as the increase in the chances of having a telephone for each increase in the independent variable, other independent variables held constant at their means.

The most consistently powerful correlates of telephone subscription were having a servant, having a head of the household who had a job as a manager, and having a notable in the home. "Managers," of whom there were only seven (three bankers, three superintendents, and one postmaster), were especially likely to have telephones, adding .03 to the probability of subscribing. (Doctors did *not* stand out even though we know they were quick to subscribe, largely because they had business, not residential, telephones in their home-offices.) Having a notable in the home also upped the chances of subscribing by .03. There was a nonsignificant tendency in these equations for marriage to promote telephones by almost the same amount. Other equations indicated that the marriage effect was stronger when the sample was restricted to male heads only; thus, wives added to the odds of having a telephone. The modest age effect suggests that the chances of subscribing increased as heads' ages increased, but that the effect tapered off, peaking at about age 65.

In all, the results suggest that a home telephone was an exceptional possession in 1900. Some managers had it, probably for business reasons, and otherwise only a few elite families did.

TABLE F—I

LOGIT REGRESSION OF THE PROBABILITY OF A RESIDENTIAL
TELEPHONE SUBSCRIPTION, 1900
(derived estimates)

	Random Sample	Total Sample
N	325	373
Overall proportion	.036	.051
Constant	−.195**	−.187***
Household Head Married	.023	.013
Servants (0/1/2+)	.014**	.013***
Occupation of Household Head		
Doctor	.016	.011
Manager	.033***	.029***
Blue Collar Worker	−.009	.000
Homeowner	.013	.000
Age of Household Head	.005	.005*
Age Squared/100	−.004	−.004*
Town		
Antioch	.008	.010
Palo Alto	−.001	.005
San Rafael	0	0
Notable in Household	—	.028***
L	−39.0	−45.0
−2 × L	25.4	60.1
pseudo-R^2	.072	.139

*$p < .10$
**$p < .05$
***$p < .01$

Notes: "Total Sample" includes the random sample plus a supplementary sample of town notables (see text). All variables coded 0/1, except servants, age, and age-squared.

ANALYSIS OF DATA: 1910

As noted above, coding residential telephones in Palo Alto presented a problem. Some error no doubt persisted despite the corrections (see note 8 above). In the regression equations in Table F–2 I used a dummy variable to take that into account. Although, in principle, a logit model is more appropriate for analysis of such a dichotomy, in practice, the results using logit and using ordinary least squares (OLS) were essentially interchangeable. For simplicity, OLS results are presented in Table F–2.

The equation for the random sample is the most efficient of many tested for 1910. The effects for "Male Head" and "Wife" together indicate that female-

TABLE F—2

REGRESSION OF THE PROBABILITY OF A RESIDENTIAL
TELEPHONE SUBSCRIPTION, 1910
(OLS: *unstandardized regression coefficients*)

		Random Sample	Total Sample
	N	355	406
	Overall proportion	.248	.278
Constant		−.03	−.08
Male Head of Household Present		.20***	.22***
Wife or Female Head of Household Present		.18**	.21***
Adult Sons at Home (0/1/2+)-		−.06	−.03
Adult Daughters at Home (0/1/2+)		.14***	.12**
Servants (0/1/2+)		.31***	.30***
Occupation of Household Head			
Doctor		.33**	.31**
Manager		.19*	.12
Blue–Collar		−.16***	−.18***
Ethnicity			
U.S., U.K., or Ireland		−.02	−.01
French or German		.18***	.15***
Other		−.16	−.14
Town			
Antioch		.01	.00
Palo Alto		.01	.00
San Rafael		−.02	.00
Has Business Telephone in Palo Alto		−.32***	−.34***
Notable in Household		—	.07
	R^2	.292	.329

*$p < .10$
**$p < .05$
***$p < .01$

Notes: "Total Sample" includes the random sample and the supplementary sample of notables. All variables coded 0/1, except the three noted as coded 0/1/2+. In estimating categorical variables of three or more values (Ethnicity, Town), SYSTAT estimates for the excluded categories are not zero, but −1 times the sum of the coefficients for the included categories. "Has Business Telephone in Palo Alto" corrects for coding problems.

headed households were less likely than others to subscribe, but that having a wife increased subscription rates for male-headed families. The presence of adult daughters also increased the odds of subscription. Similarly, the more servants, the more likely subscription. Doctors were likely to subscribe, whereas blue-collar households were unlikely to have telephones at home.[10] Finally, knowing the total number of employed persons in the household does not increase our ability to predict subscriptions.[11]

The effect of ethnicity is hard to interpret. I could not explain the advantage of the French and the Germans, who numbered 31 in the sample, largely German. It may represent "taste." The specific town is once again nonsignificant. Although some interaction effects with the "Town" variable can be extracted, they largely revolve around a few cases. Finally, having a notable in the household no longer distinguished telephone subscribers; that is, their advantage in subscribing is fully explainable by other traits.[12]

The "false negatives" in this model—those households the model suggests ought not to have telephones but did—can frequently be explained by the identity of other people living in the household. For example, a widow in Antioch who had a telephone was living with her engineer son. When I ran the equation only for those 277 households in the random sample headed by someone under 60 years of age (i.e., excluding households headed by the elderly), the fit improved to $R^2 = .32$ from .29. In a few cases, "false positives" can be explained by the presence of a business telephone. Other prediction errors are less systematic.

In 1910, telephone subscription, being more common than in 1900, was more systematically associated with wealth, household structure, and occupation.

ANALYSIS OF DATA: 1920–1936

We drew the samples for these years from city directories for Antioch and Palo Alto and voting registration lists for San Rafael. As noted earlier, these sources' coverage and detail are inferior to those of the census, but they do permit us to track key household characteristics.[13]

The first equation in Table F–3 best summarizes, after considerable experimentation, the patterns over the sixteen-year period; the other columns provide the same equation specifically for each year. The notes to the table describe a few of the more unusual coding procedures needed to use the directory and voting list samples.

The same approximate model fits well for all three years, although it fits worst for 1920 (when many households in Antioch lacked occupational descriptions and the sampling frames were the weakest).[14] The effects of town are shown in the Town × Blue-Collar interactions: Town differences in white-collar workers' propensity to subscribe were negligible, but blue-collar workers in Antioch were especially unlikely to subscribe (more on this in the following discussion). Adding the supplementary sample of notables to the random sample and a dummy variable for them to the equation increases the R^2 to .23 but does not substantially alter the other estimates in the model. Analysis of residuals identified certain high-leverage cases, most of which were "Foremen," but dropping these households did not substantially change the findings. (Later in this section, I review the misclassified cases.) A final note about the robustness of this model concerns the sampling frame. In most of the directories, households could be randomly sampled by address. In the 1919 Antioch directory and the voter registration lists used for San Rafael, however, the listings are

TABLE F–3

REGRESSION OF THE PROBABILITY OF A RESIDENTIAL
TELEPHONE SUBSCRIPTION, 1920–1936
(OLS: unstandardized regression coefficients)

	1920–36	1920	1930	1936
N	944	305	313	326
Overall proportion	.47	.45[a]	.55	.41
Constant	.15	.10	.25	.09
1920	−.02***	—	—	—
1930	.09	—	—	—
1936	−.07	—	—	—
Household Head Married[b]	.19***	.11*	.21***	.23***
Household Head Widowed[b]	.33***	.45	.15	.43***
Number of Family Adults[c]	.12***	.15***	.13***	.09**
Occupation of Household Head				
Professional	.12**	.05	.13*	.18**
Blue-Collar	−.15***	−.07	−.22***	−.16***
Foreman[d]	.30**	.56	.42***	−.32
Laborer[d]	−.18***	−.28**	−.09	−.16*
Town				
Antioch	−.01	.01	.02	−.06
Palo Alto	.02	−.03	.01	.10
San Rafael	−.01	.02	−.03	−.04
Interaction terms with town				
Blue-Collar × Antioch	−.17***	−.08	−.27**	−.14*
Blue-Collar × Palo Alto	.11	.09	.23	−.02
Blue-Collar × San Rafael	.06	−.01	.04	.16
R^2	.21	.14	.30	.23

*$p < .10$
**$p < .05$
***$p < .01$

Notes:
[a]Higher than the estimate reported elsewhere in the study of .39, because it excludes 55 cases in Antioch for which no occupation was reported, only 13 percent of whom had telephone subscriptions.

[b]Usually determined by the listing of names in the source. For example, a Mrs. Smith heading a household was assumed to be widowed.

[c]Determined by counting the number of persons with the same last name listed at same address.

[d]Foremen and Laborers are also included in the dummy variable for Blue-Collar, so estimates for them must include the latter term as well.

by name. In these sources, households with more adults had a greater chance of being sampled. I recalculated the model, weighting against this bias with the reciprocal of the number of adults in the household. The results were essentially the same.

The three-year model shows the following: From 1920 to 1930, households increased their subscription rates, but many fewer subscribed by 1936. The initial increase is underestimated due to the Antioch cases in 1920 that were missing occupational information (see note *a* to the table). Married couples were more likely to subscribe than were single household heads (excepting widow-headed households, but there were only 36 of them in all the three years). As the number of adults increased, so did the likelihood of subscription; and, as was true in the 1900–1910 data, this cannot be explained in terms of extra earners. The number of adults, rather than employed adults, fits best. If the 1900–1910 pattern holds, these extra adults were mostly women.

Occupational effects were stronger in this era. Professional families were especially likely to have residential telephones, whereas blue-collar households— except the 13 foremen's families and except those living in Palo Alto—were especially unlikely to have them. Laborers, whose numbers increased in Antioch in particular after 1920, rarely had telephones. Put another way, community had no effect on telephone subscription for white-collar or nonworking residents, but the three towns differed in the subscription rates for blue-collar workers. Those in Palo Alto were more likely to subscribe and those in Antioch were unlikely to subscribe compared to San Rafael blue-collar households. Palo Alto's blue-collar families showed about the same levels of subscription (and changes in subscription) as their white-collar neighbors; Antioch's blue-collar households had low and steadily declining levels of subscription from between 1920 and 1936; San Rafael's blue-collar households were in-between.

It is difficult to read these patterns from the regression equations. Table F–4 presents the estimated probabilities of telephone subscription for different groups, years, and towns, calculated from the year-specific equations in Table F–3 and setting the values of Married, Widowed, and Adults at their year-specific means.

Two particular questions arise: Why did blue-collar Antiochians' subscriptions decline even between 1920 and 1930? Why were blue-collar Palo Altonians so like white-collar ones? The former result may be partly a methodological artifact: Over 40 percent of the 1920 Antioch sample were missing data on occupation; they were probably disproportionately blue-collar and disproportionately without telephones. But both questions can be answered more substantively, first, in terms of the kinds of blue-collar workers each town had and second, perhaps in terms of a contextual effect. Palo Alto's blue-collar workers tended to be craftsmen and sometimes independent contractors; Antioch's were more often simple laborers and, as the years passed, factory workers. (Also, between 1920 and 1936, Antioch's labor force experienced an influx of Italian immigrants and in the 1930s of "Oakies.") This subtle within-class difference in the composition of the towns probably explains most of the interaction effect. But there may also be an additional contextual effect. Blue-collar families in Palo Alto may have been encouraged by the telephone subscription of similar households and those in Antioch discouraged from doing so by the relative absence of tele-

TABLE F–4

PROBABILITY OF A RESIDENTIAL TELEPHONE SUBSCRIPTION,
BY OCCUPATION, 1920–1936
(as estimated from Table F–3)

	1920	1930	1936
Professional			
Antioch	.55	.82	.60
Palo Alto	.51	.82	.75
San Rafael	.55	.78	.62
White-Collar[a] or No Occupation			
Antioch	.50	.69	.42
Palo Alto	.46	.68	.57
San Rafael	.50	.64	.43
Blue-Collar			
Antioch	.35[b]	.20	.13
Palo Alto	.47	.69	.39
San Rafael	.41	.46	.44
Laborer (where $N > 5$)			
Antioch	.07[b]	.10	0[c]
Palo Alto	—	—	.22
San Rafael	.14	.37	.27

[a]Other than professional.
[b]The figures for 1920 Antioch blue-collar and laborer are probably inflated because of the numerous missing cases.
[c]Regression estimate is −.04.

phones in the working-class community. (In the analysis of change over time by occupation reported in Chapter 5, pp. 148–49, town differences persist even when detailed occupational classifications are held constant.)

Analysis of residuals showed no clear pattern. Some households were, in effect, misclassified by occupation. For example, an "Employee at Fireboard" in 1920 was coded as a laborer but turned out to be an "accountant" on the 1924 voter registration lists. A few households lacking a residential telephone had a business telephone in the head's name. Although the addition of other variables or finer distinctions to the model add to the explained variance (as does the inclusion of a dummy variable for "Foreman"), they do not alter the general structure of the results.

GEOGRAPHIC ANALYSIS

The great majority of households were listed with addresses, and I coded those into several geographical sections for each town. In analyzing location, however, I used only the dichotomy of town center versus elsewhere.[15] Using essentially the same equation for all five years to hold constant household structure and

TABLE F-5

ADJUSTED PROBABILITY OF A RESIDENTIAL TELEPHONE
SUBSCRIPTION, WHITE-COLLAR HOUSEHOLDS

	1900	1910	1920	1930	1936
Antioch					
Center	.02	.21	.52	.71	.40
Periphery	.08	.41	.52	.81	.54
Palo Alto					
Center	.00	.32	.36	.72	.49
Periphery	.04	.38	.61	.78	.58
San Rafael					
Center	.00	.37	.58	.92	.37
Periphery	.03	.56	.44	.78	.51

Notes: These estimates are based on regressing, among white-collar households only, residential telephone listing on household type (five categories), on three specific white-collar occupational categories (doctor, manager, and for 1900–1910, proprietor, for 1920–36, professional) and a dummy variable for central address. For calculating the estimates, household type was set to "married couple" and occupation to white-collar "other."

occupation yields the results in Table F–5. These are the probabilities that a white-collar household subscribed to the telephone by whether the household was in a central or peripheral location, adjusted for other factors. The differential is usually in favor of peripheral households, averaging 7 percentage points (10 without San Rafael).

A more detailed examination, using the specific equations in Tables F–1 through F–3, shows that the geographic effect was negligible in 1900 but substantial in 1910. The coefficient for Center × White-Collar is significant for the latter year (partial $b = -.15, p < .05$). (Excluding the Mayfield residents listed for Palo Alto in 1910, who were not, strictly speaking, Palo Altonians until 1926, strengthens the effect.) For the period from 1920 to 1936, the overall effect of Center × White-Collar is not significant (partial $b = -.06, p < .20$). Town analysis indicated that the effect was negligible in San Rafael, for reasons I have not been able to discern. Adding that qualification into the equation yields a significant effect for most of the sample (partial b for Center × White-Collar $= -.13, p < .05$; partial b for San Rafael × Center × White-Collar $= +.16, p < .07$). The center-periphery effect, all else equal, was greatest in Palo Alto (and, again, stronger when Mayfield residents were excluded). It was greatest in 1920 and declined to 1936. Examination of other possible explanations for the effects, such as home ownership, substantiated the interpretation that location was the factor.

In 1910 in all three towns and then afterwards in two of the three, white-collar households residing away from the town center were likelier to have telephones than those living in the town center.

Analysis of Advertisement Data

CODING

We* coded advertisements from the *Antioch Ledger, Palo Alto Times,* and San Rafael's *Marin Journal,* canvassing two issues of each newspaper for each even-numbered year: the first Monday on or after 20 January and the first Monday on or after 5 May.** We coded three categories of advertisements: (1) classifieds — rentals available, jobs wanted, lost-and-founds, and household goods for sale; (2) "Cards" — notices of professional services; and (3) "Boxed" ads — advertisements of consumer goods or services, set off in their own box, for a local dealer. We excluded simple product advertisements, such as "Buy Uneeda Biscuits," unless a specific retailer was mentioned.[1] And we excluded advertisements on pages of the newspapers designated for out-of-town readers (e.g., "the Oakley" page of the *Antioch Ledger*). We read the newspapers out of chronological order so as not to confound historical time with coding experience.

The coders recorded each unique advertisement,[†] noting its product or service and other relevant information, including whether there was any mention of a

---

*Specifically, Barry Goetz and Kinuthia Macharia.

**The dates covered the years available for the newspapers: 1890 on for the *Journal*, 1894 on for the *Times*, and 1906 on for the *Ledger*. The *Ledger* shifted from a weekly to a triweekly to a daily during this period, but these shifts seem not to have affected our findings.

[†] That is, they did not include advertisements that appeared in a collective block.

telephone, usually a telephone number but occasionally something more, such as a picture of a telephone. In all, the data are based on about 8600 coded advertisements from 134 issues of the newspapers.

ANALYSIS*

From the codings of these ads, we generated, for each year and for each town, summary counts of both the number of ads of particular types and the proportions of each type that included telephone references. (Some combinatorial categories were too rare, year by year, to be useful.) Reliability is a general concern in this process. There were a few systematic differences between the two coders, each of whom coded different issues, but none affected the analyses reported here. These are "noisy" data, with considerable year-to-year fluctuations. Yet, fortunately, the coding of telephone references was reliable.[2]

Because the data are erratic, based as they are on only two issues in each year, I smoothed the percentage estimates for the graphic presentations in Figures 13 and 14. I used a three-point (i.e., six-year) moving average involving two steps, first obtaining a running median for three observations and then "hanning" that series with a weighted moving mean of three observations.[3]

The figures in Chapter 6 display the historical patterns for a few key types of advertisements. I also explored, with regression analysis, the factors that affected those trends, in particular the role of telephone diffusion. The analysis was difficult, in large part because of strong collinearity among population, number of telephones, and year. I set population aside under the assumptions, first, that population growth was antecedent to telephone diffusion and second, that the latter would be the more proximate cause of ads with telephone references. Table G–1 reports the results of dynamic modelling based on the following operational definitions and procedures: The units of analysis are town-years (e.g., Antioch 1922) with enough coded ads to justify taking a percentage. The dependent variables are logit transformations of the unsmoothed percentages of ads with telephone references, in each year for each town.[4] (The percentages in Figures 13 and 14, on the other hand, are smoothed.) I dropped data points based on fewer than 10 ads per issue for "goods and services" and fewer than 5 ads per issue for the other categories.[5] Only two measures are shown in Table G–1, ads for consumer goods and services and professional cards, because the others had relatively few observations and unstable estimates. The independent variables are the logit-transformed percentages lagged one period (i.e., two years); a three-year moving average of the number of residential telephones, logged and lagged one period;[6] a similar transformation for the number of business telephones; and control variables.

The results in Table G–1 show little "inertia" in the form of the lagged dependent variables and strong effects for the (lagged) number of telephones. The number of residential telephones helps explain the percentage of consumer

*The statistical work was done with the technical assistance of William Barnett.

TABLE G–I

REGRESSION OF THE PERCENTAGES OF ADVERTISEMENTS WITH
TELEPHONE REFERENCES, 1890–1940
(unstandardized regression coefficients)

		Logit Percentage of Ads with Telephone	
		Goods and Services	Professional Cards
N	61	61	44
R^2	.86	.89	.71
Constant	−3.95*	−2.70*	−5.55*
Logit Percentage, Lagged	.01	.14	−.16
Log Residential Telephones, Lagged[a]	.61*	.36*	—
Log Business Telephones, Lagged[a]	—	—	.77*
Years since 1890	−.01	.01	.03
Antioch	.01	−.26	2.32*
Palo Alto	.46*	.46*	1.58*
San Rafael	0	0	0
Palo Alto 1908 and 1910[b]	.50	.36	−.84
Palo Alto 1896[c]	—	−1.97*	—

*$p < .01$

Notes:

[a]Telephones were logged, then smoothed, and then lagged one period (two years). Almost all the data were derived from counts of telephone directories; three values in earliest years were estimated from company statistics.

[b]Dummy variable for two data points. Conforms to period when residential telephones were undercounted in Palo Alto. The two points have high leverage.

[c]Another high-leverage case. Conforms to lowest estimates for telephones (lagged) in entire sample.

ads two years hence and the number of business telephones helps explain the percentage of cards with telephone numbers two years later. (For classifieds and other small ads, however, telephone effects were nonsignificant, unstable, and interacted with town.) Also, the *Palo Alto Times* had a greater percentage of ads with telephones, other factors notwithstanding, and the *Antioch Ledger* had a greater percentage of cards with telephone numbers, all else equal.

APPENDIX H

Statistical Analyses for Chapter 7

OUT-OF-TOWN ADVERTISEMENTS

Appendix G describes the general procedure for sampling and coding newspaper advertisements. For the analysis of localism, coders noted whether the store or service-provider in the advertisement was located in-town, in an adjacent town, further away but still in the vicinity, or in a more distant location. (San Francisco, for example, was regarded as "distant" for all three towns.) Out-of-town advertisers' notices were not counted if they appeared on pages devoted specifically to out-of-town news — i.e., pages addressed to audiences in outlying towns. In-town advertisers' notices *were* counted on those pages.[1] Also, mail-order ads were not coded. These coding decisions may have moderately underestimated the proportion of ads directed to town residents by merchants elsewhere, but probably did not affect the general historical trend.[2] For this analysis, I collapsed "in-town" and "adjacent" into one category. (Categorizing "adjacent" locations with more distant ones makes no difference to the substantive conclusions.)

Figure 15 displays the data for those types of ads numerous enough to calculate percentages: boxed ads for consumer products and services (separating products from services results in similar but "noisier" trend lines) and professional cards placed by doctors or lawyers. For Palo Alto, there were enough rental classifieds to analyze, too. The numerator in these percentages is the number of ads placed by a business located beyond the immediately bordering communities. The denominator is the total number of coded ads of that type. If the denominator was fewer than 10 consumer ads or two cards or rentals, I set the percentage to missing data, which explains the shortness of a few series. The smoothing in Figure 15 involved two passes, one using a running median and the second a weighted moving average.

In another analysis of the same data, I regressed the *un*smoothed percentages on functions of year. The most efficient and interesting models are in Table H–1. The dependent variables are the percentages of ads that were from out-of-town businesses, transformed into logits. (Regressions done on the absolute number of extralocal ads show a similar declining trend.)

For consumer goods and services, there is a simple linear decline in out-of-town advertising. Adding interaction terms or measures of telephony contributes nothing. (An estimate of the number of automobiles in town can replace "Years since 1890" in that equation, but it fits less well and the coefficient is negative [$b = -.08$], suggesting, oddly, that the more automobiles in town, the proportionately *fewer* advertisements from out-of-town businesses.)[3] For legal and medical cards, a simple model shows a rapid decline in the percentage of

TABLE H–I

REGRESSION OF THE PERCENTAGES OF ADVERTISEMENTS FROM OUT-OF-TOWN, 1890–1940
(unstandardized regression coefficients)

| | | Logit Percentage of Ads from Out-of-Town | | |
		Goods and Services	Medical and Legal Cards	
	N	65	52	52
	R^2	.27	.40	.41
Constant		-1.41^{**}	.71	1.81
Logit Percentage, Lagged[a]		$.24^*$	.13	.17
Antioch		.38	.21	$-.90$
Palo Alto		$-.21$	-1.02	$-.95$
San Rafael		0	0	0
Coder[b]		$-.26$	$-.14$	$-.15$
Years since 1890		$-.03^{***}$	$-.21^{***}$	$-.03$
Years, Squared		—	$.002^*$	—
Town Business Telephones, Logged and Lagged[c]		—	—	$-.65^{**}$

$^*p < .10$
$^{**}p < .05$
$^{***}p < .01$

Notes:
[a] The dependent variable lagged one period (two years).
[b] In some analyses, there was a significant difference between the two coders, so this is introduced as a control.
[c] The number of business telephones in town, logged, smoothed, and lagged one period (two years).

out-of-town notices at first, followed by a leveling out after 1920. An estimate of business telephones can replace the "Years, Squared" variable and yields an intuitive interpretation: The more town doctors and lawyers with business telephones in the earliest years, the lower the percentage of such notices placed in the newspapers by their out-of-town competitors.[4]

In all, extralocal advertising became proportionately less common over the years, and the technologies—except for the relationship between local business telephones and nonlocal professional cards—played an insignificant role in this trend.

LEISURE EVENTS

To assess changes in the location of leisure pursuits—at least, the more organized, public ones—Barry Goetz coded stories and notices from samples of the local newspapers. As with the advertisement data (see Appendix G), he read January and May issues of even-numbered years, but he scanned two issues per month instead of one.* Goetz looked through each issue for all stories about leisure events that had already happened (e.g., "Debutante Ball a Big Smash," or "Roberts Family Tours Yosemite"), announcements of coming events (e.g., "Roxie to Show Pickford Film"), social columns, personal notes, lodge and club reports, and the like. He ignored "hard news" stories, syndicated features, duplications, and obvious advertisements, but did count ads announcing shows. (As with the advertising analysis, the sequence of newspapers coded was random with respect to chronological order.) He coded each qualifying item on various characteristics, including location. "In-town" meant that Antioch, Palo Alto, or San Rafael were the locales. (Stanford, Mayfield, and South Palo Alto were also included as in-town for Palo Alto.) Adjoining and nearby towns were considered in the "vicinity."** "Elsewhere" encompassed more distant places, notably the cities of San Francisco, Oakland, San Jose, Richmond, and Berkeley. (Goetz also coded the stories for which people did or were expected to attend—town residents, out-of-towners, or both—but the basic patterns were the same as those using the location measure.)

In all, Goetz coded over 9700 such items in 272 newspapers for Antioch from 1908 to 1940, Palo Alto from 1894 to 1940, and San Rafael from 1890 to 1940. William Barnett and I aggregated them into the largest and most meaningful categories—for example, "organized sports sponsored by schools," "socials open to the general public," and "lectures presented by voluntary associations." The figures presented in the text are the average number of events per newspaper issue that year.

*The first Monday after 20 January, the first Friday after 1 February, the first Monday after 1 May, and the first Friday after 15 May—or the nearest following days for newspapers not published on Mondays or Fridays.

**For Antioch, these were Pittsburg, Oakley, Brentwood, Knightsen, Byron, Sherman Island, Port Chicago, and most distant, Concord, Walnut Creek, and Martinez—essentially east Contra Costa Country. For Palo Alto, "vicinity" included East Palo Alto, Menlo Park, Ravenswood, Runnymeade, Redwood City, Mountain View, San Mateo, Woodside, Santa Clara, Los Altos, and Sunnyvale. For San Rafael, "vicinity" included all of Marin county, except Novato at the north end.

These are somewhat "noisy" and erratic data, in part because many event categories had low frequencies. For example, Antioch averaged 2.2 "performances for which admission was charged" per issue of the *Ledger* between 1908 and 1940. Split-half reliability ranged widely around .6 for different measures.[5] Not much weight, then, should be put on small variations. More importantly, the reports were subject to substantive distortions based on editorial policy. Establishing a sports or a women's section, adding pages, expanding wire-service use, printing a "lodge" column—these and other changes in the publication procedures of the newspapers no doubt altered the coverage of leisure events. Given the complexity of these distortions, there is no simple way to correct for them.[6]

The lines in Figures 16 and 17 are based on smoothing the data. In addition, I analyzed the *un*smoothed data using dynamic regression models. The dependent variables in these models are the average number of leisure events per newspaper issue, logged. For all leisure activities totalled, I distinguished three categories of location: in-town, vicinity, and far. For specific types of events, I simply dichotomized the frequencies of events into in-town versus elsewhere. In addition, I calculated the percentages of all reported activities that were in-town. (Since these percentages required a minimum number of reports per year for stable estimates, some years are missing data.) For regression purposes, the percentages are logit-transformed. The independent variables include: the dependent variable lagged one period (two years), town (represented as dummies for Antioch and Palo Alto), and a complex of highly collinear measures comprising years since 1890, an estimate of the number of residential telephones, logged and lagged one period (that is, a smoothed estimate of the residential listings in the town's telephone directory two years before), and an estimate of the number of automobiles in town, logged and lagged.[7] Controlling for specific town, the measures of years, telephones, and automobiles correlate at from .92 to .97 for the period after 1901 used in the regression analysis.[8] Almost all the equations reported below use a single indicator of the three, because adding an additional predictor did not significantly increase explained variance. (I.e., one of the measures tends to capture the entire effect of the three.)

I tested various equations, including nonlinear models, and tested for robustness by examining and excluding "influential" (high-leverage) cases. The cleanest results come from simpler models and from excluding data prior to 1902. (The numerical estimates for the leisure measures and for the automobile and telephone measures were erratic and "influential" in the years before 1902. Including them, however, does not substantially alter the conclusions.)

Table H–2 summarizes the results for all leisure activities combined and for a few of the more common subcategories. The top panel presents results using the absolute counts of activities as the dependent variables, and the bottom gives the results for the (logit-transformed) percentages. Both unstandardized and standardized regression coefficients are displayed. Note that in general, the lag parameters are weak, testifying to the "noisy" nature of the data. Still, a respectable amount of variance is explained. The results for the absolute counts (top panel) show that the number of both in-town and out-of-town events increased over time, whether the historical trend is indicated by year, automobiles, or telephones, and that the increases were substantial (as suggested by statistical

TABLE H–2

REGRESSIONS OF LEISURE ACTIVITIES, BY LOCATION, 1902–1940

(unstandardized regression coefficients; entries in parentheses are standardized coefficients)

Dependent Variables	N	R^2	Constant	Lag[a]	Antioch	Palo Alto	Years[b]	Log Autos[c]	Log Telephones[c]
			Absolute Counts of Reported Events (Logged) per Issue						
All Activities									
In-Town	56	.56	−19.8**	.06* (.42)	11.8** (.37)	6.6** (.65)			5.4*** (.33)
Vicinity	56	.52	−1.9	.07** (.29)	−2.5 (−.20)	−2.0 (−.17)	.24*** (.46)		
Far	56	.27	1.8	.02 (.09)	2.6** (.35)	−.62 (−.09)		0.5** (.33)	
Organized Sports									
In-Town	56	.50	−2.5**	.06 (.24)	0.9* (.32)	0.8** (.32)			0.5*** (.48)
Elsewhere	56	.60	−5.4***	.07** (.27)	2.3*** (.55)	0.9*** (.24)			1.0*** (.60)
Paid Performances[d]									
In-Town	56	.65	−0.8**	.08 (.30)	.21* (.22)	.25** (.28)			.18*** (.51)
Elsewhere	56	.46	−.11	.06 (.22)	−.14** (−.28)	−.03 (−.06)	.01*** (.48)		
Organization Meetings									
In-Town	56	.72	−11.3***	.10*** (.42)	4.8*** (.37)	3.9*** (.32)			2.1*** (.43)
Elsewhere	56	.40	.45	.06* (.23)	−.45 (−.10)	−1.2** (−.28)		0.4*** (.43)	

Dependent variable	N	R²							
Socials[e]									
In-Town	54[f]	.43	-.42**	.03 (.10)	.28* (.57)	.20*** (.44)			.09*** (.49)
Elsewhere	54[f]	.40	-.14	.09*** (.46)	.02 (.06)	-.03 (-.10)			.04** (.30)
In-Town Events as Percentages of Total (Logit Transformed)									
All Activities	56	.38	.71***	.26* (.26)	.02 (.02)	.40** (.38)		-.06** (-.29)	
Organized Sports (A)	53	.47	3.48***	-.13 (-.18)	-2.0*** (-.45)	.23 (.08)		-.52*** (-.68)	
Organized Sports (B)[g]	53	.55	-5.09*	-.22* (-.31)	.14 (-.01)	.40 (.14)		-1.45*** (-1.6)	2.11*** (1.1)
Paid Performances (A)	44	.32	2.11**	.15 (.15)	.92* (.30)	1.1** (.38)	-.04 (-.30)		
Paid Performances (B)	43[h]	.45	6.99***	.04 (.04)	-.45 (-.15)	1.7*** (.61)			-.92*** (-.72)
Organization Meetings	56	.31	1.02**	-.04 (-.04)	.78** (.34)	1.4*** (.63)	-.02 (-.16)		
Socials	30[i]	.57	7.66***	-.16 (-.16)	.74 (.22)	3.25 (1.0)			-1.1*** (-.81)

$*p < .10$ $**p < .05$ $***p < .01$

Notes:

[a]Dependent variable lagged one period (two years).

[b]Years since 1890.

[c]Estimate of number of automobiles and telephones, logged, smoothed, and lagged one period (two years).

[d]Spectator performances that required paid admission, such as plays, movies, and other shows.

[e]Dances, teas, mixers, and other activities largely meant for socializing.

[f]Excludes two "influential" cases, San Rafael in 1928 and 1930.

[g]Increase in explained variance significant, although beta-weights for automobiles and telephones are out of range.

[h]Excludes "influential" case, Palo Alto in 1902.

[i]Low N because of many years with too few socials; this equation also excludes an "influential" point, San Rafael in 1940.

significance and high betas). It is mildly interesting that the best predictor of in-town activities was in all cases the number of residential telephones.

The results using the percentages of activities that were local (the bottom panel of Table H–2) indicate that, for the most part, nonlocal activities became more frequent faster than did local ones. ("Vicinity" activities account for that growth.) However, this was not so for organizational meetings. Automobiles were most implicated in the relative expansion of nonlocal events overall and for sports, whereas telephones most highly correlated with the location of performances and socials. The results for organized sports, equation (B), are intriguing because they imply that automobiles increased the proportional growth of out-of-town contests and telephones increased the in-town ones. There are nonsignificant and unstable tendencies in a similar vein for other categories of events. (To mitigate the collinearity a bit, I also tested the models of Table H–2 for only the years 1928 to 1940. The results are largely the same, excepting one equation: For the percentage of all activities that are in town, telephones have a positive effect and automobiles a negative one. A similar result occurred for the percentage of meetings that were local and a similar tendency for the percentage of sports events.[9]) Still, the fundamental conclusion is that, excepting the large category of organization meetings, nonlocal leisure events expanded faster than local ones, but both kinds grew in number.

Moving back to the larger picture and simpler analyses, the percentage of all leisure activities that were local declined from 69 percent to 61 percent between the 1900s and the 1930s.[10] Organizational meetings hardly changed, from 74 percent to 73 percent in-town, but the proportion of local sports events dropped steadily from 74 percent to 35 percent of all reported events. Paid performances were most localized in the 1910s (84 percent were in-town) and least in the 1930s (68 percent), and socials followed a similar pattern (83 percent to 70 percent.) Excepting club meetings, reported leisure events out-of-town—although nearby—expanded more than did in-town events over these decades.

MARRIAGE RECORDS

Our sampling procedure was to record at least 20 marriage certificates for each town for each fifth year from 1900 on. In some instances, we recorded all the licenses; in others, we had to pool years to attain a minimum number. The denominator for the percentages shown in Figure 18 is the number of marriage licenses in which at least one of the parties listed an address in the town. The numerator is the number of licenses in which *both* parties listed in-town addresses.

For Antioch, Barbara Loomis examined marriage licenses issued in the county seat of Martinez. By combining years, all Antioch percentages are based on at least 20 certificates each. Barry Goetz tracked down the Palo Alto marriage licenses in the county seat of San Jose. Many weddings performed on the Stanford campus involved students whose homes were elsewhere, but we counted only those campus weddings that included at least one Palo Altonian. Excepting 1910, all the points are based on at least 29 weddings conducted in Palo Alto (or Stanford). John Chan tallied the San Rafael licenses. All San Rafael percentages are based on at least 20 weddings. Note that we have no record of residents' weddings that were held out of town. There seems no reason why wedding habits would have changed over the years to systematically affect the location of the ceremony in a way correlated with our estimates.*

SAN FRANCISCO NEWSPAPER CIRCULATION

We obtained the circulation figures of San Francisco newspapers in our three towns from the Audit Bureau of Circulation, Schaumburg, Illinois. (Susan M. Kidder, then of the University of Chicago, recorded most of the figures. Edward Yelin gathered some supplemental data.) Circulation data for this period are not fully compiled and are somewhat ambiguous. For example, it is not clear what boundaries of Antioch, Palo Alto, and San Rafael are referred to in the Audit Bureau files: are they city limits or trading areas? Still, there does not appear to have been any change in the statistical procedures between the 1910s and 1940 that would affect our conclusions.

In the figures, "Sunday" refers to the summed circulation in each town of the Sunday *San Francisco Chronicle* and the *Examiner,* including both mailed and delivered issues. "Morning" refers to the daily *Chronicle* and *Examiner* (then a morning newspaper) added together. "Evening" refers to the *Call-Bulletin* for the years after 1928 and to the *Bulletin* and *Call and Post* added together for the years before that. (One other newspaper, the morning *News,* had substantial circulation, but only after 1921. I did not include it because its time span was limited. However, per capita subscription to the *News* increased slightly for Palo Alto and San Rafael over the years. Including the *News* would have shifted the "morning" lines for those two towns from a slight decline after the early '20s to essentially straight lines. For Antioch, the per capita *News* circulation was small and largely unchanging.) The population estimates used as denominators are the same as those calculated for Chapter 5. In the case of Palo Alto, the population includes Palo Alto city, Mayfield before 1926, and Stanford University.

Linear regression analyses indicate that for Antioch, all three types of newspapers lost per capita circulation, from a rate of $-.002$ copies per year for the Sunday editions to $-.005$ for the evening press, all statistically significant. (With such data, statistical significance is, of course, only heuristic.) For Palo Alto and San Rafael, evening papers declined, morning papers were unchanged, whereas Sunday subscription rates increased at about .001 and .002 per capita per year, respectively.

We were unable to get any serious circulation data for the towns' own newspapers covering the years before 1945. Even contacting the newspapers themselves was of no avail. For the *Palo Alto Times,* however, we had a few crude estimates of circulation based on occasional notes in the newspapers. (Not shown in Figure 19 is an estimate of .17 for 1896 and .14 for 1908. The 1930 and 1934 estimates are of paid circulation, whereas the number of free "shoppers' editions" were considerably higher.) For Antioch, we had three circulation claims in the 1930s, and for the *San Rafael Daily Independent,* we had two estimates. These figures

---

*Such a bias would involve couples increasingly (or decreasingly) choosing to marry in locations other than either of their home towns and that this change was especially true for intertown (or intratown) couples.

need to be taken with some skepticism, as they were not "audited." Also, their growth reflects subscription outside the town, so that, to the extent that outside population grew faster or slower than town population, the per capita estimates are in error.

VOTER TURNOUT

I fitted the available data on voter turnout—garnered from day-after-election stories in the local press—to linear and quadratic functions of years since 1890 (see Table H–3). For Antioch, I modeled turnout for presidential elections and city council elections. For Palo Alto, I modeled presidential and gubernatorial election turnouts together and city council elections. For San Rafael, I was unable to obtain general turnout data and could only model city council turnout. The results indicate both greater fit and steeper accelerations for general turnouts than local ones. Measuring turnout as a percentage of the population indicates linear increases for the general elections and no time trend for the council elections in Antioch and Palo Alto. (For Palo Alto, "population" refers, of course, to just the population within the city limits.) San Rafael city elections somewhat resemble the other cities' general elections, but unfortunately I cannot compare them to general votes in that city.

TABLE H–3

REGRESSION OF VOTER TURNOUTS, 1890–1940

| | | Number of Voters | | | |
| | Antioch | | Palo Alto | | San Rafael |
	Presidential	Council	General	Council	Council
N	10	15	17	22	17
R^2	.97	.81	.96	.78	.87
Constant	935	1378	811	252	852
Years since 1890	−78	−92	−104	−4	−33
Years Squared	2.1	1.8	5.5	1	1.8

| | | Voters as Percentage of Population | | | |
| | Antioch | | Palo Alto | | San Rafael |
	Presidential	Council	General	Council	Council
N	10	15	17	22	17
R^2	.83	.01	.77	.09	.63
Constant	37	17	16	18	10
Years since 1890	−1.3	0.1	0.7	−0.1	0.6
Years Squared	0.3	—	—	—	—

Notes

(*A list of selected source abbreviations is found on page 379.*)

1. TECHNOLOGY AND MODERN LIFE

1. Knights of Columbus, "Elaborated Questions for Adult Education Committee," in file, "1926–27 K.C. Adult Education Committee," Archives of the San Francisco Archdiocese (courtesy of Lyn Dumniel).

2. How can we tell which era's technological developments were more consequential? It seems largely a subjective judgment, but at least one numerical guide can provide an estimate: the number of new patents for inventions issued each year per capita. That number was typically higher in the years before World War II than it has been in the years since. Looking at every tenth year, there were between 321 and 382 patents per million Americans from 1900 to 1940, compared to 263 in 1960, 317 in 1970, 292 in 1980, 342 in 1988 (U.S. Bureau of the Census, *Historical Statistics of the United States, 1790–1970,* 957–59, and idem, *Statistical Abstract 1990,* table 886). It is also reasonable to assume that more of the earlier patents were fundamental innovations.

3. The advertisement appears in the N. W. Ayer Collection at the National Museum of History, Smithsonian Institution, Washington, DC.

4. Telegraph: quoted by Marvin, *When Old Technologies Were New,* 189; radio: Douglas, *Inventing American Broadcasting,* 306. In the early twentieth century many major thinkers considered the new media as solutions to the lack of moral and political consensus (Czitrom, *Media,* pt. 2).

5. The automobile quotation is provided by J. Smith, "A Runaway Match," 583. Smith describes the positive and affectionate view of the automobile in early films. See also Dettlebach, *In the Driver's Seat.* On attitudes of planners and farm women, see, for example, Foster, *From Streetcar to Super Highway,* and Interrante, "You Can't Go to Town in a Bathtub."

6. Presbyterians and the bicycle: R. A. Smith, *A Social History of the Bicycle,* 71–75. *Ambersons:* Tarkington, *The Magnificent Ambersons,* 275. Family life:

see Lynd and Lynd, *Middletown*, 137, 153, 263–69, and passim. Colleges: J. N. Mueller, "The Automobile," 111ff. On telephones and rudeness, see, e.g., Antrim, "Outrages of the Telephone" (1909). On the telephone and neighborhood, see, e.g., McKenzie, "The Neighborhood" (1921).

7. The literature on modernization theory and research is enormous. Examples of the application of this approach to American history include R. Brown, *Modernization*, and Wiebe, *The Search for Order*. For critical overviews of this theoretical tradition, see, e.g., Tilly, *Big Structures;* Bender, *Community and Social Change;* Zunz, "The Synthesis of Social Change"; and Conzen, "Community Studies."

8. I have discussed these ideas in further detail in Fischer et al., *Networks and Places,* chaps. 1 and 10.

9. Tilly, *Big Structures,* 11–12, 48ff; Rutman, "Community Study," 31.

10. Some of the arguments in this section were initially pursued in Fischer, "Studying Technology and Social Life."

11. For overviews of the study of technology and society, see two recent texts, Westrum, *Technologies and Society,* and McGinn, *Science, Technology, and Society*. On the work of Ogburn, see, e.g., Ogburn, *Social Change* (1922), "The Influence of Invention" (1933), and "How Technology Causes Social Change" (1957); and Westrum, *Technologies and Society,* chap. 3. The "oblivion" quotation is from Westrum, "What Happened to the Old Sociology of Technology?" See also R. Merrill, "The Study of Technology." Today's sociologists touch on the subject tangentially for the most part, their major interest being the organization of work and production. Braverman, *Labor and Monopoly Capital,* is a popular example of the study of technology as the study of work. For the work of historians, see, e.g., Hounshell, *From the American System to Mass Production,* and especially the journal *Technology and Culture*. The trolley example is from McKay, *Tramways and Trolleys*. Boorstin, *The Americans,* is a bold and exceptional example of a historian's exploration of the social consequences of technology. A manifesto for a new sociology of technology is Bijker, Hughes, and Pinch (eds.), *Social Construction*. A recent critique of this movement is Woolgar, "The Turn to Technology."

12. Recent general treatises on the subject of housework technologies include Cowan, *More Work for Mother,* and Strasser, *Never Done*. Examples of specific empirical studies include studies of air conditioning, garbage disposals, and electrical sockets.

13. See R. Merrill, "The Study of Technology," 576–77; Kranzberg and Purcell, *Technology and Western Civilization,* 4–6; Winner, *Autonomous Technology,* 8–12; Hughes, *American Genesis,* 5.

14. Giedion, *Mechanization Takes Command,* 3.

15. LaPorte, "Technology as Social Organization."

16. The distinction between impact and symptomatic analysis is at least implicit in Stack, *Communication Technologies;* Winner, *Autonomous Technology;* Williams, *Television, Technology, and Cultural Form;* and Daniels, "The Big Questions."

17. See Ogburn, "How Technology Causes Social Change," "National Policy and Technology," and *Social Change,* 200ff. For an intriguing application

of this approach, see Ogburn's projections in *The Social Effects of Aviation*. See also: White, *Medieval Technology* and "Technology Assessment"; and Winner, *Autonomous Technology*, chap. 2.

18. I am thinking here in particular about retrospective technology assessment. See, e.g., Segal, "Assessing Retrospective Technology Assessment"; and Coates and Finn, *A Retrospective Technology Assessment*.

19. See, e.g., Bijker, Hughes, and Pinch, *Social Construction*.

20. See, e.g., St. Clair, "The Motorization and Decline of Urban Public Transit"; Yago, *The Decline of Transit*. For challenges to these claims, see, e.g., Bottles, *Los Angeles and the Automobile;* Foster, *Streetcar;* and Davis, *Conspicuous Production,* including 248, n.3, for further citations.

21. Daniels, "The Big Questions," 3, 6. See also responses to Daniels's argument following the article; and see Schmookler, "Economic Sources of Inventive Activity," for a similar statement specifically regarding economic change.

22. McKay, *Tramways*. On France and the telephone, see, e.g., Attali and Stourdze, "The Birth of the Telephone"; Holcolmbe, *Public Ownership* (1911); and Bertho, *Télégraphes et Téléphones*.

23. Kern, *The Culture of Time and Space*: "vast extended," 318; "habit of mind," 91. Here Kern quotes the journalist Casson, *The History of the Telephone* (1910).

24. An extended excerpt may more fairly represent Kern's style:

> In comparison with written communication or face-to-face visits the telephone increased the imminence and importance of the immediate future and accentuated both its active and expectant modes, depending on whether one was placing or receiving a call. A call is not only more immediate than a letter but more unpredictable, for the telephone may ring at any time. It is a surprise and therefore more disruptive, demanding immediate attention. The active mode is heightened for the caller who can make things happen immediately without enduring the delay of written communication, while the intrusive effect of the ringing augments the expectant mode for the person called by compelling him to stop whatever he is doing and answer. (*Culture,* 91)

25. Touring: see Belascoe, *Americans on the Road*, chap. 2; idem, "Cars versus Trains." Farmers: see Murphy, "How the Automobile Has Changed the Buying Habits of Farmers" (1917), 98; Wik, *Henry Ford and Grass-Roots America;* and M. Berger, *The Devil Wagon in God's Country*.

26. Meyerowitz, *No Sense of Place,* 115.

27. D. H. Fischer, *Historians' Fallacies,* 177–78.

28. "Symptomatic": Williams, *Television,* chap. 1; "technological politics": Winner, *Autonomous Technology,* 226; and "rationalization": ibid., chap. 5.

29. Mumford, "Two Views of Technology and Man," 4–5.

30. Leo Marx, *The Machine in the Garden*. The assembly line was most dramatically used by Charlie Chaplin in *Modern Times*. On uses of the engine and automobile as symbols, see Tichi, *Shifting Gears,* and Lewis (ed.), "The Automobile and American Culture." On the telephone, see Brooks, *Telephone*.

31. Stack, *Communication Technologies,* 77.

32. Borgmann, *Technology and the Character of Contemporary Life*. "Center and illuminate," 4. Other authors in the same vein include Lewis Mumford, Marshall McLuhan, and Daniel Boorstin.

33. By the "myth of cultural integration," we mean the notion that all aspects of a society need to be harmonious, closely and functionally integrated. See Archer, "The Myth of Cultural Integration"; see also Boudon, "Why Theories of Social Change Fail."

34. On trains and cars, see Moline, *Mobility and the Small Town*. On housework, Cowan, *More Work for Mother*, especially chap. 4. On communication, Willey and Rice, *Communication Agencies and Social Life*.

35. On interaction, see Willey and Rice, *Communication Agencies;* touring, Belascoe, *Americans on the Road;* and blacks, Preston, *Automobile Age Atlanta*.

36. The washing machine example is from Lynd and Lynd, *Middletown*, 174–75.

37. Sales of video cartridges grew from about 10 million in 1980 to about 75 million in 1983 and then fell below 15 million in 1986 (*New York Times*, 17 September 1986). Although there was later a resurgence of home video games (*New York Times*, 9 March 1989), we remain far from the original projections. Those who predicted that home computers would sweep the nation in the 1980s were eventually disappointed.

38. The basic document of the school of social constructivism is Bijker, Hughes, and Pinch (eds.), *Social Construction*. On radio, see Douglas, *Inventing American Broadcasting*, and Marchand, *Advertising the American Dream*, 88–94. Marvin, *Old Technologies*, uses this orientation in analyzing debates around electric lighting and telephony in the nineteenth century. My comments here draw also on suggestions by Everett Rogers.

39. Marvin, *Old Technologies*.

40. Cowan, "The Consumption Junction."

41. These paradoxical outcomes are examples of "collective irrationalities," a class of social phenomena that have been discussed by rational-choice scholars such as Thomas Schelling and Mancur Olson.

42. "Intuitions": *Technology and the Character of Contemporary Life*, 11. For example, Borgmann makes claims about public opinion—that people view technology as a solution to eternal human problems (7), machines as "characteristic of our era" (8), and so forth—but ignores survey literature on what people actually think about science and technology. He makes debatable historical assertions, such as the claim that technology dissolved the "tradition of cooking" (59; some historians date the development of household cuisine in the United States to the home economics movement of the early 1900s); social psychological assertions, for example, that touring Glacier Park in a motor home is no richer an experience than viewing a film of it (56); and assertions about the history of technology, such as the declaration that science is "necessary" to technological advance (50), a claim that many historians of technology would dispute. Borgmann, like some other intellectuals, dismisses "mere empiricism" because he is confident in his own anecdotal and subjective experience of the world. One lesson we learn from social studies, however, is the variability of subjective knowledge and thus the necessity for more objective research.

43. Bartky, "The Adoption of Standard Time."

44. Some excellent work has been done, for example, on the role of the telegraph in business and administration by scholars such as Pred, *Urban Growth and the Circulation of Information,* and DuBoff, "The Telegraph and the Structure of Markets in the United States."

45. This position is stated most baldly in *The Division of Labor,* but elsewhere as well. A similar line of argument is implicit in Tonniës, Simmel, and perhaps Marx.

46. Social problems: Abler and Falk, "Public Information Services and the Changing Role of Distance in Human Affairs." Literary: Wisener, " 'Put Me on to Edenville,' " 23. Several contributors to Pool (ed.), *The Social Impact of the Telephone* note the same point; see especially 146–51. Herbert Dordick ("The Social Uses of the Telephone") suggests that the Bell System, by its lack of interest in such issues, is to blame.

47. Pool, *Forecasting the Telephone.* See also Marvin, *Old Technologies.*

48. On the theoretical underpinnings of the claim that telephones altered American urban form, see, e.g., Abler, "Space-Adjusting Technologies"; Falk and Abler, "Intercommunciations, Distance, and Geographical Theory"; Pool, "The Communication/Transportation Tradeoff"; and Salomon, "Telecommunications and Travel Relationships." Pool, *Forecasting,* 41–45, includes predictions along these lines. Gottman, "Metropolis and Antipolis," and Moyer, "Urban Growth and the Development of the Telephone," 342–69, both in Pool's *Social Impact,* review the arguments and some evidence.

49. See, e.g., McLuhan, *Understanding Media,* 238; Dordick, "Social Uses for the Telephone"; and Meyerowitz, *No Sense of Place,* 161. Marvin, *Old Technologies,* 100, points out that in the early years experts debated whether the telephone did more to empower the masses or more to coordinate the forces of repression. Authors in the Marxist vein, such as Martin, "Communication and Social Forms," contend that the telephone, like other tools developed by capitalism, generally advantages the rulers over the ruled, although they agree that insurrectionary uses of the telephone can occur in the interstices of the social order. In a recent international comparison, Buchner, "Social Control and the Diffusion of Modern Telecommunications Technologies," found that (pre-1989) Marxist regimes seemed to disfavor the development of telephones as compared to television, which he interprets as resistance to the possibly rebellious features of telephony.

50. On AT&T, see Marchand, "Creating the Corporate Soul," 12. Other commentators include McLuhan, *Understanding Media,* 225, and Abbott, *Seeking Many Inventions.*

51. "Wider distances": Kern, *Time and Space,* 215. See also, e.g., McKenzie, "The Neighborhood" (1921), 391; Calhoun, "Computer Technology, Large-Scale Social Integration, and the Local Community"; P. Berger, *The Heretical Imperative,* 6–7; Strasser, *Never Done,* 105.

52. The "antidote" comment is from Barrett, "The Telephone as a Social Force" (1940). Marvin, *Old Technologies,* chap. 5, reviews the discussion at the turn of the century on media and cross-cultural contact. AT&T engineer John J. Carty hailed the brotherhood of man in "Ideals of Telephone Service" (1922),

11. Casson, *History of the Telephone* (1910), 199, predicted one big family. On shallower communities and placelessness, see Abbott, *Seeking Many Inventions*, 74; Westrum, *Technologies and Society*, 276; and Meyerowitz, *No Sense of Place*.

53. Willey and Rice, *Communication Agencies and Social Life* (1933). "Back fence": Moyer, "Urban Growth," 365.

54. On tenseness, see, e.g., Kern, *Time and Space*, 91, and Brooks, *Telephone*, 117–18. Aronson, "The Sociology of the Telephone," 162, on the other hand, sees "an increased feeling of psychological and even physical security."

55. The quotation is from Marvin, *Old Technologies*, 6. See the volume for more along these lines; also Pool, *Forecasting*, 139–42. The social psychological arguments can be found in, e.g., Meyerowitz, *No Sense of Place*; Mitchell, "Some Aspects of Telephone Socialization"; and Abbott, *Seeking Many Inventions*, 167.

56. The quotation is from Jackson, *Crabgrass Frontier*, 280–81. See also Marvin, *Old Technologies*; Keller, *The Urban Neighborhood*. On liquor delivery, see, e.g., Duis, *The Saloon*, 220.

57. Willey and Rice, *Communication Agencies*, 199.

58. Pool, *Forecasting*, 43, provides the 1902 citation and the specific 1908 quotation from Carty. Recent repetitions include Asmann, "The Telegraph and the Telephone," 275; Dordick, "Social Uses of the Telephone—An U.S. Perspective," 227; and Abbott, *Seeking Many Inventions*, 70. Regarding the historical timing, the first acknowledged skyscraper, of ten stories, appeared in Chicago in 1885. Although that was several years after the first telephone exchange, it and other tall buildings appeared before the telephone was common, even in business.

59. The Heilbroner quote is from Rae, *The Road and the Car in American Life*, 370. See Steiner, "Recreation and Leisure Time Activities" (1933), 944; M. Berger, *Devil Wagon*, for samples of this view. Willey and Rice, *Communication Agencies*, argue for increased localization and Jackson, *Crabgrass Frontier*, for the retreat into parochial bastions.

60. On dangers to the family, see discussions in, e.g., Flink, *The Automobile Age*, 159 ff.; Brownell, "A Symbol of Modernity," 38–39; and, of course, Lynd and Lynd, *Middletown*. On privatism, see, e.g., Ariès, "The Family and the City in the Old World and New," 38; and Jackson, *Crabgrass Frontier*.

61. On the automobile as an agent of liberation, see, e.g., Wik, *Henry Ford*, 25; M. Berger, *Devil Wagon*, 65–66; and Scharff, *Taking the Wheel*. On the enslavement side, see, e.g., Hawkins and Getz, "Women and Technology"; McGaw, "Women and the History of American Technology"; and Cowan, *More Work*.

62. Several books on the history of the automobile include sections on "social impact," and almost inevitably the key—if not the only—source is *Middletown*. Aside from the problem of relying on one source, the Lynds' evidence on some issues, notably on sexuality, is quite slim and often little more than contemporary opinion or hearsay. See also Jensen, "The Lynds Revisited," on the Middletown methodology.

63. Boorstin, *The Americans*, 72–73. See also Daniels, "The Big Questions," for skepticism about the ability of technology to alter social patterns.

64. Modell, "Review Essay," 141–42. For a recent debate about whether one can or ought to apply theoretical abstraction to the history of technology, see

Buchanan, "Theory and Narrative in the History of Technology," and the subsequent comments.

2. THE TELEPHONE IN AMERICA

1. For example, the books by Pound, *The Telephone Idea* (1926), and Casson, *History of the Telephone* (1910), were paid for by AT&T, according to Danielian, *A.T.&T.*, chap. 13. See discussion in Chapter 3 of AT&T's general effort to plant stories in the press.

2. On district telegraph systems, see Tarr et al., "The City and the Telegraph"; Schmidt, "The Telephone Comes to Pittsburgh." On Dundas: Town Files of Bell Canada. Dordick, "The Origins of Universal Service," usefully contrasts the ways telegraphy and telephony were initially established.

3. Hounshell, "Elisha Gray and the Telephone."

4. Reproduced in AT&T, *A Capsule History of the Bell System,* 11.

5. Crandall, "Has the AT&T Break-up Raised Telephone Rates?" 40–41.

6. On the early evolution of the Bell System, see especially Garnet, *The Telephone Enterprise;* Lipartito, *The Bell System and Regional Business.*

7. Letter of 2 February 1880, in "Measured Rate Service," Box 1127, AT&THA. In 1884 Hall recalled the situation in 1879: "In Buffalo we had so many subscribers that the service became demoralized. Our switchboards were not equipped to handle the necessary amount of business . . . growing under the high pressure of competition" (Hall, "Notes on History of the Bell Telephone Co. of Buffalo, New York," [1884], 9). "Most far-seeing:" Lipartito, *The Bell System,* 252, n. 39.

8. See Chapius, *100 Years of Telephone Switching;* Mueller, "The Switchboard Problem."

9. See Paine, *Theodore N. Vail* (1929). Accounts of Vail appear in most telephone histories.

10. Los Angeles: "Telephone on the Pacific Coast, 1878–1923," Box 1045, AT&THA; Boston: Moyer, "Urban Growth," 352.

11. Letter (no. 75968) to Vail, 13 February 1884, in "San Francisco Exchange," Box 1141, AT&THA.

12. Bell had planned to switch from flat-rate to measured-rate (per-call) charges. With the support of city hall, subscribers organized a boycott. Eventually, Bell agreed to delay measured service for five years, place its wires underground, and pay the legal costs. See MacMeal, *The Story of Independent Telephony,* 111.

13. Los Angeles rates: "Telephone on the Pacific Coast, 1878–1923," Box 1045, AT&THA. Wage data come from the United States Bureau of the Census, *Historical Statistics of the United States,* tables D735–38.

14. On the debate over rates, see "Measured Rate Service" and other files in Box 1127, AT&THA. The quotation about "superfluous business" is from a Vail letter to Hall dated 7 February 1880.

15. Aronson, "Bell's Electrical Toy," 19.

16. Biographical notes on Hall were culled from press releases and clippings provided by Mildred Ettlinger at AT&THA. See also Lipartito, *The Bell System,* for an account of Hall's work in the South.

17. Patten (*Pioneering the Telephone in Canada,* 1ff) points out that Canadian telephone executives also had backgrounds in telegraphy.

18. On doctors and telephony, see Aronson, "*Lancet* on the Telephone," and Aronson and Greenbaum, "Take Two Aspirin and Call Me in the Morning." Histories of telephony often note the early role of doctors. See also chap. 5.

19. Letter to Thomas Sherwin, 11 July 1891, in "Classification of Subscribers," Box 1247, AT&THA.

20. Circular by Vail, 28 December 1883, and attachments, in Box 1080, AT&THA.

21. Mueller, "The Switchboard Problem."

22. Responses to 28 December 1883, circular by Vail, Box 1080, AT&THA.

23. Anderson, *Telephone Competition in the Middle West,* 13–14.

24. The California cases are in "PT&T News Bureau Files," Telephone Pioneer Communications Museum of San Francisco, Archives and Historical Research Center (hereafter TPCM). See also Olsen, "But It Won't Milk the Cows," and Atwood, "Telephony and Its Cultural Meanings."

25. See, e.g., the rationale for not pursuing wider markets as expressed in the United States Bureau of the Census, *Special Reports: Telephones and Telegraphs 1902* (1906), chap. 10. The stress on high quality was indeed important. In the South, for example, local managers were discouraged from building simple, low-cost systems, because of the national company's insistence on maintaining quality levels needed for the long-distance network (Lipartito, *The Bell System,* chap. 5).

26. In the New York–New Jersey data over 80 percent of telephones were located in businesses. In the same year, in Kingston, Ontario, about 70 percent of telephones were located in businesses, although many officially listed as residential were really used for business by people such as physicians (Pike, "Kingston Adopts the Telephone"; Pike and Mosco, "Canadian Consumers and Telephone Pricing").

27. Allred et al., *Rural Cooperative Telephones in Tennessee* (1937), 20. See also, e.g., Olsen, "But It Won't Milk the Cows"; People's Mutual Telephone Company, *It's for You;* and Atwood, "Telephony and Its Cultural Meanings."

28. The 1902 figure is from the United States Bureau of the Census, *Telephones... 1902* (1906), 26. See especially Gabel, "Where Was the White Knight When the Competition Needed One?" on the independents' efforts to develop long distance. See also Mueller, "From Competition to 'Universal Service' "; and Bornholz and Evans, "The Early History of Competition in the Telephone Industry."

29. J. I. Sabin letter to J. E. Hudson, 15 April 1895; and letter from St. Louis manager to Hudson, 2 June 1899, both in Box 1184, AT&THA.

30. In addition to the general sources, see, on Bell's competitive strategy, Bornholz and Evans, "Early History"; Mueller, " 'Universal Service' "; and Lipartito, "System Building at the Margin."

31. On Bell's anticompetitive tactics, see, e.g., MacMeal, *Independent Telephony;* Rippey, *Goodbye, Central,* 108–10 (which reports the grocery and line-cutting incidents); Gary, "The Independents and the Industry" (1928); and McDaniel, "Reminiscences." On bribery, see also Bean, *Boss Reuf's San Francisco,*

chap. 8. On the Kellog case, see Brooks, *Telephone,* 113–4. For a catalog of Bell tactics in Canada, see Martin, *"Hello, Central,"* 29–30.

32. In 1893 net income was 12 percent of the book value of Bell property and declared dividends were 7 percent; in 1907 they were 6 percent and 4 percent respectively (United States Bureau of the Census, *Historical Statistics,* table R17). Bornholz and Evans, "Early History," estimate that Bell returns on investment were 46 percent between 1876 and 1894 and 8 percent between 1900 and 1906.

33. Lipartito, "System Building"; Armstrong and Nelles, *Monopoly's Moment;* Cohen, "The Telephone Problem and the Road to Telephone Regulation."

34. Correspondence in "Party Line Development," Box 1284, AT&THA, provides much of the Bell perspective. Percentages of party-line customers are in a letter from J. P. Davis to President Cochrane, 2 April 1901, in "Telephone Statistics, 1877–1902," Box 1312, AT&THA. The Davis quotation is from a letter to President Hudson, 6 February 1899. The Vail quote is from New York State, *Report of the Committee,* 472. See also, e.g., Mahon, *The Telephone in Chicago,* 26–27; Masters, Smith, and Winter, *An Historical Review of the San Francisco Exchange.* The habit quotation is from Hibbard, *Hello-Goodbye,* 210.

35. Martin, "Communication and Social Forms," 287–319, and Martin, *"Hello, Central?"* 107–08, discusses how operators in Canada bent the rules. A couple of our interviewees who ran or had a relative who ran a rural switchboard acknowledged that listening in was common.

36. New York: Kingsbury, *The Telephone and Telephone Exchanges* (1915), 478; national: *AT&T Annual Report,* 1910; Chicago: Rosher, "Residential Telephone Usage among the Chicago Civic-Minded," 25–26, and Mahon, *The Telephone in Chicago.*

37. See, in particular, MacMeal, *Independent Telephony,* 171–72; Hibbard, *Hello-Goodbye,* 205ff; and Mahon, *Chicago,* 25.

38. Purchases: Temin, *The Fall of the Bell System,* 11. See, in particular, Loeb, "The Communications Act Policy toward Competition"; Bornholz and Evans, "Early History"; Lipartito, "System Building"; United States House, "Federal Communications Commission" (1934); Faris, "Some Lessons Learned from the Past" (1926), 17; and Gary, "Independents."

39. Vietor, "AT&T and the Public Good," 40.

40. See, e.g., United States House, "Federal Communications" (1934); Federal Communications Commission, *Proposed Report* (1938) and *Investigation of the Telephone Industry* (1939); and Danielian, *A. T. & T.* (1939).

41. Gherardi and Jewitt, "Telephone Communication System of the United States" (1930), 34.

42. Most of these data come from Gherardi and Jewitt, "Telephone Communication System" (1930). The party-line estimate is reported in "Sales Activities in the Residence Market," 33, in *Bell System General Commercial Conference, 1930,* microfiche 368B, IBIC. For a detailed look at the telephone instruments themselves, see Haltman, "Reaching out to Touch Someone?"

43. Rates are from Gray, "Typical Schedules of Rates for Exchange Service" (1924). Comparative data are from the United States Bureau of the Census, *Historical Statistics,* 168–71.

44. On the stability of rates, see Williams and Hansen, "Money Disbursements of Wage Earners and Clerical Workers" (1941), 35ff.

45. The surprise over the increase in subscriptions was expressed in Helms, "The Relation between Telephone Development and Growth and General Economic Conditions" (1924).

46. L. B. Wilson, "Sales Activities" (1928).

47. Reports of tribulations and extra exertions during the Depression can be found in various internal sources, including many issues of *Telephony*, such as 12 September 1931, 14–19, and 10 June 1933, 114–15.

48. Canada Department of Trade and Commerce, "Telephone Statistics for 1930"; see also Pike and Mosco, "Canadian Consumers."

49. Ayer Collection, National Museum of American History.

50. These statistics and those in the next paragraph come from annual editions of "The World's Telephones," in *Bell Telephone Quarterly*, and from Chapuis, *100 Years*, 43, 139–41, and 279–83.

51. I calculated these ratios for countries with their GNPs, circa 1927, listed in Mitchell, *European Historical Statistics*, 817–39. The exchange rates, for mid-1927, to convert currencies to dollars came from the United States Bureau of the Census, *U. S. Statistical Abstract 1929*, and the telephone statistics are from "The World's Telephones," *Bell Telephone Quarterly* 8 (July 1929), 218–30. The key problem with this analysis is that for both per capita wealth and per capita telephones, the United States is an extreme outlier, much higher than any other country. In regression analyses the U.S. case has very high statistical "leverage." Still, graphs and regressions using logged values all indicate that, at the least, telephones were not much more numerous in the United States than in European countries once relative wealth is taken into account. In corroboration of the point an AT&T executive worried to his colleagues, in 1928, that telephone diffusion in the United States was, by standards of wealth, falling behind that of Europe (Page, "Public Relations and Sales," 2). On the party-line issue, L. B. Wilson, "Promoting Greater Use of Exchange Service," 45, quoted Vice President Gherardi as noting that most American telephones were on party lines, while all those in Great Britain were on individual lines, and that the American pattern was "looked upon, not with admiration, but with astonishment."

52. Using 1985 data from the United States Bureau of the Census *Statistical Abstract of the United States 1988*, tables 1396 and 1389, I regressed telephones per capita on gross domestic product per capita for the United States, Canada, and the 12 members of the European Economic Community (EEC). Whether raw figures or natural logs were used and whether the United States was or was not included in the estimations, the figure for the United States — 76 per 100 — is almost exactly what would be predicted by gross domestic product. (Indeed, since the cost of living in the United States is lower than in some other Western countries, the United States should have exceeded its expected rates of telephone diffusion.) Chapius, *100 Years*, 43, displays the relationship between GNP per capita and telephones per capita (both logged) in 1978 for many more countries. His graph shows that the United States was slightly above the regression line; that is, it had slightly more telephones than expected. But other nations, such as New Zealand, Great Britain, Greece, and Spain, had many more telephones than would be predicted. (Saunders,

Warford, and Wellenius, *Telecommunications and Economic Development*, 4, present other data for circa 1980 that would lead to the conclusion that the United States had more telephones than expected. The numbers in their table, both for telephones and GNP for the United States, seem to be wrong; they do not match comparable data in the United States Bureau of the Census, *Statistical Abstract 1990*, 840, 844, and imply that the United States had a per capita GNP substantially below that of a few European countries.)

53. See Nye, *Electrifying America*, 259ff, 140, 180–81, 387.

54. Hughes, "The Evolution of Large Technological Systems."

55. The literature on the automobile is, by comparison to that of the telephone, vast. This is a small inventory. Basic sources on automobile history include the works of renowned automobile historian John B. Rae, *The American Automobile* (1965), *The Road and the Car in American Life* (1971), and *The American Automobile Industry* (1984), and of a younger, more critical, dean, James J. Flink, *Americans Adopt the Automobile, 1895–1910* (1970), *The Car Culture* (1976), " 'The Car Culture' Revised" (1980), and a summary volume, *The Automobile Age* (1988). A dissenting interpretation of the industry's formation is Donald Davis, *Conspicuous Production* (1988). Other studies on the industry include Epstein, *The Automobile Industry* (1928); Hounshell, *Mass Production;* Hugill, "Good Roads and the Automobile in the United States"; S. Miller, "History of the Modern Highway in the United States"; and Yago, *The Decline of Transit*. Studies of the automobile in rural America include M. Berger, *Devil Wagon;* Wik, *Henry Ford;* and Interrante, "You Can't Go to Town in a Bathtub." Studies of how people used the automobile and how it functioned in society include, aside from the chapter in Lynd and Lynd, *Middletown,* Belascoe, *Americans on the Road;* Brilliant, "Some Aspects of Mass Motorization in Southern California"; Brownell, "A Symbol of Modernity"; Interrante, "The Road to Autopia"; Lewis (ed.), "The Automobile and American Culture," special issue of *Michigan Quarterly Review* (Fall, 1980); Moline, *Mobility and the Small Town;* Nash, "Death on the Highway"; Preston, *Automobile Age Atlanta;* Tobin, "Suburbanization and the Development of Motor Transportation"; and Willey and Rice, *Communication Agencies* (1933). Some international comparisons are available in Bardou et al., *The Automobile Revolution;* and D. Davis, "Dependent Motorization."

56. Foster, *Streetcar,* 60–62.

57. Model-T: Clymer, *Henry's Wonderful Model T,* 109–27; Automobile Manufacturers' Association, *Automobile Facts and Figures,* 4; families: Lebergott, *The American Economy,* 290.

58. For the statistics, see Heer, "Trends in Taxation and Public Finance" (1933); and the United States Bureau of the Census, *Historical Statistics,* tables Q82, Q97. On road improvement, see McShane, "Transforming the Use of Urban Space."

59. A 1982 Federal Highway Administration study estimated that the general tax subsidy of the automobile amounted to 11.8 cents per mile driven (Lazare, "Collapse of a City," 274).

60. Page, "Public Relations," 2–3.

61. Among the sources on streetcars are McKay *Tramways;* Foster, *Streetcar;* Cheape, *Moving the Masses;* St. Clair, "Motorization and Decline"; and Bottles, *Los Angeles.*

3. EDUCATING THE PUBLIC

1. Monaco, "The Difficult Birth of the Typewriter."

2. The phrase, "theatre of science," comes from Armstrong and Nelles, *Monopoly's Moment*, chap. 3; see also Marvin, *Old Technologies*, esp. 209–16; and Aronson, "Bell's Electrical Toy." Flink, *Americans Adopt the Automobile*, and Corn, *The Winged Gospel*, describe how the automobile and airplane were popularized. Quebec and Hamilton: Collins, *A Voice from Afar*, 77–79. PT&T: *Pacific Telephone Magazine*, December 1907, 13, and September 1916, 20. Hearst: Letter from N. C. Kingsbury to G. E. McFarland, August 1915, in "AT&T Long Lines," TPCM.

3. Hibbard, *Hello-Goodbye*, 7. Lodi: PT&T News Bureau, TPCM.

4. In the Ayer Collection, National Museum of American History, Washington, DC. Emphases in original.

5. Vail testimony: 9 December 1909, in New York State, *Report of the Committee*, 398; Young, "The Elements of a Balanced Advertising Program," 91.

6. Letter from Michaelis and Ellsworth to F. P. Fish, 3 October 1906, and letter from Publicity Bureau to J. D. Ellsworth, "Statement of Work in Hand," 10 December 1907, in "Advertising and Publicity—Bell System—1906–1910, Folder 1," Box 1317, AT&THA. In 1909, after Vail's changes, internal correspondence involving Ellsworth discussed newspaper publishers' increasing resistance to "free publicity"—and AT&T's apparent immunity from the publishers' censure (letter from P. Dudley to J. D. Ellsworth, 22 September 1909, and letter from Ellsworth to Vail, 24 September 1909, in ibid.) On publicity, see Griese, "AT&T: 1908 Origins of the Nation's Oldest Continuous Institutional Advertising Campaign."

7. E.g., Atwood, "Telephony."

8. Considerable details on Bell system publicity and advertising are available in various files labeled "Advertising and Publicity," "Publicity Conference," "Advertising Conference," and the like in boxes 23, 1310, and 1317, AT&THA (as well as in other archives). For example, "Publicity Conference—Bell System—1914," Box 1310, details Ellsworth's suggestions to local publicity men. It also documents the continuation of free publicity. See also Steel, "Advertising the Telephone"; Hower, *The History of an Advertising Agency;* Griswold, "How AT&T Public Relations Policies Developed"; T. T. Cook, "The Advertising of [AT&T]"; and W. J. Phillips, "The How, What, When and Why of Telephone Advertising," talk given on 7 July 1926, in "Advertising," TPCM. Some of the ads Ayer placed can be found in N. W. Ayer & Son, *Fifty Years of Telephone Advertising*, and in the Ayer Collection, National Museum of American History. These ads largely promote the industry, rather than basic sales.

9. Eureka: A. E. McLaren, "Telephone History Notes," 1934, in "Eureka," PT&T News Bureau, TPCM. Illinois Bell: Mahon, *The Telephone in Chicago*, 101–03. All-employee sales efforts are described, for example, in *Pacemaker*, 1928, TPCM; Lauderback, "Introduction," 1; Summerfield, "Kansas Company's Campaign Record."

10. See, e.g., on the 1910s "Publicity and Advertising Conference—Bell System—1916," Box 1310, AT&THA; 1924: see letter from B. Gherardi, AT&T,

to G. E. McFarlane, PT&T, 26 November 1924, in "282 Conferences," TPCM; late 1920s: Mahon, *Chicago,* 89; *General Commercial Conference, 1928,* Microfiche 368B, IBIC; Lauderback, "Introduction," and "Residence Exchange" (much of that material also appears in Lauderback, "Bell System"); Depression: see, e.g., Mahon, *Chicago,* 101; Fancher, "Every Employee is a Salesman for American Telephone and Telegraph."

11. Aronson, "Bell's Electrical Toy"; Walsh, *Connecticut Pioneers in Telephony,* 27; Glauber, "The Necessary Toy," 70–86; Tarr, "The City and the Telegraph."

12. *Printers' Ink,* "Results by Telephone," 35, and "Broadening the Possible Market," 20; Mock, "Fundamental Principles of the Telephone Business: Part V: Advertising and Publicity," 21–23; Valentine, "Some Phases of the Commercial Job," 40; and H. B. Young, pages 91 and 100 in "Publicity Conferences—Bell System—1921–34," Box 1310, AT&THA.

13. On the search for uses, see generally *Telephony, Pacific Telephone Magazine,* and *Printers' Ink.* On broadcasts, see Marvin, *Old Technologies,* 209–23; Walsh, *Connecticut Pioneers,* 206; Dilts, *The Telephone in a Changing World,* 32–33.

14. The advertisements are from "Advertising and Publicity—Bell System—1906–1910, Folders 1 and 2," Box 1317, AT&THA.

15. 1878: Walsh, *Connecticut Pioneers,* 47; Illinois: Central Union Telephone Company, *Instructions and Information for Solicitors,* 13–14; the 1910 ads accompany a letter from E. Harris to J. D. Ellsworth, 29 January 1910, in "Advertising and Publicity—Bell System—1906–1910, Folders 1 and 2," Box 1317, AT&THA; see also "Advertising," TPCM; Reynolds, "Selling a Telephone"; Haire, "The Telephone in Retail Business," and "Bell Encourages Shopping by Telephone."

16. Instructions on sales emphasis appear in the various "Advertising and Publicity Conference" folders at AT&THA, TPCM, and IBIC. On *Pacemaker* (at TPCM), see, e.g., volume 2, April, 1929.

17. 1905: Steel, "Advertising the Telephone," 15; 1920s: "Advertising," folder, TPCM, and interviews with retired telephone men conducted by John Chan. Extension statistics: For example, Lauderback ("Residence Exchange Service Sales," 33) reported that on 1 January 1929 there was one extension per 12 Bell residential subscribers nationwide. He thought the increase of 70,000 extensions in 1929 compared unfavorably to the 630,000 refrigerators installed in 1928 and the 80,000 luxury cars sold in 1929. Lauderback also reported that individual-line subscriptions grew from 31 percent to 37 percent of all nonrural lines between 1926 and 1930, an unimpressive figure given the economic boom of those years.

18. Fishing ad: "Advertising," TPCM. On long-distance sales policy, see correspondence between A. W. Page and PT&T President H. D. Pillsbury, March and April 1929, TPCM.

19. The U.S. government helped the industry teach the use of the dial telephones to the general public (Haltman, "Reaching Out," 340).

20. Canada: Collins, *A Voice,* 137; California: San Rafael telephone directory, 1884, in PT&T News Bureau files, TPCM; Indiana: "Scrapbook, Southern Indiana Telephone Company," MIT. See also: *Telephony,* "Educating the Public"; "Advertisements," TPCM; *Pacific Telephone Magazine,* passim, and specifically on dial phones, 16 March 1923, 16.

21. Ad and pledge: "Advertising and Publicity—1906–1910," AT&THA. On laws see *Telephony,* "Swearing over the Telephone"; article reprinted from *Hous-*

ton Post, in *Pacific Telephone Magazine* 3 (August 1909): 6; Marvin, *Old Technologies,* 321–32. In general, industry publications were replete with items and complaints on this matter.

22. Contest essay: see "Courtesy Article," 2 February 1910, and related correspondence of J. D. Ellsworth, 1909, in "Advertising and Publicity—1906–1910, Folders 1 and 2," AT&THA.

23. *Voice Telephone Magazine,* "Telephone Etiquette."

24. Arnold, *Party Lines, Pumps and Privies,* 150. One Bell survey done in 1945 found that "in one area more than half of the rural customers . . . did not object to 'listening in' by others on the same line" (Hanselman and Osborne, "More and Better Telephone Service for Farmers," 226). Another done in Canada reported that customers wanted fewer stations on their lines largely, it seems, to prevent a weakened signal (Carruthers, "Rural Telephone Development").

25. Indiana: Arnold, *Party Lines,* 151. Rakow, "Gender, Communication, and Technology," 230. On efforts to enforce time limits, see, e.g., *Telephony,* "Limiting Party Line Conversations," and idem, "Limiting 'Talkers' in Jamestown, N.Y." A former telephone operator told Lana Rakow that she had to enforce a ten-minute limit on some who abused the line ("Gender, Communication, and Technology," 210).

26. The sort of basic instruction and blatant efforts at instilling etiquette noted here have seemingly tapered off over the years. Perhaps the industry succeeded in its social control, perhaps it gave up efforts at social control, perhaps it solved many problems technically (by eliminating operators through automatic dialing, increasing calling capacity on trunk lines, and so on), or perhaps younger generations were more comfortable with the telephone, or some combination of these. Nevertheless, some teaching of telephone courtesy continues, at least for children. A 1986 book on telephone etiquette describes a "Gallery of Phone Monsters," including "Tie-Uppasaurus" and "Grumple" (Weiss, *Telephone Time*).

27. The initial quotation is from an anonymous letter endorsed by Kingsbury to J. D. Ellsworth, in "Advertising and Publicity—Bell System—1912–1913," Box 23, AT&THA. The Kingsbury quotation is from Kingsbury, "Advertising Viewed as an Investment"; see also Kingsbury, "Results from the American Telephone's National Campaign." General studies of AT&T's public relations include: Federal Communications Commission, *Proposed Report* and idem, *Investigation;* Danielian, *A.T.&T.,* chap. 5; N. Long, "Public Relations of the Bell System"; Griswold, "AT&T Public Relations"; Griese, "AT&T: 1908 Origins"; Marchand, "Creating the Corporate Soul."

28. See Federal Communications Commission, *Proposed Report,* chap. 17; N. Long, "Public Relations."

29. In a letter to Fish dated 3 October 1906, the Bureau's Michaelis and Ellsworth chided a Bell publicist in northern California for believing that the only "way to get help from the newspapers is by buying them. . . . We were impressed by the amount of favorable matter which the newspapers would print for nothing if the matter was properly prepared." A report from the Bureau to J. D. Ellsworth, by then at AT&T, dated 10 December 1907, lists the contents and fates of scores of stories. Both items appear in "Advertising and Publicity—Bell System—1906–1910, Folder 1," Box 1317, AT&THA.

30. For example, a long article in a 1937 *National Geographic* (Colton, "The Miracle of Talking by Telephone") reads as a lengthy press release. Also, as noted earlier, a handful of books were quietly subsidized by AT&T (see Danielian, *A.T.&T.,* chap. 13).

31. See Atwood, "Telephony," for example, on newspapermen as telephone entrepreneurs. Greene, "Operator—Could You Please Ring?" reports an Idaho case in which a small-town publisher established the local telephone system and then sold out to PT&T. PT&T free service: letter from E. C. Bradley to T. N. Vail, 3 August 1908, in "Advertising and Publicity—Bell System—1906–1910, Folder 2," Box 1317, AT&THA. Sometimes local managers felt blackmailed into advertising. For example, in 1910 PT&T obtained the agreement of the *San Francisco Chronicle* to pay full rates for its telephones, but only in return for the placement of ads in the newspaper representing an equivalent charge for PT&T. This, however, was preferable to "having any trouble with a paper like the *Chronicle....*" (E. C. Bradley to J. D. Ellsworth, 24 June 1910, in "Advertising," TPCM). Southwestern: "Meeting of Publicity Men of AT&T and Associated Companies," 26 June 1914, Toronto, in "Publicity Conference—Bell System—1914," Box 1310, AT&THA. Portland: letter from C. H. Casey to H. D. Pillsbury, 10 January 1927, in "Advertising," TPCM. *Moody's:* Letter from H. W. Pool to Ellsworth, 7 September 1909, in "Advertising and Publicity—Bell System—1906–1910, Folder 2," Box 1317, AT&THA. These practices also occurred in Canada (Martin, *Hello, Central,* 36, 46–47) and are generally discussed in Federal Communications Commission, *Proposed Report.*

32. On clergy, see, e.g., the case of Rev. J. Whitcomb Brougher, 8 November 1914, in "Correspondence—Miscellaneous," TPCM. On prominent men, see letter from J. D. Ellsworth to J. A. Nowell, 18 September 1933, in "Advertising," TPCM. On other activities, see Mahon, *Chicago,* 96–97; Federal Communications Commission, *Proposed Report,* chap. 17, as well as sources cited in note above.

33. The ads come from the Ayer Collection, National Museum of American History, and from "Publicity Conference—Bell System—1921–1924," Box 1310, AT&THA. See Marchand, "Corporate Soul." On boosting employee morale, Griswold, "AT&T Public Relations." Marchand points out that Bell's own employees were an intended audience. AT&T hoped to bolster their morale and thereby improve their performance and enhance public encounters.

34. See, e.g., discussions in "Publicity Conference—Bell System—1914" and "Advertising Conference—Bell System—1916," both in Box 1310, AT&THA; correspondence between H. B. Thayer and G. E. McFarland, September 1914, in "Correspondence—H. B. Thayer," TPCM.

35. See "Correspondence—E. S. Wilson, V.P., AT&T," TPCM.

36. "Casual Phoning," reprinted in the *San Francisco Chronicle,* 7 February 1989.

37. "Newspaper Clippings, Ads, 1919–1922," file, Bell Canada.

38. Pope, *The Making of Modern Advertising;* S. Fox, *The Mirror Makers;* and Schudson, *Advertising,* 60ff. The stove example is from Busch, "Cooking Competition."

39. Edwards, "Tearing Down Old Copy Gods."

40. Bell Telephone Company of Canada, "Selling Service on the Job."

41. See comments, especially by Vice Presidents A. W. Page and Bancroft Gherardi, during *General Commercial Conference, 1928* and *Bell System General Commercial Conference, 1930,* both Microfiche 368b, IBIC. One hears many echoes of "comfort and convenience" at lower Bell levels during this period. The direct-mail messages were reported in Schoonmaker, "When Sales Lagged, Direct Mail Saved the Day."

42. The ads are from various collections in AT&THA, TPCM, and Bell Canada archives. In 1931, a sales manager from New England said that his men, in trying to sign up non-users, had found that "the low cost of daily service, the protection that a telephone affords, the convenience of telephone service and the desire for possession are among the most effective appeals" (Harrell, "Residential Exchange Sales," 83).

43. The emphasis is in the original. These newspapers are sources used in the community studies described in Chapter 5. A more complete statistical analysis of these and other sources is presented in C. Fischer, " 'Touch Someone'."

44. Illinois booklet: Central Union Telephone Company, *Instructions.* Note that Central Union had been, at least through 1903, one of Bell's most aggressive solicitors of business. 1931 memorandum: Illinois Telephone and Telegraph Company, *Sales Manual;* 1935 memorandum: Ohio Bell Telephone Company, "How You Can Sell Telephones," 1935 (Bell Canada archives file, "Salesmanship").

45. Sociologists Robert Pike (personal communication, 10 February 1987) and Michèle Martin ("Communication and Social Forms"), using Canadian archives, have detected similar shifts toward sociability themes. Atwood, ("Telephony") notes that social themes began appearing in rural Iowa ads earlier, circa 1911.

46. Worries: Marvin, *Old Technologies;* A. G. Bell: quoted by Pool, *Forecasting,* 103.

47. 1881 announcement by National Capitol Telephone Company, reported in a letter to Bell headquarters dated 20 January 1881, Box 1213, AT&THA. Gossip: Judson, "Unprofitable Traffic." He also wrote, "The telephone is going beyond its original design, and it is a positive fact that a large percentage of telephones in use today on a flat rental basis are used more in entertainment, diversion, social intercourse and accommodation to others, than in actual cases of business or household necessity" (p. 645). Canadian directory: Martin, "Communication and Social Forms," 353.

48. Hall's philosophy is evident in the correspondence over measured service before 1900 (Box 1127, AT&THA). Decades later, he pushed it in a letter to E. M. Burgess, Colorado Telephone Company (30 March 1905, Box 1309, AT&THA), even arguing that operators should stop turning away calls made by children and should instead encourage such "trivial uses." Another, more extreme populist was John I. Sabin of PT&T and the Chicago Telephone Company (see Mahon, *Chicago,* 29ff). The tone of these men's letters and accompanying comments indicate that they spoke as voices of a minority.

49. A. W. Page, "Public Relations and Sales." See also comments by Vice President Bancroft Gherardi and others in same conference and related ones of the period, such as Lauderback, "Introduction."

50. Tomblein, "Population Changes in Small Communities and in Rural Areas"; Holland, "Telephone Service Essential to Progressive Farm Home"; see also R. T. Barrett, "Selling Telephones to Farmers by Talking about Tomatoes."

51. On the telegraph background of Vail and other early telephone leaders, see, e.g., Rippey, *Goodbye, Central;* Patten, *Pioneering the Telephone in Canada.* The leaders of the small mutuals, however, were typically businessmen and farmers. See, e.g., Wisconsin State Telephone Association, *On the Line,* for histories of Wisconsin telephone companies. Yet even there, the statewide leader of independent telephony was originally a teacher of telegraphy (Barsantee, "The History and Development of the Telephone in Wisconsin.")

52. A. W. Page in L. B. Wilson (Chair), "Promoting Greater Use of Toll Service," 53.

53. Hibbard, *Hello-Goodbye,* 205–6.

54. Martin, "Communication and Social Forms," 397.

55. I refer here to Rakow, "Gender, Communication, and Technology"; Arnold, *Party Lines;* and the oral histories we conducted.

56. At the same 1928 conference that Page used to announce the comfort and convenience theme, Vice President Bancroft Gherardi said, "People think as a matter of course of turning to the telephone for business purposes, but they don't do it for social purposes," suggesting that the subscriber practices their line personnel grappled with—"idle gossip"—had not yet impressed the top management (in L. B. Wilson, "Sales Activities," 55).

A subtler economic argument is that the old practicality sales pitches had exhausted their appeal at a 42 percent market penetration level. The telephone industry turned to social conversations as a new use to attract the other 58 percent. Though plausible if one assumes that the industry expected significantly more diffusion (a debatable assumption; see Chapter 4), this explanation does not account for the continuity of the sociability theme into the Depression, when market penetration dropped sharply.

57. Rate schedules varied greatly from place to place. Also, the proportion of telephones on flat versus measured service did not substantially change around the time of the advertising change. For example, in San Francisco, the proportion of telephones on measured rates actually declined, from 79 percent in 1914 to 55 percent in 1924 ("Bell Telephones in Major Cities, 1924," in "Correspondence—DuBois," TPCM). As a different sort of evidence, a recent California survey suggests that customers on flat-rate or measured-rate (or even lifeline) services make equal proportions of social calls (Field Research Corporation, *Residence Customer Usage and Demographic Characteristics Study,* 41–43). Although company concerns about these economic matters no doubt influenced their orientation toward sociability, they do not seem to have been crucial. (Regarding this and other arguments, C. Fischer, " 'Touch Someone' " provides more extensive detail.)

58. Martin, "Communication and Social Forms," 376, notes that warnings against long talks appeared next to party-line instructions in Canadian directories before 1920.

On local variations in party- versus individual-line telephones: The variation among cities in 1929 was reported by Gherardi and Jewitt, "Telephone Communication System." Data from 1930 indicate that the proportion of nonrural residential telephones on four-party lines varied from none in New York City and Washington, D.C. to over 50 percent in the Midwest (Lauderback, "Residence Exchange," 34).

On change over time: In 1901, in the five cities with the most Bell subscribers, an average of 31 percent of all telephones were on party lines (letter from J. P. Davis to A. Cochrane, 2 April 1901, in Box 1312, AT&THA). For those same five cities in 1929, the average was 36 percent (Gherardi and Jewitt, "Telephone Communication System"). In 1929 40 percent to 50 percent of Bell's main telephones in almost all major cities were still on party lines, little changed from 1915 (ibid; in San Francisco, the proportion of stations on party lines increased from 55 percent in 1914 to 72 percent in 1924 ["Supplemental Telephone Statistics," in "Correspondence—DuBois," TPCM]). Between 1926 and 1930, however, the proportion of Bell residential telephones on party lines in all nonrural exchanges declined from 69 percent to 63 percent (Lauderback, "Residence Exchange," 34). Given the problems of comparability in much of the available data, the best estimate is that party-line telephones grew as a proportion of nonrural residential telephones until about the mid-1920s and then started to decline. Party-line telephones, nevertheless, remained the vast majority of subscriptions through the 1920s. No noteworthy change in that proportion coincides with the shift in advertising.

One could argue that even though the proportion of individual-line households was static or declining between 1900 and the 1930s, the absolute numbers were increasing, thus explaining the shift to the sociability theme. However, this seems a stretch, and there is no narrative material to support such a theory.

A related issue was the tying up of toll lines between exchanges, especially those between small towns, but this, too, does not explain the timing of the shift toward sociability. The toll-line bottleneck was resolved much later, when it became possible to have several calls on a single trunk line. Nor does the increased emphasis on selling long-distance account for the shift, despite the use of sociability in long-distance ads, because the thematic shift also appeared in basic service marketing.

59. See Singer, *Social Functions of the Telephone,* esp. 20; Mayer, "The Telephone and the Uses of Time"; and Wurtzel and Turner, "Latent Functions of the Telephone."

60. 1906 quote: *Printers' Ink,* "Advertising the Automobile," 47. 1914 quote: J. H. Newmark (Advertising Manager, Oakland Motor Company), "Have Automobiles Been Wrongly Advertised?"; and idem, "The Line of Progress in Automobile Advertising." On automobile advertising, see Flink, *America Adopts* and *Car Culture;* Goist, *From Main Street to State Street.*

61. Utility: G. L. Sullivan, "Forces that are Reshaping a Big Market." Newmark, "The Line of Progress," 97, wrote in 1918 that it "has taken a quarter century for manufacturers to discover that they are making a utility." Diffusion: An AT&T researcher estimated that in 1927, 59 percent of American households had automobiles, as opposed to only 39 percent that had telephones ("Charts on Operation and Engineering, Yama Conference, 1927," in file "Conferences, Presidents," TPCM).

62. Scharff, *Taking the Wheel,* 123ff.

4. THE TELEPHONE SPREADS: NATIONAL PATTERNS

1. Households without telephones tend to be poor, black, Southern, rural, and to have many children, or only a single male resident. See Wolfe, "Characteristics of Persons with and without Home Telephones"; Mahan, "The Demand for Residential Telephone Service"; Kildegaard, "Telephone Trends"; and Taylor, *Telecommunications Demand*, 74–76. On usage, see, e.g., Mayer, "The Telephone and the Uses of Time"; Field Research Corporation, *Residence Customer Usage;* and Brandon, *The Effect of the Demographics of Individual Households on Their Telephone Usage.*

2. For a basic review, see Rogers, *Diffusion of Innovations.*

3. 1927 exchanges: L. Wilson, "Sales Activities," 7. 1930 states: Untitled report, Chief Statistician's Division, AT&T (27 July 1932), AT&THA.

4. Four states: AT&T, *Events in Telecommunications History,* 5–6; Census Bureau: quoted by A. B. Foote in U.S. Bureau of the Census, "Statistics of Manufactures: 1890," 3.

5. Bell's agents in the South seemed hesitant and the company eventually transferred its southern franchises to northern "carpetbaggers." Kenneth Lipartito attributes the South's slow start to a regional tradition of weak entrepreneurship (Lipartito, *Bell System*). On the general patterns of franchising, see, e.g., Langdale, "The Growth of Long-Distance Telephony in the Bell System," and Garnet, *The Telephone Enterprise.*

6. U.S. Bureau of the Census, *Special Reports: Telephones and Telegraphs 1902.* On errors in the count, see the two subsequent reports, *Special Reports: Telephones: 1907* and *Telephones and Telegraphs . . . 1912.* Independent telephone businessmen claimed that the census continued to undercount their customers (e.g., *Telephony,* "Report of Secretary Ware"). On cities: Among the six largest, New York City was credited with 26 telephones per 1000 residents, less than the national average estimated at 29 per 1000. Chicago had 33; Philadelphia, 34; St. Louis, 32; Boston, 53; but Baltimore, 29 (U.S. Bureau of the Census, *Telephones 1902,* 27). It was not until the 1920s that the large metropolises clearly exceeded the hinterland in telephones (*Bell Telephone Quarterly,* "The World's Telephones, 1927").

7. In 1902 there was a modest statistical association between telephone competition in a state, measured by the number of telephone systems per capita, and telephone diffusion, measured by telephones per capita ($r = .30$; without Iowa, an "influential" case, $r = .16$). In the next census, covering 1907, when rural telephony was better measured, the correlation peaked ($r = .62$; without Iowa $r = .53$), followed by declining levels of association in subsequent years (1912: $r = .40, .29$; 1917: $r = .41, .31$; 1922: $r = .25, .16$; and $r < .20$ afterwards), as would be expected with the decline of real competition after World War I. Moreover, competition, as measured by number of lines, was associated with reduced profits for the telephone companies, implying that competition lowered telephone rates. (The partial correlation between number of telephone systems and profits—measured as the difference between gross revenues and gross expenditures in each state—was $r = -.34$ once population is controlled. See below for more on telephone rates.) The importance of competition may be yet understated, because the measure does not account for the threat of competition. Also,

this analysis controls for nothing but population (see note 9 on analysis of rates), and, by using ratio variables, can only be suggestive. The 1902 data, for 44 states, come from United States Bureau of the Census, *Telephones 1902.*

8. U.S. Bureau of the Census, *Telephones 1902*, 26. On competition and telephone diffusion, see also G. W. Brock, *The Telecommunications Industry;* Federal Communications Commission, *Investigation,* 132–3. On competition in the Midwest: 59 percent of all independent commercial systems in 1902 were in the North Central division of the census, versus 34 percent of the estimated population; three-fourths of all recorded rural lines were in Illinois, Indiana, Iowa, and Missouri (U.S. Bureau of the Census, *Telephones 1902*, 8, 16). On "telephone crusades," see Casson, *History of the Telephone,* 216–18. Lipartito, *The Bell System,* chap. 6, also points to the midwestern providers' access to several regional electrical manufacturers as low-cost sources of telephone equipment.

9. The 1902 census report (U.S. Bureau of the Census, *Telephones 1902*) provides a rare glimpse of the rate structures across states, albeit indirectly, by reporting the gross revenues and gross expenditures per telephone in each of 44 states (p. 29). There is a small negative correlation of telephones per 1000 residents with gross revenues per instrument ($r = -.10$ when New York, highest in revenues and high in diffusion, is dropped) and a larger one with gross profit per instrument ($r = -.16$ overall, $-.30$ without New York).

10. Cover of *San Francisco Telephone Directory,* Pacific States Telephone and Telegraph Company, October 1900 (TPCM).

11. The quote and a rich description of Sabin's "raising the devil" is provided by Hibbard, *Hello-Goodbye,* 196–223. See also *Telephony,* "John I. Sabin." In 1901, Sabin moved to the presidency of Chicago Bell, where this same strategy led to internal disruption, technical complications, and a quick departure in 1903 (see Garnet, "The Central Union Telephone Company," 9ff). For a taste of Sabin's marketing, see Sabin, "Special Reports."

12. See Lipartito, *Bell System,* on timidity. The 1930s study examined only white, native-born, families not receiving government relief, with both a husband and wife, living in "villages" in the Southeast versus those in the Middle Atlantic and North Central regions, and it controlled for occupational type and income. See Kyrk et al., *Family Expenditures for Housing and Household Operation, Urban and Village Series,* 161ff.

13. The discussion that follows is based in part on Fischer and Carroll, "Telephone and Automobile Diffusion," and also on subsequent data-analysis summarized in the text and Appendix B.

14. This point summarizes unpublished analyses by Glenn Carroll and myself, aided by Sylvia Flatt.

15. B. G. Hubbell, testimony in New York State, *Report of the Committee,* 791.

16. See *Telephony,* passim.

17. The statistics come from various sources. The 1920 through 1940 estimates of farms with telephones are provided in the agricultural censuses of those years. Other sources estimate 41 percent for 1925 (Builta, "How Can Publicity Assist") and 21 percent for 1935 (Independent Telephone Institute, "The Farm Telephone Story"). The 1912 estimate is from the U.S. Department of

Agriculture, "Income Parity for Agriculture." The 1902 estimate is from the U.S. Bureau of the Census, *Telephones 1902*. Estimates for nonfarm households were then calculated by combining these data with approximate estimates of total household rates and the number of residential telephones (U.S. Bureau of the Census, *Historical Statistics*, 783).

18. On Colfax County: Olsen, "But It Won't Milk the Cows." Bell's position on serving rural areas is well-documented in the independent telephony literature; see also, e.g., Martin, "Communication and Social Forms," 152–53; Lipartito, *Bell System*, chap. 6. On country tinkerers: for example, a case in Chico, California, is reported in the PT&T News Bureau Files, TPCM. Similar stories were reported elsewhere, e.g., Atwood, "Telephony."

19. The quotation is from E. Arnold, *Party Lines*, 144. This account captures the stories of many rural operations. See also, e.g., Atwood, "Telephony"; Greene, "Operator—Could You Please Ring?"; Wisconsin State Telephone Association, *On the Line;* Spasoff and Beardsley, *Farmers' Telephone Companies;* and other references in Appendix A.

20. Sidwell, "Some Phases of Rural Telephony"; Dobbs, "Are Country Lines Profitable?"

21. Letter from Bush to Bradley, 21 November 1908, in "Rural Telephone Service, 1907–1910," Box 1363, AT&THA. A similar evaluation came from Buffalo in 1907: Pickernell to Hall, 16 December 1907, in ibid.

22. Minutes of "Advertising Conference—Bell System—1916," Box 1310, AT&THA.

23. Wonbacher, "Proper Development of the Rural Telephone," 242.

24. Dobbs, "Are Country Lines Profitable?" 248, 298.

25. Barker, "Operating Rural Lines." *Telephony* is the basic source on independents' relations with farmers. The various histories of independent companies are also useful.

26. The higher cost of rural lines was, for example, the explanation for the reluctance to pursue rural customers used by AT&T President Walter S. Gifford in 1944 in "An Address," as well as by many others in the telephone industry. See Fischer, "The Revolution," esp. p. 12 and note 32, for a discussion of costing out rural lines. A Congressional Budget Office study in 1984 concluded that rural costs for telephone service were, in total, *not* higher than urban costs (C.B.O., *Local Telephone Rates*). Although conducted long after our period, the logic of the 1984 study still applies. One can also see consideration of rural telephony's external advantages in the evaluations made by managers of the government systems in the Canadian Prairie provinces. Although they reluctantly agreed that the rural telephones were unprofitable, the managers nevertheless stressed their view that farm telephones undergirded the larger systems (Chant, "The Value of Rural Development to Exchange and Toll Earnings"; Armstrong and Nelles, *Monopoly's Moment*, 289–90).

27. On spending studies: see Kyrk et al., *Family Expenditures For Housing and Household Operation*, Farm Series, 41–60, and Kyrk et al., *Family Expenditures*, Village Series, 41–68. Rankin, *The Use of Time in Farm Homes*, is one study, from 1928, showing that farmers were likelier to have telephones than townsfolk. See also Farrell, *Kansas Rural Institutions: IX. Rural Telephone Company;* Robertson

and Amstutz, *Telephone Problems in Rural Indiana.* On 1940s spokesmen: see, e.g., California Public Utilities Commission, "Development of Telephone Service in Rural Areas."

28. Atwood, "Telephony," 326ff; Umble, "The Telephone Comes to Pennsylvania Amish Country."

29. Crude estimates from a later period suggest as much. See Kyrk et al., *Family Expenditures,* Urban and Village Series, and Kyrk et al., *Family Expenditures,* Farm Series.

30. Any number of publicists' stories or advertisements directed to farmers listed such uses. See also, U.S. Bureau of the Census, *Telephones 1907.*

31. Colfax County: Olsen, "But It Won't Milk the Cows"; business ends: Borman, "Survey Reveals Telephone as a Money Saver on Farm"; Robertson and Amstutz, *Telephone Problems,* 22; Farrell, *Kansas Rural Institutions,* 10, 27.

32. 1904: *Telephony,* "The Farmer and the Telephone"; magazines: Weinstein, "The Telephone in Popular Journals: 1908–1913;" Federal Investigations: U.S. Bureau of the Census, *Telephone 1907,* 77–78; U.S. Senate, *Report of the Country Life Commission;* and Ward, "The Farm Women's Problems"; and the academic was Card, "Cooperative Fire Insurance and Telephones," 305; see also J. Pope, "Rural Communication." There is a tidbit of survey data. The Country Life Commission surveyed 55,000 farmers and found that 69 percent were dissatisfied with "social intercourse," but 73 percent, the highest level in the survey, were satisfied with their postal and telephone service, asked together. (Larson and Jones, "The Unpublished Data from Roosevelt's Commission on Country Life.")

33. For general sources on automobiles in rural America, see M. Berger, *Devil Wagon;* Interrante, "You Can't Go to Town"; Moline, *Mobility and the Small Town;* and Wik, *Henry Ford.* Local studies include, e.g., C. Larson, "The Automobile in North Dakota"; Preston, *Dirt Roads to Dixie.*

34. Data on the diffusion of cars in rural America come from the histories cited above and the Censuses of Agriculture. On sharecroppers, see Federal Writers' Project, *These are Our Lives.*

35. Auto budget: Kyrk et al., *Family Expenditures for Automobile and Other Transportation,* and Monroe et al., *Family Expenditures for Automobile and Other Transportation;* Country Life: U.S. Department of Agriculture, *Social and Labor Needs of Farm Women,* and Ward, "The Farm Women's Problems."

36. Nye, *Electrifying America,* chap. 7; Hurt, "REA," reports that in 1925, 90 percent of French farmers had electricity, compared to 3 percent of (nonsuburban) American farmers.

37. The decline in rural telephone percentages is even more noteworthy because it came when many marginal operators left their farms; that alone should have increased the percentage of telephone subscribers among the remaining farmers. The data on farm telephones and automobiles comes from the agricultural censuses of 1920 through 1940, and the percentages are of farms, rather than households. The estimate for nonfarm households with telephones was calculated by subtracting farm households from total households in U.S. Bureau of the Census, *Historical Statistics,* 783. The estimate for nonfarm households with automobiles was calculated using the figures (the "total" column) provided by

Lebergott, *The American Economy,* 290. The electricity data for both farm and nonfarm are households from the U.S. Bureau of the Census, *Historical Statistics,* 827. (Note that agricultural census reports have different estimates for electricity in 1920 and 1930, 7 percent and 13 percent, respectively.)

38. For example, an AT&T analyst estimated in 1930 that 40 percent of farms had telephones, based on the assumption that nothing much had changed over the previous decade (Wallace, "Service Outside the Base Rate Area"; see also Gherardi and Jewett, "Telephone Communication System," 33, for similar overestimates in Iowa). Also, the independents' journal, *Telephony,* made little, if any, note of losing customers in rural areas until the 1930s.

39. Independent Telephone Institute, "The Farm Telephone Story," 20. For the sources of other explanations, see C. Fischer, "Technology's Retreat."

40. On spending patterns, see Blalock, "Streamlining Rural Telephone Service," 468–69; Kyrk et al., *Family Expenditures,* Farm Series, 41; Allred et al., *Rural Cooperative Telephones in Tennessee,* 5; Kirkpatrick, "The Farmer's Standard of Living" passim; United States Bureau of Labor Statistics, "Family Spending and Saving in Wartime," 75; and D. Brown, *Electricity for Rural America,* 70.

41. An Illinois farm official claimed that in 1929 two bushels of corn could pay for a month's telephone service, but by 1932, it took six bushels (Mathias "The Rural Telephone Situation"). A retired PT&T executive told us that when Bell absorbed rural lines in California's Marin County, it had to repair and upgrade the plants, which drove up costs. Lipartito makes a similar point about consolidation in the South (*Bell System,* 331). For estimates of telephone rates, see U.S. Department of Agriculture, "Income Parity for Agriculture"; Federal Communications Commission, *Preliminary Studies on Some Aspects of the Availability of Landline Wire Communications Service;* Spasoff and Beardsley, *Farmers' Telephone Companies;* Mulrooney, "Some Aspects of Rural Telephone Service"; C. Mason, "America's Farm Telephone Program Is Well on Its Way"; Farrell, *Kansas Rural Institutions;* Sichter, "Separations Procedures in the Telephone Industry."

42. C. Mason, "America's Farm Telephone Program," 86–88.

43. Among those who argued for the collapse explanation are Shenk, "Small Exchanges Struggle for Survival"; Robertson and Amstutz, *Telephone Problems in Rural Indiana;* and Parker, "Cooperative Telephone Association 1936." Iowa: Iowa State Planning Board, "Telephone Communication in Iowa." On Bell areas, see AT&T, "Selected Statistics Relative to Residence and Farm Markets" (1932); AT&T, "Families and Residence Telephones" (1940), 6; AT&T, "Rural Population and Telephones" (1941); and Tomblein, "Population Changes in Small Communities." On numbers of small companies, see U.S. Bureau of the Census, *Census of Electrical Industries 1932;* and idem, *Census of Electrical Industries 1937.* Between 1922 and 1937, the number of small companies dropped by 11 percent, but the number of subscribers they served dropped by 43 percent. Also, some reduction in the number of companies resulted from consolidation rather than disbandonment.

44. See Federal Communications Commission, *Preliminary Studies;* also U.S. Department of Agriculture, "Income Parity for Agriculture"; Spasoff and Beardsley, *Farmers' Telephone Companies;* and arguments presented in California Public Utilities Commission, "Development of Telephone Service in Rural Areas."

Lipartito, *Bell System,* chap. 9. The report of the Independent Telephone Institute ("The Farm Telephone Story," 20) quotes one industry man as saying that the major companies were "so busy developing the larger centers . . . that they have neglected the farmers. They have been skimming the cream."

45. C. Fischer, "Technology's Retreat." The report here also relies on further, unpublished analyses.

46. It is clear, for example, that Bell and other companies simply did not consider blacks, renters, or the poor to be worthwhile customers for telephone service (see discussions in various *Bell System Commercial Conferences* and *Advertising Conferences,* microfiche 368B, IBIC; and Box 1310, AT&THA).

47. Bare, "Service Outside the Base Rate Area," 32.

48. See, e.g., D. Brown, *Electricity for Rural America,* and idem, "North Carolina Rural Electrification"; Childs, *The Farmer Takes a Hand;* and Ellis, *A Giant Step,* among others. The delivery systems of electricity and telephony are similar; electric companies also resisted serving farmers and were skeptical that they had any need for the service.

49. On the beginnings of European systems, see Holcombe, *Public Ownership* (1911); see also Chapuis, *100 Years of Telephone Switching;* and Bertho, *Télégraphes et Téléphones.* Statistics come from "The World's Telephone Statistics," appearing annually in *Bell Telephone Quarterly,* 1922 to 1937.

50. On the Prairie experience, see, e.g., Armstrong and Nelles, *Monopoly's Moment* (the Alberta quotation is from p. 291); Cashman, *Singing Wires;* and Collins, *A Voice from Afar.* Statistics on Canadian telephony can be found in Canada, Department of Trade and Commerce, "Telephone Statistics" (1925–1940), and Urquhart and Buckley, *Historical Statistics of Canada,* 352–54. Whereas American farmers did not regain their 1920 level of telephone adoption until 1950, Prairie farmers reached their 1924 level by 1940.

51. Bell ad, 4 January 1902, from Bell Canada archives, quoted by Pike, "Adopting the Telephone," 33.

52. On monopolistic pricing, see G. Brock, *Telecommunications Industry.* On extras, see, for example, Louderback, "Bell System Sales Activities." Louderback warned that current nonusers would, if enrolled, probably disconnect in bad times, so it was better to concentrate on current users. On estimates, see letter from S. H. Mildram to W. S. Allen, 22 May 1905, "Estimated Telephone Development, 1905–1920," in Box 1364, AT&THA. In this analysis, Mildram assumed that telephone diffusion would top out in 1920 at one per five Americans and even that "may appear beyond reason." (One per five Americans was not reached until 1945, in actuality, but was doubled by 1960 and tripled by 1975.) Mildram also assumed that growth in business telephones would make up for an absence of residential telephones in the bottom half of the population. On zoning, see Pool, "Foresight and Hindsight," 143–44; Rubin, "Urbanization." On lone voices, see Hibbard's colorful descriptions of Sabin's marketing efforts to all sorts of San Franciscans and the derision it raised (*Hello-Goodbye,* 196–223). E. J. Hall's correspondence from 1880 on shows that he was an aggressive proponent of rate reductions that would popularize the telephone, but also shows some frustration at his failure to convince others to agree. See letter from Hall to Vail, 26 October 1880, in Box 1127 and letter to E. M. Burgess, 30 March 1905, in Box 1309, AT&THA.

53. On party lines as competitive strategies: John Sabin's 10-party lines ploy in San Francisco was noted in Chapter 2 (letter to J. E. Hudson, 15 April 1895, Box 1278, AT&THA). The same files include similar reports from the Hudson River Telephone Company (17 October 1895) and the exchange manager in St. Louis (6 February 1899). "Lower strata": Young, "The Elements of a Balanced Advertising Program." See also Page, "Public Relations and Sales." AT&T analyses showed that by 1930, adoption of automobiles, electricity, gas service, and later, radio, had outstripped telephone subscription and that the discrepancy was especially great in Bell's territories ("Charts on Operation and Engineering," Yama Conference, 1927, in "Presidents' Conferences," TPCM; and AT&T, "Selected Statistics Relative to Residence and Farm Markets," [November, 1932], and Untitled Report [July, 1932]). This embarrassment was even publicized by the *Literary Digest* in "More Automobiles Now than Telephones" (1925).

54. Expectations: Flink, *Americans Adopt the Automobile,* 36, and passim; see also Rae, *The American Automobile,* 57–61. Flink (*The Automobile Age,* 131) notes that the automobile industry did not consider "illiterates, immigrants, and Negroes" to be a market. Low-priced cars: Epstein, *The Automobile Industry.* Marketing: Clymer, *Henry's Wonderful Model T.*

55. The 1923 variation is documented in Gray, "Typical Schedules." Estimates used for Figure 6 are as follows: In 1895, the average monthly residential rate for Bell telephones was $4.66 (AT&T, *Annual Report 1910;* U.S. Bureau of the Census, *Telephones 1902,* 53). Earnings for 1895 alone cover all nonfarm employees (U.S. Bureau of the Census, *Historical Statistics,* table D–735). The 1909 estimate of $2.00 is also from the 1910 AT&T Annual Report. Earnings for 1909 on cover manufacturing employees only (Bureau of Census, *Historical Statistics,* table D-740). The 1915 "median minimal rate," $1.50, is from Gray, "Typical Schedules," figure 3, as is the 1923 median minimal rate (the low estimate for that year). The 1923 high estimate also includes the "typical" two-party residential rate of $3.25 (ibid, figure 5). The 1933 range is based on a consensus in the literature that the nominal rates did not fall, but earnings did. On comparable estimates, see, e.g., Pike and Mosco's estimate ("Canadian Consumers") that in the 1880s the average telephone rent in Hamilton, Ontario, was equal to 10 percent of manufacturing workers' wages. Also, in 1922, a "typical middle-class" household budget included about $2.00 a month for telephone (Horowitz, *The Morality of Spending,* 92).

56. Comparison based on United States Bureau of Labor Statistics, "The Cost of Living in the United States," Table G.

57. The Ford prices appear in Clymer, *Henry's Wonderful Model T.* See also Epstein, *The Automobile Industry,* and Roos and Szeliski, "Factors Governing Changes in Domestic Automobile Demand," on automobile prices, especially p. 42 of Roos and Szeliski on drops in prices expressed in dollars per pound of closed vehicle. Williams and Hanson, "Money Disbursements."

58. Many of the sources used in Appendix D are also summarized in Horowitz, *The Morality of Spending.*

59. Cowan and Cowan, *Our Parents' Lives,* 65.

60. The 1907 census of telephones indicates that about 4 percent of all Bell telephones were pay telephones (U.S. Bureau of the Census, *Telephones 1907,* 21). Statistics for PT&T's major cities in 1915 show that 1 percent to 4 percent

of its telephones were pay stations (in "Correspondence—DuBois," TPCM). In 1930, pay telephones formed about 2 percent of all AT&T connections (Gherardi and Jewitt, "Telephone Communication System," 12). Statistics for the three towns studied in depth in Chapter 5 show that in the late 1930s pay telephones equalled about 4 percent of all residential telephones in San Rafael and Palo Alto, 12 percent in Antioch (Untitled folder, San Jose Forecasting Office, PT&T).

61. Field Research Corporation, *Residence Customer Usage,* 3–5.

62. 1918 budget: Horowitz, *The Morality of Spending,* 92ff; subsistence and minimal adequacy budgets: Ornati, *Poverty Amid Affluence,* 21, 52.

63. See Wolfe, "Characteristics of Persons with and without Home Telephones"; Mahan, "The Demand for Residential Telephone"; Taylor, *Telecommunications Demand,* 4–6; Kildegaard, "Telephone Trends." However, among families with telephones, income does not seem to determine use, except perhaps inversely (Mayer, "The Telephone and the Uses of Time," 232; Field Research Corporation, *Residence Customer Usage,* 3–5).

64. Robert and Helen Lynd reported in *Middletown,* for example, that half of their 1924 sample of working-class families owned automobiles, and payments were at the expense of housing, clothing, and other necessities (p. 235ff). Their calculations have been questioned, however. See Jensen, "The Lynds Revisited"; Flink, *Automobile Age,* 131–35. Jessica B. Peixotto also challenged these sorts of claims in her studies of spending in the San Francisco area. Commenting on the 17 of 82 typographers' families with automobiles in 1921, she claimed that their expenditures were financed by "small economies" and did not undercut savings. Commenting on the 55 of 96 Berkeley faculty with automobiles, she described the automobile as "a relatively new type of expenditure that custom is rapidly ranging in the class of necessities though comfortable conservatives still regard it darkly as a luxury." (Peixotto, "Cost of Living Studies, II. How Workers Spend," 200ff; and idem, *Getting and Spending,* 196). On the automobile's popularity with small earners: see Flink, *Americans Adopt the Automobile,* 72; Moline, *Mobility and the Small Town.* On worries: Flink, *The Car Culture,* 140ff; Horowitz, *The Morality of Spending.*

65. Calculated from Williams and Hanson, "Money Disbursements," 139 and 258. "Lowest income" here means those in the $200 to $300 "expenditure class."

66. The advertising critique appears in many places. See, e.g., Flink, *The Car Culture.* The Lynds' *Middletown* is another classic example. The Bell System's analysis appears, in particular, in *General Commercial Conference, 1928* (microfiche 368b, IBIC). In 1929, a Pennsylvania Bell official documented the surpassing of telephones by automobiles and claimed it was "evidence of the very extensive advertising which has been done by the automobile industry and the very limited advertising which we have been doing" (Young, "The Elements of a Balanced Advertising Program").

67. Unfamiliar: McHugh, "Introductory Review of Sales," 23; rates: L. B. Wilson, "Commercial Results"; Canvass: Harrell, "Residential Exchange Sales," 67–68. Harrell estimates that 15 percent of families in his New England region fell below $1300. National census data, although not specifically appropriate, suggest figures closer to 30 percent or 40 percent (U.S. Bureau of the Census, *Historical Statistics,* 299). According to Michèle Martin, Bell Canada lines would

even pass through low-income neighborhoods without serving the residents (*"Hello, Central?"*, 116).

68. Telephone: Kyrk et al., *Family Expenditures for Housing and Household Operations*, Urban and Village Series, 66, table 57; automobiles: Kyrk et al., *Family Expenditures for Automobile*, 67–73, table 50. Wage-earners tended to own cheaper and older cars, to drive them less, and to spend less money on them than white-collar workers. However, since white-collar families attributed more of their automobile costs to business purposes, in net, expenditures for "family use" of cars were the same for both classes.

69. Pike and Mosco, "Canadian Consumers"; Pike, "Kingston Adopts the Telephone"; Pike, personal communication, 10 February 1987.

70. This point was suggested by Mark Rose.

71. Williams and Hanson, "Money Disbursements," 97; see also R. O. Eastman, Inc., *Zanesville and 36 Other American Communities*, 81–82; Rose, "Urban Environments and Technological Innovation"; and analysis from the U.S. Bureau of Labor Statistics data in Appendix E.

72. Spending study: On industry attitudes: About 1930, one AT&T official argued that telephone development advanced with the "homogeneity" of the population. Puzzling, though, was his use of San Francisco as an illustration, since it was anything but homogeneous (Andrew, "The New America"). Also, a New York manager reported in 1930 on a novel campaign to sell telephone subscriptions in "lower quality" Bronx neighborhoods, using men of "foreign extraction" in some, that met with some success. Another reported some success using special salesmen to sell to "national groups," thus blacks on commission to sell to the "better classes of negro [*sic*] families," Italians to Italians, and so forth. The context suggests that these approaches were unusual (*Bell System General Commercial Conference, 1931*, Microfiche 368b, IBIC, 77–78). On blacks and telephones today, see Wolfe, "Characteristics of Persons"; Mahan, "The Demand for Residential Telephone Service," chap. 5; and Taylor, *Telecommunications Demand*, 74–76.

5. THE TELEPHONE SPREADS: LOCAL PATTERNS

1. Recalled in the *Antioch Ledger,* 17 and 27 March, 1938.

2. The telephone histories of many communities are available in the various industry archives, though usually only as unpublished or privately published company annals. Most are brief and dwell on personalities and colorful anecdotes. Useful sources include Rippey, *Goodbye, Central; Hello, World;* Mahon, *The Telephone in Chicago;* and Cashman, *Singing Wires.* See also, Atwood, "Telephony and Its Cultural Meanings"; Farrell, *Kansas Rural Institutions;* Greene, "Operator—Could You Please Ring?"; Howe, *This Great Contrivance;* Masters et al., *An Historical Review of the San Francisco Exchange* and *An Historical Review of the East Bay Exchange;* Robertson and Amstutz, *Telephone Problems in Rural Indiana;* and Wisconsin State Telephone Association, *On the Line.* MIT archives contain several independent company histories, and AT&T Historical Archives contain several histories and memoirs of local Bell companies. Bell Canada and TPCM have dossiers on many local exchanges.

3. See W. Barnett, "The Organizational Ecology of the Early Telephone Industry," for a statistical analysis of early telephone company birth and death rates.

4. A major scandal occurred in San Francisco, where the supervisors, in effect, auctioned a franchise for personal gain. PT&T was quite generous to local politicians (see Bean, *Boss Reuf's San Francisco*, chap. 8). New England Bell had numerous politicians and "phantom workers" on its payroll (Weinstein, "The Telephone in Popular Journals"). For a detailed account of rate regulation during this period, see "The Telephone on the Pacific Coast," Box 1045, AT&THA. See also Chapter 2.

5. Palo Alto is much better documented than our other towns. We relied largely on issues of the *Palo Alto Times,* the files of the Palo Alto Historical Society (PAHS), and, within their files, the following most often used books and pamphlets: Wood, *History of Palo Alto;* Palo Alto City Council, *Annual Reports;* Wyman, "A Study of Contemporary Unemployment and of Basic Data for Planning a Self-Help Cooperative in Palo Alto, California 1935–6," and Gullard and Lund's recent *History of Palo Alto.*

6. Starr, *Americans and the California Dream,* 319.

7. The figure presents selected statistics describing the three towns. The data are largely drawn from census data, newspaper reports, and State of California Controller's Reports.

8. *Times,* 27 October 1913; Starr, *California Dream,* 321.

9. See, e.g., Gullard and Lund, *History of Palo Alto,* chap. 11.

10. Among the many sources consulted on San Rafael history the primary ones are the *Marin Journal, San Rafael Daily Independent,* and *Marin Independent-Journal,* the files of the Marin County Historical Society, the Marin County Library's California Room, and the following texts: Toogood, *A Civil History of Golden Gate National Recreation Area;* B. Brown, *Marin County History;* Donnelly, "A History of San Rafael"; Wilkins, *Early History of Marin;* J. Mason, *Early Marin;* Meret, *A Chronological History of Marin County;* and Keegan, *San Rafael.*

11. Keegan, *San Rafael,* 80.

12. The sources consulted on Antioch history include the *Antioch Ledger* from 1908 on; Benyo, *Antioch to the Twenties; History of Contra Costa County;* Hohlmayer, *Looking Backward;* Bohakel, *Historic Tales of East Contra Costa County;* Dillon, *Delta Country;* Emanuels, *Contra Costa;* Thompson, "The Settlement Geography of the Sacramento-San Joaquin Delta, California"; Purcell, *History of Contra Costa County;* Wiltsee, "The City of New York of the Pacific."

13. *Antioch Ledger,* 6 April 1909.

14. Based on comparing samples of the 1910 census to the 1937 city directory (see Appendix F).

15. *Antioch Ledger,* 22 December 1927.

16. We estimated the number of household telephones in each town by drawing samples from telephone directories for every other year, and we estimated annual population largely from the decennial censuses. At some points, however, artifacts distorted the time trends. PT&T had lists (found at TPCM) providing the total number of telephones in each exchange up through 1928, but we needed the number of telephones within each town, which covered somewhat

less territory, and we needed that number to be differentiated by business versus residence. Therefore, we (largely, Kinuthia Macharia) estimated the number of residential telephones by using telephone directories for each even-numbered year from 1900 through 1940. (The books were made available to us by PT&T.) For the early years, we counted all the listings, categorizing them as business or residential telephones. (Residential listings usually had an *r* after the subscriber's name.) For later years, we randomly sampled six pages in each directory, counted telephones, and weighted back to a full-directory estimate. I then interpolated for the odd-numbered years. Because of sampling errors and seasonal variation—the directories were for different months of the year and subscription fluctuated seasonally—these estimates were still quite noisy. To smooth them, I combined the series based on simple interpolation with a series of estimates calculated by using a regression equation. That equation estimated telephones as a function of year, season, and the official PT&T exchange statistics previously mentioned. The Antioch and San Rafael numbers estimated in Figure 9 are essentially a result of this procedure. For Palo Alto, the effort to get estimates was more difficult, given the complications of Stanford University and of Mayfield, the town that Palo Alto annexed in 1926. This caused discontinuities in various statistical series. Using our rough directory counts, I estimated several series for Palo Alto with regression equations and then combined simple interpolation of our counts with these smoothed equations.

Obtaining the per capita estimates also required developing annual approximations for population even though the census provided one for only every decadal year. We had other population proxies for intervening years, such as the number of registered voters and population estimates provided by town officers, but they were often unreliable or at best partial (e.g., town clerks' estimates of population for decadal years were often far off the figure the census eventually delivered). Using linear interpolations for nine years is clearly suspect, so I tried several procedures. For Palo Alto the estimate of population I used combines (1) an equation fitting the first reported population of the city and census figures to a nonlinear function of year and the geographical size of the city and (2) up to 1926, an equation that fitted census counts and state controller estimates of Mayfield's population to a nonlinear function of year. For San Rafael, I combined (1) a series based on simple interpolation with (2) a series based on a complex function of year and (3) adjustments for a few disjunctive periods. For Antioch, the final population estimate is essentially the average of three procedures: (1) simple interpolation; (2) an equation fitting decadal counts to functions of year; and (3) an equation fitting census estimates onto a nonlinear function of year and counts of registered voters for decadal years and four years after. The justification for averaging the results of these different estimation procedures is simply that any one is errorful and by triangulation we may minimize idiosyncratic errors.

17. *Palo Alto Times,* 16 February 1893, and 2 June 1893.

18. Pioneer settler: *Times,* 7 December 1894; fat man: *Times,* 5 July 1901. Parkinson's biography appears in Wood, *History of Palo Alto,* and his many activities are documented in the pages of the *Times.* See also, Palo Alto Historical Society folder, "Telephone"; "City Files," in Telephone Pioneer Communications

Museum "Local Telephone History Recited," *Times*, 8 April 1932; *Stanford Daily*, 14 April 1932.

19. *Times*, 6 October 1893, and 2 October 1896. There is some inconsistency in the archives about dates and pace of development. PT&T accounts claim that the exchange moved in 1896 due to the press of business, e.g., "Local History of the Telephone," *Times*, 18 August 1930.

20. Ads: *Times*, 12 May 1893, and 1 December 1893; merchants: *Times*, 23 January 1896, and 20 March 1896; regulation: Miller, "Bulletin No. 144," PAHS files; competitor and canvasser: *Times*, 26 March 1897, and 6 August 1897; subscribers: *Times*, 23 April 1897, and 3 December 1897.

21. Offices: "Local History," *Times*, 18 August 1930, and 27 March 1900; Menlo Park: PAHS, "Telephone"; 500 subscribers: "The Pacific Telephone and Telegraph Company," mimeo, n.d., TPCM; $2.50: telephone bill in PAHA, "Telephone"; poles and wires: G. Miller, *Palo Alto Community Book*, and *Times*, 13 March 1903, and 25 March 1905. On competitors: In 1901 an outsider showed up in town, called a meeting and gathered petitions for a new telephone system (*Times*, 20 December 1901). Nothing came of it. In 1905 agents of San Francisco's Home Telephone Company scouted Palo Alto for possible expansion (*Times*, 25 June). On complaints and the meeting: *Times*, 8 May 1903, 17 August 1903, and 29 June 1905.

22. Improvements: *Times*, 7 October 1907, and 7 June 1908; strike: *Times*, 23 and 24 April 1907; customer approval: *Times*, 5 February 1908.

23. Ads: *Times*, 18 June 1908, 27 August 1912, 10 July 1912, and 13 March 1914; rates: *Times*, 28 June 1910, and 12 July 1910.

24. Expansions: see, e.g., *Times*, 22 June 1912; Stanford: *Times*, 20 June 1914; Camp Fremont: *Pacific Telephone Magazine*, 18 June 1918, and 18 November 1918; strike: see, e.g., *Times*, 30 June, 2 July, and 18 July, 1919.

25. Police: *Times*, 9 September 1922; automatic-dial service: *Times*, 28 April 1932.

26. The basic source on San Rafael's early telephone history is a set of press releases and notes in "PT&T News Bureau Files," TPCM. See also, "San Rafael Telephone Service Began in '84," *Marin Independent-Journal*, 22 September 1964. Information on Bogle comes from Cross, *Banking in California*, 522.

27. Billing: *Marin Journal*, 10 December 1908; upgrading: 17 December 1908, and 10 August 1909; franchise: 8 November 1917; advertising: John Chan's survey of ads in the local press.

28. Fools: *Contra Costa Gazette*, 29 August 1903; all-night service: *Gazette*, 4 June 1904; 1907: *Ledger*, 9 February, 8 June, and 7 December, 1907, and *Pacific Telephone Magazine*, December; 1908: 9 February, 14 March.

29. *Ledger*, 1909: 20 March and 14 August, 1909; 1910: 19 February, 2 April, 4 June, 13 August, 12 and 19 November, 1910; pigeons: Antioch Historical Society, newsletter, n.d.

30. *Ledger*, 7 March 1908. The first local advertisements explicitly mentioning the telephone in San Leandro, California (just south of Oakland) were also by liquor dealers (Dierkx, "San Leandro"). On liquor shopping by telephone, see Duis, *The Saloon*.

31. *Ledger*, 14 January 1911; 13 and 20 April 1912; 29 November 1913; 27

February 1915; 11 November 1916; and 17 August 1918. Farm systems: *Ledger,* 11 February and 14 June 1913, 24 January 1914, and 11 November 1916.

32. The figure of 363 comes from the exchange statistics found at TPCM. Antioch tax rolls recorded a total of 233 telephones as property of PT&T. Our own estimate for 1920 is 315, based on listings.

33. *Ledger,* 11 July 1939.

34. *Ledger,* 7 April 1906.

35. *Marin Journal,* 14 March, 11 June, 3 September, 1903.

36. Antioch: *Ledger,* 13 April 1907, and 8 June 1907. Harger, "The Automobile in Western Country Towns," claims that the Winter of 1906–1907 saw a reversal in rural Americans' attitudes toward the automobile.

37. *Ledger,* 6 April 1907; *Marin Journal,* 3 September 1908.

38. Caravans: e.g., *Ledger,* 2 January 1915; *Times,* 28 September 1907. Sisters: *Ledger,* 4 October 1913.

39. Palo Alto speeders: *Times,* 5 May 1933, and 6 May 1934; Wood, *History of Palo Alto,* 134. Police: see, e.g., *San Rafael Daily Independent,* 6 January 1938, and Palo Alto, *Annual Reports.*

40. *Times,* 5 January 1907; Birge Clark, Memoirs, in "Veterans—Miscellaneous," PAHS.

41. *Marin Journal* and *San Rafael Daily Independent,* passim, 1925–1931.

42. Horton, "Against All Odds."

43. Andrew Latham, "Hotel Raphael," pamphlet file, Marin County Civic Library.

44. Strike: *Ledger,* 10 January 1920; jitneys: *Ledger,* 23 February 1928; see also, Thompson, "Settlement Geography."

45. Fourth Street: *Daily Independent,* 7 January 1925; homeowners: *Daily Independent,* 9 August 1934.

46. See, e.g., Berger, *Devil Wagon;* Wik, *Henry Ford;* Carr, "How the Devil-Wagon Came to Dexter"; West, *Plainville, U.S.A.;* Sundstrom, "The Paving of America."

47. See, e.g., Flink, *Americans Adopt the Automobile,* 70ff, 107; Bardou, et al., *The Automobile Revolution,* 22; Harger, "The Automobile in Western Country Towns," 1280.

48. On early press coverage, see Flink, *America Adopts,* 134ff, who claims that coverage of automobile developments was "well out of proportion" to their importance and may have been a major spur to diffusion. On variations in public concern about accidents: Brilliant, "Some Aspects of Mass Motorization," especially 202; J. Mueller, "The Automobile"; and Nash, "Death on the Highway."

49. Los Angeles: Bottles, *Los Angeles;* Scharff, "Reinventing the Wheel"; on general traffic controversies and the merchants, see, e.g., Brownell, "A Symbol of Modernity"; Brilliant, "Some Aspects of Mass Motorization"; and Moline, *Mobility and the Small Town,* 132–41. On planning: Foster, *Streetcar.*

50. For example, in 1912 an official of the National Bureau of Statistics blamed the high cost of living on the enormous increase in "expenditures for unproductive consumption." He listed many such purchases, including the telephone, trolley rides, and card parties, but the most expensive item was certainly the automobile. (Literary Digest, "Fallacies as to the High Cost of Living." See also

Horowitz, *The Morality of Spending,* and Flink, *The Car Culture,* 140ff.) The Lynds' critique of the automobile in Middletown was in this tradition (see Lynd and Lynd, *Middletown,* and Jensen, "The Lynds Revisited").

51. Sharpless and Shortridge, "Biased Underenumeration in Census Manuscripts."

52. The difference between the notables and the total sample is actually even sharper. Some of the notables we sampled from the newspapers also came up in the random household sample. The estimates in the text lump these twice-chosen families with the random sample. If instead we group together all the notable households, then 23 percent of notables had telephones versus a mere 1.5 percent of the remaining, once-chosen, randomly-selected households.

53. *Our Society Blue Book,* 196–98.

54. Although only about half of the originally sampled households could be found in the later city directories or voting lists, this exercise did permit us to estimate the proportions of original subscribers who held on to their telephones and the proportions of original nonsubscribers who got telephones in the intervening time. The following table provides estimates of these probabilities. (The general pattern remains if one includes all households whether traced or not or if one uses more stringent matching criteria.) The displayed estimates are based on regressions to control for the town, but table analysis shows the same pattern.

	Probability of Subscribing Four Years Hence				
	1900–04	1910–14	1920–24	1930–34	1936–40
Original Subscribers	(7 of 9)	.61	.89	.85	.87
Original Nonsubscribers	.18	.23	.33	.08	.14

I cannot explain with certainty the low estimate for resubscription in the 1910–14 period, .61. A small part of it is due to measurement problems in the Palo Alto sample (see Appendix F). Perhaps the changing regulatory regime between 1910 and 1914 affected rates and procedures against the interests of current subscribers. Or, more generally, telephony may have been less entrenched—reliable, useful, necessary—in the pre-war period.

In any event, after World War I, renewal rates stabilized at between 85 percent and 90 percent. The historical pattern revealed in Figures 9 and 10—accelerating adoptions to 1930 and then a drop—must therefore be attributed to the changing probabilities that nonsubscribers would sign up.

55. In technical terms, year did not interact significantly with household traits; see Appendix F.

56. There was a slight tendency (significant at about $p = .11$, when added to the model in Table C1) for the blue-collar disadvantage to widen after 1920. Put another way, white-collar workers advanced a bit more rapidly in telephone subscriptions after 1920 than did blue-collar workers.

57. Figure 11 displays the results of the following operation: For each year, I regressed (using ordinary least squares estimates) residential telephone subscription on a seven-category occupational classification scale (professional and managerial; proprietors; other white-collar; skilled blue-collar; semi-skilled; laborers and transportation workers; and all other categories, such as retired, student,

or "own account"), a five-category typology of households based on the adults present (only a single male; only a single female; unmarried adults; married couple; married couple and [an]other adult[s]), and dummy variables for the towns. The displayed results are the expected values for the major categories, setting towns and household types at their mean values for each year.

58. The same equation and procedure described in the preceding note for Figure 11 were used to estimate the effects in Figure 12. The 27 cases that were missing or that did not fit the household categories were dropped.

59. The issue of who was an earner is not clear-cut. No doubt, some married women worked informally or in a family business but preferred not to be recorded as such. Nevertheless, the data consistently demonstrate that the number of earners is not a significant predictor of telephone subscription, whereas the number of adults, especially female adults, is.

60. Cowan, "The Consumption Junction."

61. This point was suggested by Mark Rose, personal communication.

62. Moyer reports that in Boston in 1910 there were 14.3 telephones, business and residential, per 100 residents in the city center, 4.7 in the near suburbs, and 6.4 in the far suburbs. If one heavily discounts the city center number for largely reflecting business telephones, the pattern of distribution is consistent with our findings here for towns that were not metropolitan business centers. (See Moyer, "Urban Growth," 360.)

6. BECOMING COMMONPLACE

1. Wisener, " 'Put Me on to Edenville,' " 23.

2. Aronson, "Lancet"; Aronson and Greenbaum, "Take Two Aspirin." The quotation is from p. 16 of the latter.

3. Menlo Park doctor: Palo Alto Times, 12 May 1893; drugstores: Times, 7 August 1903; telephone consultation and delay: Telephony, "Doctors and Rural Telephones," 492; safety and fees: Aronson and Greenbaum, "Take Two Aspirin," 24ff; lowered death rates: Literary Digest, 1924, quoted by Aronson and Greenbaum, "Take Two Aspirin," 4.

4. Nebraska: note in Telephony 107 (7 July 1934), 20; fire alarms: Palo Alto Times, 8 February 1901, and Antioch Ledger, 26 July 1913; police encouragement: Times, 3 September 1922; pants: Ledger, 14 September 1932.

5. Roland Marchand, Advertising, has described how the telephone was often highlighted and used as a symbol of modernity in advertisements during the 1920s. For example, a 1917 ad in the Ladies' Home Journal shows a woman speaking on the telephone and instructs the reader as to which brand of cold cream to order (Ladies' Home Journal, April 1917, 74). In a 1935 Journal ad, a woman says over the telephone: "Of course I'll go—Listerine got rid of my sore throat." (Ladies' Home Journal, April 1935, 46. Steve Derné surveyed ads in the Journal.) Most major companies used telephones in their advertisements at some time (R. Kern, "In Advertising, the Telephone Is the Message").

6. Three businesses: Times, 27 August 1912; California Market: Times, 23 March 1905; Solares Cleaners: Times, 17 August 1916; telephone directory: Times, passim; Viera: Ledger, 17 February 1906, and 19 February 1916.

7. I explored the possibility that it had something to do with the regional nature of the *Marin Journal,* but if one looks only at advertisements placed by persons or businesses with addresses in town, the pattern looks much the same.

8. A rough canvass of Dun & Bradstreet business listings suggests that San Rafael businesses were less likely to have telephones than Palo Alto businesses. (There were too few Antioch listings in the Dun & Bradstreet reports to analyze.) This finding is based on credit reports for our three towns available at the Dun & Bradstreet Corporation Library, New York City.

9. The books are, in chronological order: Benham, *Polite Life and Etiquette* (1891); Dale, *Our Manners and Social Customs* (1891); Cooke, *Manual of Etiquette* (1896); Sherwood, *Manners and Social Usage* (1897); A. White, *Twentieth-Century Etiquette* (1900); Holt, *Encyclopedia of Etiquette* (1901); Sangster, *Good Manners for All Occasions* (1904); Ayer and Berghe, et al., *Correct Social Usage by Eighteen Distinguished Authors* (1906); "A Woman of Fashion," *Etiquette for Americans* (1909); Glover, *"Dame Curtsey's" Book of Etiquette* (1909); F. Hall, *Good Form for All Occasions* (1914); Eichler, *Book of Etiquette* (1921); Holt, *Encyclopedia of Etiquette,* supra, 1923 edition; Post, *Etiquette* (1923); Wade, *Social Usage in America* (1924); Fishback, *Safe Conduct* (1938); Harriman, *Book of Etiquette* (1942); Post, *Etiquette,* supra, 1942 edition; M. Wilson, *The New Etiquette* (1947); Fenwick, *Vogue's Book of Etiquette* (1948); and Post, *Etiquette,* supra, 1955 edition.

10. White, *Twentieth-Century,* 146; 1906: Ayer and Berghe, et al., *Correct Social Usage,* 385.

11. "A Woman of Fashion," *Etiquette,* 57–59, 70–71; Glover, *"Dame Curtsey,"* 40; Hall, *Good Form,* 53–54. The ad was enclosed with a letter from E. Harris to J. D. Ellsworth, 29 January 1910, in "Advertising and Publicity, 1906–1910, folder 2," Box 1317, AT&THA.

12. 1921 book: Eichler, *Book of Etiquette,* 24, 49, 59; 1923 volume: Holt, *Encyclopedia,* 1923 edition, 70; Washington Society: Wade, *Social Usage,* 231; and Post, *Etiquette,* 1923 edition, 128.

13. Fishback, *Safe Conduct,* 105; Wilson, *New Etiquette,* 240; Fenwick, *Vogue's,* 361; Harriman, *Book of Etiquette,* 78, 67.

14. White, *Twentieth-Century,* chap. 16, 146; "A Woman of Fashion," *Etiquette,* 169–73; Fishback, *Safe Conduct,* 71, 105.

15. Fenwick, *Vogue's,* 360, 14; Wilson, *New Etiquette,* 164.

16. White, *Twentieth-Century,* 145.

17. Miss Manners, "Calling for a Few Simple Phone Rules."

18. Rakow, "Gender, Communication, and Technology," 168.

19. Palo Alto Business and Professional Women's Club *News Notes* (PAHS), Winter 1929–30, March 1931, and April 1931.

20. *Ledger,* 24 January 1930, and 14 January 1931; *Times,* 9 September 1932, and 5 December 1934.

21. Palo Alto: *Times,* 1 November 1922; Antioch: *Ledger,* 3 November 1939.

22. Banks: *Ledger,* 20 May 1931; unions: Thompson, "Settlement Geography," 302–3.

23. Rainwater, *What Money Buys,* 48; see also Horowitz, *Morality of Spending;* Ornati, *Poverty Amid Affluence.*

24. Wurtzel and Turner, "Latent Functions," table 4.

25. Horowitz, *Morality of Spending,* 92–119.

26. Peixotto, *Getting and Spending at the Professional Standard of Living,* 175; Ornati, *Poverty Amid Affluence,* 52ff.

27. Willey and Rice, *Communication Agencies,* 199.

28. Lynd and Lynd, *Middletown,* 253; college professors: Peixotto, *Getting and Spending,* 196; typographical workers: Peixotto, "How Workers Spend a Living Wage."

29. M. Berger, *Devil-Wagon,* 212. See also Interrante, "The Road to Autopia."

30. U.S. Bureau of the Census, *Historical Statistics,* 469.

31. Between 1973 and 1988, the Roper organization presented lists of devices to samples of Americans, asking them whether they "personally pretty much think of [each object] as a *necessity* or pretty much as a *luxury* you could do without." Between 87 percent and 90 percent answered necessity to an automobile (27 percent to a *second* automobile). *Roper Reports 89–1,* 22–23, courtesy W. Bradford Ray, Roper Organization, New York.

7. LOCAL ATTACHMENT, 1890–1940

1. Quoted by Chicago sociologist, Roderick McKenzie, who added that "[t]he automobile, street car, telephone, and press, together with increased leisure time, have all contributed greatly to the breakdown of neighborhood ties.... It is an obvious fact that in isolated rural communities where the telephone has not become an instrument of common usage and where poverty restricts the use of secondary means of transportation... are to be found the best examples of the old neighborly forms of association" (both quotes in "The Neighborhood," 72–73).

2. Textbook: Layton, *The Community in Urban Society,* 98. The phrase, a "community of limited liability," was coined by Morris Janowitz, in *The Community Press in an Urban Setting.* It is one whose residents, although satisfied with and involved in the town or neighborhood, feel free to leave at any time for another similar community. I have substituted "locality" for "community," since such places seem to lack the ideal qualities of "community."

3. Wiebe, *The Search for Order;* Rossi, "Community Social Indicators," 87. For claims of declining localism, see, e.g., Boorstin, *The Americans;* Abler, "Effects of Space-Adjusting Technologies on the Human Geography of the Future"; and Stein, *The Eclipse of Community.* A few works provide good overviews on this issue, including Bender, *Community and Social Change,* which largely concludes that locality has lost its significance; Hunter, "The Persistence of Local Sentiments in Mass Society," which largely stresses the continuing importance of locality; Conzen, "Community Studies"; and Gusfield, *Community.*

4. For example, Stuart Blumin (*The Urban Threshold*) argued that, in the nineteenth century, the townsfolk of Kingston, New York, attained *greater* community consciousness as the town grew and became more integrated into national affairs. Michael Frisch (*Town into City*) has suggested that during the nineteenth century Springfield, Massachusetts, became a more substantive and active town, but also a place of weaker emotional bonding.

5. This passage alludes to a vast literature on community change. For some synthetic statements that also provide bibliographies, see, e.g., Abbott, *Seeking*

Many Inventions; Atherton, *Main Street on the Middle Border;* Bender, *Community and Social Change;* Berthoff, *An Unsettled People;* R. Brown, *Modernization;* Conzen, "Community Studies"; Hine, *Community on the American Frontier;* Rutman, "Community Study"; Stein, *Eclipse of Community;* Wiebe, *The Search for Order;* Zunz, "The Synthesis of Social Change."

6. The claim that residential mobility has declined is surprising to many, but the best evidence shows that rates of moving in the twentieth century were equal to or lower than those of the nineteenth century. Census data since World War II shows a slight trend toward Americans staying in the same home longer. See, e.g., Thernstrom, *The Other Bostonians,* 221–32; Knights, *The Plain People of Boston;* Parkerson, "How Mobile Were Nineteenth-Century Americans?," 99–109; L. Long, "The Geographical Mobility of Americans"; idem., *Migration and Residential Mobility in the United States;* U.S. Bureau of the Census, *Historical Statistics,* 96.

7. Neighborhood organizations emerged in the early twentieth century and seemed to remain at least as efficacious over the decades. See, e.g., J. Arnold, "The Neighborhood and City Hall"; Z. Miller, "The Role and Concept of Neighborhood in American Cities"; and Melvin, "Changing Contexts."

8. For a discussion of localism and the ways in which it has changed, see C. Fischer, "Ambivalent Communities: How Americans Understand Their Localities." We can distinguish localism as a property of the community (e.g., political autonomy) from localism as a property of its residents (e.g., the scope of their political interest). Another distinction can be made among levels of locality—street, neighborhood, town, region. Perhaps the neighborhood gains residents' loyalties at the expense of the town, or vice-versa. Yet another is among different domains of localism—economy, polity, private life, and so on—that need not change in concert. Finally, different types of people and places may change differently. For example, many middle-class people have become more cosmopolitan over the decades, but not working-class people.

9. Propinquity: quoted by Marvin, *Old Technologies,* 66. See also the quotation from McKenzie in note 1 of this chapter as another comment on the telephone's negative effect on localism. AT&T publicists claimed that the telephone enabled people to maintain contact with intimates scattered around the city, reducing the need to communicate with bothersome neighbors (Pound, *The Telephone Idea,* 37) and that the telephone was an "antidote to provincialism" (Barrett, "The Telephone as a Social Force," 131). Lewis Atherton, who attributed the disintegration of small Midwestern towns mostly to the rise of fraternal orders, also blamed telephone companies for undermining the separate communities' autonomy (*Main Street,* chap. 7). Today, writers such as Joshua Meyerowitz (*No Sense of Place*) and Philip Abbott (*Seeking Many Inventions,* 74, 163) blame the telephone, among other developments, for weakening local ties.

10. Heilbroner is quoted by Rae, *The Road and the Car,* 370. Michael Berger (*Devil Wagon*) quotes early observers' comments about the automobile and rural neighborhoods. Brunner and Kolb point to improved roads to explain a shift in rural folks' interests from the neighborhood to translocal concerns (*Rural Social Trends,* esp. 67–68). West agrees with town residents that the automobile undermined neighborhood solidarity and introduced cosmopolitan aspirations

(*Plainville*, 74–75, 221). Preston suggests that autos and roads altered the culture of Spartanburg, South Carolina (*Dirt Roads*, 134–36).

11. Willey and Rice, *Communication Agencies*, 202.

12. C. Fischer, "Changes in Leisure Activities in Three Towns," and "Community Cohesion and Conflict."

13. Researchers have long used local shopping as an indicator of attachment to the community. For example, see Hunter, "The Loss of Community."

14. Moline, *Mobility and the Small Town.*

15. *Times,* 3 November 1893; *Times,* 19 April 1895.

16. In 1900 an insurgent slate running for city council promised to impose license fees on outside vendors; they lost to the incumbents. In 1937 a proposed ordinance against peddling lost by more than five to one. On campaigns for fees, see, e.g., *Times,* 8 October 1894, and 22 January 1897; on the voting, see *Times,* 16 March 1900, and 12 May 1937.

17. Dispute over advertising: *Times,* 11 December 1920; bridge: *Times,* 13 July 1905; Christmas festivals: *Times,* 9 December 1922.

18. *Journal,* 19 November 1908.

19. On advertising from San Francisco, see, e.g., the *Marin Journal,* 10 August 1899, and 17 December 1908. Buy-at-home campaigns were common throughout; see, e.g., *Daily Independent,* 18 May 1932. On campaigns against peddlers, see, e.g., *Journal,* 5 February 1934. On the chain store issue, see *Independent,* 10 January 1935, and *Journal,* 5 November 1936 (the county vote was 57 percent against the fee). The growth of chain businesses is evident in the Dun & Bradstreet credit reports (Dun & Bradstreet Corporation Library, New York).

20. *Ledger,* 22 December 1906, and ad placed by the Bank of Antioch, 1 January 1925.

21. Railroad fare increase: *Ledger,* 18 April 1908; fare rebate: for example, *Ledger,* 7 January 1926.

22. The best estimate is that the early 1920s was the period of fastest growth in automobile ownership in Antioch, from about 100 to about 250 cars.

23. Moline, *Mobility,* 128–32.

24. We must rely on impressions rather than counts of national or mail-order ads, because we systematically coded only those advertisements that had addresses for retailers (e.g., a general ad for Chevrolet was not coded; one with a local dealer's address was).

25. As noted above, we did not code corporate ads unless the name of a specific retailer appeared. Also, we did not code ads on pages directed to out-of-town residents—say, the page of Oakley news in the *Antioch Ledger.*

26. Those upward ticks near the end are somewhat misleading because most of those percentages are based on quite small numbers. For example, in the 1920s the *Palo Alto Times* published an average of about two legal and medical notices per issue, compared to over seven in the 1890s.

27. A major study of villages in rural Wisconsin between 1913 and 1929 concluded that rural families increased their shopping in distant cities, but also continued to shop locally. Small hamlets varied in how well they survived the 16 years. Many stores specialized, abandoning their roles as general goods stores; others became convenience stores (Kolb and Polson, *Trends in Town-Country*

Relations). A study of rural Illinois villages concluded that smaller ones suffered commercially, but that there was also great variation among them (Bureau of Business Research, "The Automobile and the Merchant").

28. *Times,* 22 October 1921.

29. 1933: Steiner, "Recreation and Leisure Time Activities," 944; Lynd and Lynd, *Middletown;* West, *Plainville;* Moline, *Mobility,* 103–105.

30. *Palo Alto Times,* passim.

31. *Marin Journal* and *San Rafael Independent,* passim.

32. On the Meyers and the Jacobs, see *Ledger,* 24 January 1914; on Asians, see Thompson, "Settlement Geography," 164; on Azevedo and Portuguese organizations, see *Ledger,* 14 January 1965; on others, see *Ledger,* passim.

33. There were fewer participatory events reported in the press, so the numbers are less stable and must be collapsed. Through 1920, 76 percent of those reported were in-town; in the 1920s, 70 percent; and in the 1930s, 71 percent.

34. Moline, *Mobility.*

35. One report estimates that 40 percent to 50 percent of all household calls are made to a point within two miles (Mayer, "The Telephone and the Uses of Time," 226). See also Singer, *Social Functions of the Telephone,* 23.

36. Wellman, "The Community Question"; see also C. Fischer, *To Dwell among Friends;* and C. Fischer et al., *Networks and Places.*

37. For discussions of endogamy as an indicator of the dispersion of social ties, see Clarke, "An Examination of the Operation of Residential Propinquity as a Factor in Mate Selection"; and Shannon and Nystuen, "Marriage, Migration, and the Measurement of Social Interaction."

38. Moline, *Mobility,* 120. The countywide rate of endogamy decreased only from 50 percent to 47 percent between 1900–1902 and 1928–1930; the rate for the town of Oregon actually rose, from 42 percent to 49 percent.

39. For a study of auto-touring, see Belascoe, *Americans on the Road.*

40. One might use sources such as diaries and letters, reading them for appropriate references. Besides the difficulties in interpretation, however, there exist the problems of how few and how special were the people who left such documents behind. (For an ambitious yet problematic effort along these lines, see R. Brown, *Knowledge Is Power.*)

41. On the other hand, historian Lewis Atherton construed publication of local news to mean that residents could no longer learn community events by word-of-mouth, thus a sign of the disintegration of local ties (Atherton, *Main Street,* 165).

42. Lynd and Lynd, *Middletown,* 533.

43. *Ledger,* 9 April 1910.

44. A 1922 story on various social events thrown for Mrs. McDaniel listed friends whose names had often appeared in the *Ledger* (5 January 1922).

45. *Ledger,* 2 January 1931.

46. This account draws on Wood, *History of Palo Alto,* 166–71, in addition to Steve Derné's reading of the *Times.*

47. This account is based, aside from readings of the newspapers by John Chan, on Gardner, "Newspapers of San Rafael," *San Rafael Independent-Journal,* 23 March 1974.

48. We coded the newspaper reports on leisure to identify whether items appeared in the news columns, club columns, or other specialty pages. Using linear regressions, I estimated the average annual change in the number of recreational items reported in each type of space.

	Antioch	Palo Alto	San Rafael
News Pages	−.00	.08*	.39*
Club Columns	.08	.82*	.55*
Special Pages	.35*	.37*	.10*
Number of Years	17	24	26

*$p < .01$

With the exception of special pages in San Rafael's *Journal,* the estimates are consistent with the impression that, proportionately, social reporting moved from the news pages to other columns of the newspapers. That is, reports increased least rapidly in the news pages.

49. A classic study of the persistence of local newspapers in the modern metropolis is Janowitz, *The Community Press.*

50. Blumin, *The Urban Threshold;* Rosenzweig, *Eight Hours for What We Will;* Moline, *Mobility,* 105–6.

51. Applebaum, *The Glorious Fourth.*

52. See the *Ledger* between 3 July and 7 July of each year.

53. See early July issues of the *Times.*

54. *Times,* 9 December 1922.

55. *Times,* 9 December 1935.

56. *Journal,* 8 July 1915.

57. For example, one of Moline's sources reports that there was "no one in town" on 4 July 1915, a year in which there was only 1 automobile or truck for every 15 residents in the entire county (Moline, *Mobility,* 106, statistics from 17, 55).

58. See Tropea, "Rational Capitalism and Municipal Government"; Tarr, "The Evolution of the Urban Infrastructure in the Nineteenth and Twentieth Centuries."

59. The quotation is from Chambers, *Seedtime of Reform,* 202. See also, Brock, *Welfare, Democracy, and the New Deal;* Reinhold, "Federal-Municipal Relations"; M. Katz, *In the Shadow of the Poorhouse,* 208ff.

60. The sources for this account are, in addition to the *Times:* the *Annual Reports* of Palo Alto; a WPA report, "A Study of Contemporary Unemployment," in the PAHS files; and other PAHS files on "Hoboes," the "Business and Professional Women's Club," and the "Shelter."

61. See pamphlet file of MCHS, "Mayor's Committee on Unemployment Relief."

62. *Ledger,* passim, especially 22 March 1919, 4 January 1932; 6 January 1933, 10 November 1933, 15 October 1934, and 30 January 1940.

63. This history is discussed in more detail in C. Fischer, "Community Cohesion and Conflict."

64. For 1960 and 1975 data, see Morlan, "Municipal versus National Election Voter Turnout." See also Verba and Nie, *Participation in America,* 30.

8. PERSONAL CALLS, PERSONAL MEANINGS

1. Ronnell, *The Telephone Book,* 95ff, 265. This complex literary treatment draws many other implications about the telephone, almost all in a different realm of discourse than this study. See also Haltman, "Reaching Out to Touch Someone?"

2. John Chan interviewed 10 people in San Rafael; Laura Weide, 11 in Antioch; Lisa Rhode, 13 in Palo Alto; and I interviewed an elderly woman in Palo Alto, too. Our procedure varied slightly from town to town. John Chan in part used personal connections to find San Rafael old-timers. Laura Weide and Lisa Rhode relied on institutional contacts such as local ministers and historical societies to find respondents in Antioch and Palo Alto. As the interviews progressed, we focused more on interviewees' memories of the telephone and automobile in their childhoods. Each interviewer taped the conversation and then abstracted key comments in writing. I am drawing from those abstracts. Although almost all our respondents had grown up in the vicinity of the three towns, many had spent parts of their youth in outlying villages and farms. Our sample is, as one would expect, skewed toward people who had been middle-class as children and thus to families that had telephones.

This sample bias, typical of oral histories, is one reason this evidence must be taken with some reservation. (It is also a reason not to use these interviews to make comparisons among towns.) Moreover, because people's individual histories vary so much, their claims about what was generally true differ. For example, the son of a leading Antioch businessman recalled that "phones were pretty universal" around 1922, when our census data shows that only about 30 percent of Antiochians subscribed (see Chapter 5). One female respondent in Palo Alto remembered that women didn't drive, whereas other women proudly described how they and their mothers loved to drive. Furthermore, given the advanced age of the interviewees—some were nearly 100—when we talked to them, it is not surprising that they occasionally contradicted their own recollections. Finally, the accounts are tinted by retrospection. That is, judgements about the past are in part based on views of the contemporary period. For example, respondents often described "visiting" on the telephone as uncommon *relative* to nowadays.

Despite all these cautions, the senior citizens we spoke to still gave us an otherwise unattainable peek at the past. We are grateful to them for their time.

3. Sifianou, "On the Telephone Again! Differences in Telephone Behavior"; Carroll, *Cultural Misunderstandings,* chap. 6.

4. McLuhan, *Understanding Media,* 225; Marchand, *Advertising,* 12.

5. Mead, "Looking at the Telephone a Little Differently," 12–14. Robert Collins counted over 650 telephone songs written between 1877 and 1937 (*A Voice from Afar,* 141–42). On magazine articles, see Weinstein, "The Telephone

in Popular Journals." Government commissions include U.S. Senate, *Report of the Country Life Commission.* Industry commentators regularly appeared in *Telephony.* See also, Pool, *Forecasting,* 129–31.

6. The 1893 article appeared in *Cosmopolitan* and was retold by Marvin, *Old Technologies,* 201–2; "larger but shallower" is from Abbott, *Seeking Many Inventions,* 163; Westrum, *Technologies and Society,* 273; the Berger quotation is from *The Heretical Imperative,* 6–7; See also Strasser, *Never Done,* 305.

7. *Chambers Journal,* "The Telephone," 310; Professor Jacks is quoted in Lundberg et al., *Leisure: A Suburban Study,* 7 (see also another professor: Schlesinger, *The Rise of the City,* 97); Willey and Rice, *Communication Agencies,* 203; H. Smith, "Intrusion by Telephone," 34. A chronicler of Australian telephony claims that, unlike in the United Kingdom and the United States, Australians were not concerned about intrusions by telephone (Moyal, *Clear Across Australia,* 147).

8. "Twaddle" comes from Antrim, "Outrages of the Telephone," 126; see also Bennett, "Your United States," and discussion in Chapter 3, pp. 78–80. On inappropriate contacts, see, e.g., Marvin, *Old Technologies,* 67ff; Kern, *Time and Space,* 215. On privacy, see Marvin, *Old Technologies,* 128ff; J. Katz, "US Telecommunications Privacy Policy;" Pool, *Forecasting,* 139–41. Margaret Mead, however, claimed that the telephone increased privacy, as compared to conversations in the open ("Looking at the Telephone a Little Differently," 13).

9. Calls to five numbers: Mayer, "The Telephone and the Uses of Time," 228. The New York City survey is reported in Wurtzel and Turner, "Latent Functions." In that study 36 percent of the respondents said that they most missed calling friends, 31 percent calling family, and 16 percent making business calls. Moreover, 45 percent missed getting calls from friends, 37 percent from family. The California findings are from Field Research Corporation, *Residence Customer Usage,* 40–43. The leisure study is United Media Enterprises, *Where Does the Time Go?,* table 2.1. Daily or almost daily telephone conversation with friends or family was reported by 45 percent, television watching by 72 percent, newspaper reading by 70 percent, music listening by 46 percent, exercise by 35 percent, reading for pleasure by 24 percent, shopping by 6 percent, drinking by 9 percent, and sex by 11 percent. The same table also shows that 33 percent of respondents said they never write letters to friends or family, but only 3 percent never talk with them over the telephone. Single parents and teenagers were the most frequent telephone conversationalists, dual-career parents and childless couples were the least frequent (ibid, table 2.2).

A Swiss study estimated that half of calls were made to kin or friends (Jeannine et al., "Pratiques et Representations Télécommunicationelles des Ménages Suisses"). A large study conducted in Lyons, France, asked several hundred people to keep a log of their calls and to characterize each one. The researchers categorized the calls as "functional" or "relational." By frequency, relational calls (chats, checking up on family news, etc.) accounted for only about 45 percent of the logged calls. By time, these accounted for 60 percent of calls. The authors concluded that they had done away with "the myth of the convivial telephone." The results of this study may differ from the others for either cultural or measurement reasons (Claisse and Rowe, "The Telephone in Question"). Saunders et al., *Telecommunications and Economic Development,* report survey data for four less-developed nations in which the proportion of calls residential subscribers made that were to family or friends ranged from 40 percent in rural Thailand to 69 percent in urban Chile. (The figure for the United Kingdom was 74 percent.) For calls made from public telephones, the proportion going to friends and family varied from 2 percent to 76 percent depending on the country (p. 222). Note that these authors try to stress the economic value of telephones in less-developed nations. (See also Saunders and Warford, "Evaluation of Telephone Projects in Less Developed Countries.") See also Singer, *Social Functions of the Telephone,* and Synge et al., "Phoning and Writing as a Means of Keeping in Touch in the Family of Later Life."

10. Judson, "Unprofitable Traffic." The estimate of one-third of Seattle households comes from the statistic that there was one residential telephone for each nine Seattle residents in 1914 (Item 248–6, letter from F. C. Phelps, in "Correspondence—DuBois" folder, TPCM). In the Judson data, the mean call length was about 7.5 minutes and the median about 5. In a statewide study done by AT&T in recent years, the mean was 4.25 minutes and the median about 1.5. A more accurate historical comparison may be to contrast this call length with data from cities that today have flat-rate pricing, as Seattle probably did then. Their current average is about 5.4 minutes, still about 30 percent less than the Seattle figure (Mayer, "The Telephone and the Uses of Time," 228–29). Although the longer calls for 1909 may indicate more talk, it may also reflect more difficulty in being heard. In either case, if sociability were truly uncommon, we would expect to find notably briefer calls at that time.

11. This claim has been repeated to the point of exaggeration in the telephone literature. For example, in his history of telephony, Brooks writes, "By the end of the 1880s, telephones were beginning to save the sanity of remote farm wives by lessening their sense of isolation" (*Telephone,* 94). Yet, at the end of the 1880s, very few farm homes, and effectively no remote ones, even had telephones.

12. North Electric: *Telephony,* "Facts Regarding the Rural Telephone"; Ohio Company: Kemp, "Telephones in Country Homes," 433.

13. See U.S. Bureau of the Census, *Special Reports: Telephones: 1907,* 78; U.S. Senate, *Report of the Country Life Commission,* 45ff; U.S. Department of Agriculture, *Social and Labor Needs of Farm Women,* 11–14; Ward, "The Farm Women's Problems," 6–7.

14. See, e.g., farm-related citations in note 13, as well as repeated testimony from industry sources. Ann Moyal quotes rural Australian women making similar claims in the 1980s ("The Feminine Culture of the Telephone," 22–25).

15. Rakow, "Gender, Communication, and Technology." The quotations come from pp. 218, 160–61, 159, 230, and 210.

16. Rakow, "Gender, Communication, and Technology," 207.

17. I employed three different interviewers in northern California, two of whom had only this connection to the larger project. (See note 2 above.) And I tried to make sure that the questioning about telephone use was neutral, avoiding leading questions.

18. E. Arnold, *Party Lines, Pumps and Privies,* 144–53; the quotation is on p. 153.

19. See, e.g., R. Paine, "What is Gossip About?"; di Leanardo, *The Varieties of Ethnic Experience,* 194ff; Spacks, *Gossip.*

20. Marvin, *Old Technologies,* 23.

21. On AT&T studies, see Wolfe, "Characteristics of Persons;" Mayer, "The Telephone and the Uses of Time," 231; Brandon, *The Effect of the Demographics,* chap. 1; and Arlen, *Thirty Seconds,* 46–47. Australian study: reported in Steffens, "Bewildering Range of Topics," 176. The French study found, for example, that working-class women who were employed had three times as many calls a week for four times as long as working-class men who were employed (Claisse and Rowe, "The Telephone in Question"). English study: Willmott, *Friendship Networks and Social Support,* 28–29. Ontario: Synge et al., "Phoning and Writing." New York: Litwak, *Helping the Elderly,* appendix C4. In another study of the elderly, on grandparents, Cherlin and Furstenberg discovered that maternal grandparents spoke on the telephone to their grandchildren 50 percent more often than did the paternal grandparents. The authors attributed this difference to the frequency with which mothers called their parents (*The New American Grandparent,* 116). Toronto: Wellman, "Reach Out and Touch Some Bodies."

22. Sorokin and Berger, *Time-Budgets of Human Behivior,* 52.

23. In our three towns in 1910, for example, 12 percent of the households headed by single males had telephones (n = 33), as did 28 percent of those with husband and wife but no adult daughter (n = 229) and 44 percent of the households with husband, wife, and adult daughter (n = 27). The analysis of the town data from 1900 to 1936, reported in Chapter 5 and Appendix F indicates that the presence of women modestly increased the chances of having a telephone, other factors held constant. In the Dubuque data, 60 percent of single-male-headed households (n = 72) had telephones, as did 72 percent of those with husband and wife and six of the seven households (86 percent) with husband, wife, and another adult female. The logit regression analyses in Appendix C suggest that the effect of women on telephone subscription in Dubuque County was real and robust. Analysis of the 1918–1919 Bureau of Labor Statistics survey in Appendix E found that extra adults in the household depressed the odds of having a telephone, but the depression was due to additional male adults. Thus, the proportion of females was an independent predictor of telephone subscription across all the data sets.

24. Rakow, "Gender, Communication, and Technology," 142.

25. A 1930s survey of 27 "typical" Iowa farm families found that women made 60 percent of the calls, including many regarding the farm business (Borman, "Survey Reveals Telephone"). A 1940s rural Indiana survey of 166 subscribers gives the following explanation for this pattern: "Women used the phone most frequently. Many men said they did not like to use the phone, so they had women call for them" (Robertson and Amstutz, *Telephone Problems in Rural Indiana,* 18; see also Rakow, "Gender, Communication, and Technology," 169–70).

26. Moyal, "Woman and the Telephone in Australia," 288; see also, Moyal, "Feminine Culture."

27. Marchand, *Advertising;* Cowan, *More Work.*

28. See, e.g., *Printers' Ink,* "Bell Encourages Shopping by Telephone"; Shaw, "Buying by Telephone at Department Stores."

29. In material circulated to merchants in 1933, the Bell Company claimed that over 50 percent of housewives in Washington "would rather shop by telephone than in person" (*Printers' Ink,* "Telephone Company Works with Retailers on Campaign"). A 1930s Bell survey of 4500 households in one city found that 40 percent of subscribers were "willing" to buy over the telephone, but in another survey, a bare majority of 800 subscribers answered yes to the question, "Do you like to shop by telephone?" The same source reported that telephone orders represented just 5 percent of business in large department stores (Shaw, "Buying by Telephone"). Taking into account the telephone company's interest in exaggerating the figures, we are left with the conclusion that as late as the 1930s, only a minority of women shopped by telephone.

30. Fredrick, *Household Engineering,* 329. I also examined the indices to Gilbreth (of "cheaper by the dozen" fame), *The Home-Maker and Her Job;* Balderston, *Housewifery* (the 1921 edition; the 1936 one is similar); Baxter, *Housekeeper's Handy Book;* and Nisbitt, *Household Management.* See also Strasser, *Satisfaction Guaranteed,* 265–67.

31. Martin, " 'Rulers of the Wires'? Women's Contribution to the Structure of Means of Communication."

32. The time-budget forms and related materials are in Box 653, Record Group 176, Bureau of Human Nutrition and Home Economics, "Use of Time on Farms Study, 1925–1930," Washington National Records Center, Suitland, MD. Despite the title, the raw data that have survived are not from farms, but largely from a select sample of urban and suburban housewives. Barbara Loomis alerted me to these records. For summaries of the data, see Kneeland, "Is the Modern Housewife a Lady of Leisure?" and U.S. Department of Agriculture, "The Time Costs of Homemaking." This sample is obviously unrepresentative of American women generally, and the data are vulnerable to numerous errors relevant to our interests, particularly the underreporting of calls. Nevertheless, the forms provide a rare glimpse of the daily routines of upper-middle-class women over a half-century ago, in some ways more systematic and comprehensive than even the diaries relied upon by many historians.

33. Since the women often listed telephone calls with little or no explanation, I could only estimate ranges for these content categories. The low estimates assume that calls fell into the social category only if they were explicitly labeled as such (for example, "called friend"), whereas the high estimates rest on bolder assumptions (for example, an unexplained outgoing call after 6 p.m. was likely to be social). Most of the plausible biases in the data would have depressed the number of recorded social calls: The study was explicitly conducted to see how hard women were working; the instructions to respondents clearly implied that the primary interest was in tabulating homemaking activities; the sample is of affluent and busy women (many had part-time paid or unpaid jobs), just the sort who presumably were most often too rushed to chat; the evident prestige concerns in some of the responses (one woman noted that her evening's reading was in Greek, another that it was on psychology) probably led many to minimize seemingly frivolous telephone "visiting." On the other hand, perhaps the onerousness of the time-budget task led to an overselection of women with time on their hands. (Shopping by telephone was probably *not* underreported simply because it was commonplace; entries describing in-person marketing are

very frequent, averaging about one every other day.) In sum, it is reasonable to assume that even among these upper-middle-class and presumably skilled home managers, sociability was the most common use of the telephone.

34. A study of Chicago "society women" found that by 1895, one-fourth of those who were officers in reform groups had telephones, compared to less than 1 percent of Chicagoans generally; by 1905, 66 percent of women activists had telephones, versus 3 percent of Chicagoans. The author of the study argues that the club women were quick to adopt the telephone and suggests that it may have been a major factor in the increasing civic activity of Chicago women (Rosher, "Residential Telephone Usage among the Chicago Civic-Minded," 110).

35. Indiana: E. Arnold, *Party Lines,* 145; "Patty": Richardson, "Telephone Manners"; Stanford: *Palo Alto Times,* 15 November 1934. On courting by telephone, see Rothman, *Hands and Hearts,* 233ff.

36. This is, for example, Michèle Martin's explanation.

37. Rakow, "Gender, Communication, and Technology," 297; see also Moyal, "Feminine Culture"; di Leonardo, *The Varieties of Ethnic Experience,* 194ff; Ross, "Survival Networks"; Steffens, "Bewildering Range of Topics"; among others.

38. On sociability, see, e.g., Fischer and Oliker, "A Research Note on Friendship, Gender, and the Lifecycle"; Hoyt and Babchuk, "Adult Kinship Networks." See C. Fischer, "Gender and the Residential Telephone," for a fuller development of these arguments.

39. See Wellman, "Reach Out"; Cherlin and Furstenberg, *New American Grandparent* (maternal grandparents exceeded paternal grandparents in telephone contacts with grandchildren but not in how often they personally visited with grandchildren); Litwak, *Helping the Elderly,* Appendix C4 (gender differences in telephone contacts between elderly-helper pairs was greater than gender differences in face-to-face contacts); Synge et al., "Phoning and Writing" (although women were less than 1.5 times more likely than men to see their friends face-to-face, they were 2 to 3 times more likely to telephone their friends); Willmott, *Friendship Networks,* 28–29 (although women wrote kin and friends more often than men did, the difference was even greater for telephoning).

40. Some analysts of housework reason that any change that makes a gender-stereotyped female role easier to perform ensures that women will be more able, willing, and compelled to continue in that role. See, e.g., McGaw, "Women and the History of American Technology"; and Rothschild, "Technology, Housework, and Women's Liberation." For a more nuanced treatment, see Cowan, *More Work.* See also discussion in C. Fischer, "Gender and the Residential Telephone."

41. Moyal, "Feminine Culture," 25; Martin, *"Hello, Central?"* 171.

42. See, for example, Falk and Abler, "Intercommunications, Distance, and Geographical Theory"; Pool, "The Communication/Transportation Tradeoff"; Salomon, "Telecommunications and Travel."

43. Lynd and Lynd, *Middletown,* 273–75, quotations from 275n.

44. E.g., Atwood, "Telephony and its Cultural Meanings," 360–61. Similarly, Martin suggests that by enlarging women's social contacts in the home, telephoning "may have reduced women's opportunites for socializing outside" it (*"Hello, Central?"* 165).

45. Mayer, "The Telephone and the Uses of Time," 234.

46. Wurtzel and Turner, "Latent Functions," 254; Lee, "The Resilience of Social Networks to Changes in Mobility and Propinquity." William Michelson found in a Toronto survey that people tended to telephone more often during the winter than during the spring, presumably because they used the telephone to avoid making trips into the cold weather (Michelson, "Some Like It Hot").

47. The British and Chilean studies show that people with telephones or who use the telephone more have more contacts of all sorts (Britain: Clark and Unwin, "Telecommunications and Travel," and C. Miller, "Telecommunications/Transportation Substitution"; Chile: Wellenius, "Telecommunications in Developing Countries," and idem, "The Role of Telecommunications Services in Developing Countries"). The U.S. study found that people who make more long-distance calls also more often write and visit their distant associates (reported in Mahan, "The Demand for Residential Telephone Service"). These studies must be taken with some reserve, of course, since the correlation between telephoning and other contacts may be spurious, the product of personality types, for example. But Barry Wellman's Canadian data show a strong and positive partial association between telephone and face-to-face contacts, holding constant other aspects of relationships (Wellman, "Reach Out," table 8). In the study of grandparents, those who called more often also visited more often, suggesting a synergy between the two modes (Cherlin and Furstenberg, *New American Grandparent,* 115–16).

48. Westrum, *Technologies and Society,* 276.

49. Mid- to late-twentieth century Americans changed homes *less* often than did earlier generations. There may be a tendency for the moves to be of greater distance now than before, but even that is uncertain. (See L. Long, *Migration and Residential Mobility;* C. Fischer, "Ambivalent Communities.") Americans have moved farther from their jobs, however (see, e.g., Jackson, *Crabgrass Frontier,* Appendix).

50. Experimental studies indicate that voice-only communications are experienced as more psychologically distant than visual ones (Rutter, *Communicating by Telephone*). Most respondents to Synge et al.'s Ontario survey ("Phoning and Writing") said that a telephone conversation was less personal than a face-to-face one. On the other hand, many women who responded to Moyal's Australian survey felt that telephone conversations with friends were franker and more initimate than in-person ones (Moyal, "Feminine Culture," 15).

51. Willey and Rice, *Communication Agencies,* 202. Those who claim that the telephone introduces alienation and inauthenticity are numerous. Historian Susan Strasser, for example, has argued:

> [N]ineteenth century American society, at least as mobile as that of today, had no such technology [as the telephone]; many people moved thousands of miles, never expecting to see or converse with friends and relations again. When they arrived in their new homes, they established new day-to-day intimacies, free of charge: women who hung their laundry outside knew their neighbors, who genuinely filled some of the needs the telephone meets inadequately. (Strasser, *Never Done,* 305).

52. On distance between people and feelings of closeness, see Fischer et al., *Networks and Places,* chap. 9; C. Fischer, *To Dwell among Friends,* chap. 13;

C. Fischer, "The Dispersion of Kin Ties in Modern Society"; Wellman, "The Community Question." In a small survey in northern California, people generally said that they would not like their friends to live next door, but would rather have them within easy driving distance (Silverman, "Negotiated Claim").

53. Rakow, "Gender, Communication, and Technology," 159; Synge et al., "Phoning and Writing"; Moyal, "Woman and the Telephone," 284; see also Moyal, "Feminine Culture."

54. Kenneth Haltman cites several *New York Times* articles complaining about telephone intrusions and one, dated 1922, which revealed that Bell himself would never have a telephone in his study (Haltman, "Reaching Out," 343); Gilbreth, *The Home-Maker and Her Job,* 79–81.

55. Rakow, "Gender, Communication, and Technology," 175; Lynd and Lynd, *Middletown,* 275n.

56. See Katz, "US Telecommunications Privacy Policy;" Martin, "Communication and Social Forms," 128, 279, 370, 380.

57. First-hand account quoted by Umble, "The Telephone Comes to Pennsylvania Amish Country," 12.

58. Brooks, "The First and Only Century of Telephone Literature"; see also Wisener, " 'Put Me on to Edenville.' "

59. Ads from "Advertising and Publicity, 1906–1910, folder 1," Box 1317, AT&THA. Telephone as advertising symbol: Marchand, *Advertising,* 169, 190, 209, 238–47; Mennonites: Atwood, "Telephony and its Cultural Meanings," 326–47, and Umble, "The Telephone Comes to Pennsylvania Amish Country." See also Atwood, 83, on how rural telephone companies in Iowa used the modernity theme in their marketing.

60. *Chamber's Journal,* "The Telephone"; Casson, *History of the Telephone,* 231; Brooks, *The Telephone,* 117–18; see also S. Kern, *Time and Space,* 91.

61. Indiana: E. Arnold, *Party Lines, Pumps and Privies,* 146–53; Rakow, "Gender, Communication, and Technology," 231; AT&T: Mayer, "The Telephone and the Uses of Time," 232. Whether older people were more nervous around the telephone because of their cohort—the telephone was new in their youth—or because of their age—more of their friends might be ill, for example—is not evident.

62. E. Arnold, *Party Lines, Pumps and Privies,* 146–47.

63. *Sales Management,* "Average Person Finds Letter Writing More Difficult than Telephoning or Wiring."

64. Mayer, "The Telephone and the Uses of Time," 232; Singer, *Social Functions of the Telephone,* 62–63, 26, 14–15.

65. Wurtzel and Turner, "Latent Functions," 253, 256.

66. Noble, "Towards a 'Uses and Gratifications' of the Telephone."

67. Aronson, "The Sociology of the Telephone," 162.

68. Two historical studies suggest that the character of visiting in rural areas changed from overnight stays to more frequent but briefer "drop-ins" when rural folk switched from buggies to Tin Lizzies (M. Berger, *Devil Wagon,* 55; Moline, *Mobility,* 30–31). Berger argues that these increased contacts were largely impersonal (pp. 73–74). We can assume that farm families who bought cars could pursue more leisure activities and social contacts than before. On the

automobile and farm life, see also C. Larson, "A History of the Automobile," 23; Wik, *Henry Ford;* and Chapter 4 of this book. On the family, see Flink, *The Automobile Age,* chap. 9, for a summary, largely of arguments supporting the theory of weakening bonds. For examples of arguments that the automobile strengthened the family, see Belascoe, *Americans on the Road,* 84, and Allen, "The Automobile," 125. On earlier complaints about bicycles, see R. Smith, *A Social History of The Bicycle.* The concession that we may not know whether the car affected sexuality is from F. Merrill, "The Highway and Social Problems," 140–41; the judge is quoted in Lynd and Lynd, *Middletown,* 114; and the *Tennessean* quotation is from Brownell, "A Symbol of Modernity," 38–39. *Middletown* appears repeatedly in studies as the only, or the most-often, cited source on the topic of the automobile and sexuality, and the evidence there is basically of this sort of anecdotal assertion. An oft-cited statistic from *Middletown* is that 19 of 30 (but the correct figure is actually 40) young women charged with sexual crimes in 1924 had allegedly committed those crimes in an automobile—evidence too weak to prove anything except that police officers were checking cars. On the automobile and sex, see also Brilliant, "Some Aspects of Mass Motorization," and the collection edited by Lewis, *The Automobile and American Culture.* In his own essay ("Sex and the Automobile," 518–28), Lewis reviews both the material aspects of the car—including the establishment of motels—that facilitated sexual encounters and the cultural products, such as songs, that linked the automobile to sexuality.

69. On the other hand, Rothman cites the Kinsey Report findings that the greatest increase in premarital sex occurred in the generation that came of age in the late 1910s and early 1920s, which would coincide with the fastest period of automobile diffusion (Rothman, *Hands and Hearts,* chap. 7, 294–97).

70. Belascoe, *Americans on the Road;* Hyde, "From Stagecoach to Packard Twin-Six."

71. See Scharff, *Taking the Wheel;* Scharff, "Reinventing the Wheel"; Interrante, "You Can't Go to Town in a Bathtub"; M. Berger, *Devil Wagon,* 65–66; and Wik, *Henry Ford,* 25ff.

72. For example, Nash ("Death on the Highway") points out the manner in which American painters in the interwar period, such as Grant Wood, used the automobile as a symbol of modernity. (The most extreme version, of course, was the Italian Futurists' apotheosis of automobile parts.) See also Lewis, "The Automobile and American Culture," on this point. Gallaher (*Plainville Fifteen Years Later*) suggested that the automobile helped instill in Plainville residents the notion that new technologies were a source of prestige. Bardou et al., in their cross-cultural review of the automobile (*The Automobile Revolution*), conclude that its greatest cultural effect was to teach people to value progress, technology, and consumption. Warren Belascoe, however, adds a dissenting note. He claims that because automobiles allowed people to get closer to nature than did train travel, it was popularly seen as rustic and *anti*-modern (Belascoe, *Americans on the Road,* and idem, "Cars versus Trains").

73. Marvin, *Old Technologies.* See also Nye, *Electrifying America,* on reactions to other aspects of electricity.

9. CONCLUSION

1. In 1971, AT&T had assets of $53 billion, compared to Standard Oil of New Jersey's $19 billion, and net income twice that of Standard Oil (*Fortune* magazine, quoted by Temin, *The Fall of the Bell System,* 10). U.S. Bureau of the Census data lists about 800,000 employees for the giant telephone company as of 1970 (*Historical Statistics,* 785).

2. On the cross-subsidy issue: Historically, AT&T assigned the cost of everything that was not explicitly long-distance equipment to local costs and, thus, to the basic monthly charge. That is, utility commissions allowed them to charge for cost plus stipulated profit; all local plant costs were calculated into the monthly bill. During World War II, the FCC and AT&T agreed that some portion of the local system served *inter*exchange callers and that a percentage of the costs of the common plant—of the system that ran from a subscriber's house to the main switchboard—would be assigned to long-distance service and, so, paid by long-distance callers. (The technical phrase for these calculations is *separations.*) That proportion was—arbitrarily—set as the equivalent of the proportion of long-distance calls to all calls, at that time 3 percent. Starting in the 1950s, regulators steadily increased that percentage (as a multiple of, rather than equal to, the proportion of long-distance calls) until it reached about 25 percent in 1980. The effect was to have an increasing proportion of telephone capitalization costs paid by long-distance callers. This practice has been labeled *cross-subsidization* of local service by long-distance service, although there is no way to know for certain what the "true" costs of each service would have been if there had been totally separate local and long-distance telephone systems—as there were briefly in the nineteenth century. Critics charge, for example, that "conventional wisdom, assiduously propagated by the telephone companies, [holds] that long-distance revenues subsidize local service. . . . [But] the subsidy may actually go the opposite way—local users may be supporting long-distance services. . . . The local system has been specially engineered, at enormous cost, to handle long-distance service and local users have paid for most of this" (Schwartz, "It's a Bird"; cf. responses in subsequent issues of the *New York Times;* see also *Consumer Reports,* "Will Your Phone Rates Double?"). This complaint echoes charges made by small telephone companies in the 1920s and 1930s that connection with the Bell System required expensive upgrading of their equipment that would otherwise not have been necessary.

Still, the general assumption of economists and regulators is that "true" subsidies have run toward local users, so that by 1980, local service was "underpriced" and long-distance was "overpriced" from a market point of view. This would explain why local rates have risen faster than interstate rates in recent years (or, put in real dollars, local rates have dropped more slowly), as a greater accounting separation has been forced between the competitive interexchange business and the monopolies of local exchange service. Competition has forced long-distance rates down toward their "true" costs, leaving less excess with which to subsidize local service. (Critics would contend, however, that the "Baby Bells" have merely restored the old system of subsidizing in the other direction at a regional level.) Ultimately, of course, what charges are "best" depends on nonmarket concerns as well, such as equity and the common good (i.e., the positive externalities of providing cheap service to small-scale users). On these matters, see also Temin, "Cross Subsidies in the Telephone Network after Divestiture;" Sichter, "Separations Procedures in the Telephone Industry"; Crandall, "Has the AT&T Break-up Raised Telephone Rates?"; Vietor, "AT&T and the Public Good."

3. On the complexities of this history, see, e.g., Temin, *The Fall of the Bell System*, and von Auw, *Heritage and Destiny*, as well as sources cited in the previous note. Regarding moves to restructure local service, see Andrews, "Regulators Moving to Break Local Telephone Monopolies."

4. Displeased: Temin, *The Fall of the Bell System*, 365n. Subscriptions: In 1980, 93.0 percent of American households had telephone service; in March 1984 and March 1985, 91.8 percent did; by 1989, it was back to 93.0 percent (U.S. Bureau of the Census, *Statistical Abstract, 1990*, 550). Crandall, "Has the AT&T Break-up Raised Telephone Rates?" argues that basic local charges had not increased by 1987 and that increases in the *relative* cost of local versus long-distance service was a result of concurrent changes in the industry, not specifically of divestiture.

5. In 1980 dollars, the monthly charge for individual residential service was $16 in 1955 and $9 in 1980 (Congressional Budget Office, *Local Telephone Rates*, Appendix B). Off-peak pricing: von Auw, *Heritage and Destiny*, 164. On rural developments: By 1988, 95 percent of American farms and 90 percent of the rural population generally had telephones; 9 out of 10 had single-party lines. See general histories in Rural Electrification Administration, *Twenty-five Years of Progress;* Fullarton and Goss, *NTCA: Our First Twenty-five Years;* Renshaw, "Experience of the United States in Bringing Modern Telephone Service to Unserved Areas." As an example of technological developments, see Boyd et al., "A New Carrier System for Rural Service." Fuhr, "Telephone Subsidization of Rural Areas in the USA," reviews the economic aspects of rural telephone expansion and presents the statistics on farm subscriptions.

6. The more recent figures are from U.S. Bureau of the Census, *Statistical Abstract, 1991*, tables 919 and 1289.

7. Conversations: cf. U.S. Bureau of the Census, *Historical Statistics,* 783, and idem, *Statistical Abstract, 1990,* 552. The estimating procedure changed in 1988. On continuing increases in calling, see T. Hall, "With Phones Everywhere, Everyone is Talking More." This expansion affected business and residential lines equally. Experts quoted by Hall attributed the 24 percent increase in calls between 1980 and 1987 to answering machines and cellular telephones.

8. As of 1988, L. L. Bean, for example, received 70 percent of its orders by telephone (T. Hall, "With Phones Everywhere").

9. Occasionally, newspapers feature stories on communities without telephone service—e.g., Johnson, "Where Phone Lines Stop, Progress May Pass." People who eschew telephones are usually worth a paragraph or two, as well. In 1988 only families with annual incomes under $12,500 had rates of telephone subscription below 90 percent. Among the poorest, those under $5000, 72 percent had telephones (Fuhr, "Telephone Subsidization," 188).

10. By 1988, 28 percent of households had answering machines. One academic gave to a journalist the following analysis of the answering machine: "[It] raises incredible metaphysical questions of location and identity. . . . [It] allows us to cultivate calculative responses to the callers. It deadens our humanity." See S. Davis, "After the Beep"; Miss Manners, "You Don't Have to Talk to a Machine" and idem, "Proper Use of the Telephone."

11. Avital Ronnell, who, in *The Telephone Book,* wrote about the psychology of the telephone, said of the answering machine: "The telephone is an extremely aggressive instrument. . . . The answering machine acts as a tool of delay and deferral. It defuses the power of the telephone, by permitting us not to answer" (in S. Davis, "After the Beep").

12. The computer industry, in its efforts to "educate the public," has searched for and advertised nonwork and nonschool uses of the home computer. It has, to date, largely failed to persuade consumers of such uses. As one industry observer wrote in 1990, the "essential nature of the home-PC market is this: home PC's are mainly used by Mom and Dad to stretch their workdays" (Seymour, "The 'Home-PC Market' Myth Rises Again"). The president of a software marketing firm told a journalist in 1985: "Our surveys over the past two years found that a preponderance of Americans cannot imagine any way they could use a personal computer, either at home or at work . . . ." The industry is waiting for the "Great Idea, the breakthrough that will make the computer an indispensable part of the American home, like a washer or a toaster . . . ." (Liberatore, "Home Computers"). A 1989 U.S. Census Bureau survey found that only 9 percent of American adults used a computer at home. Most of them also used one at the office, suggesting that the home machine was largely being used to "stretch the workday." (A total of 28 percent used a computer either at work or at home.) Fifteen percent reported having a computer in the home, but that was probably inflated by counting game-playing computers (see Kominski, *Computer Use in the United States, 1989*). Telecommuting was, at the end of the 1980s, uncommon and was not growing particularly fast (Gerstein, "Working at Home in the Live-in Office"). One should recall that many workers have always engaged in "homework," be they seamstresses in putting-out systems, real estate agents, writers, or whatever.

Some commentators have pointed to the French "Minitel" as a harbinger of the telecommunications future. It is a system of computer terminals that the communications ministry placed, free of charge, in several million homes in lieu of telephone directories. (The telephone subscriber punches up people's names to find their numbers instead of using a book.) Minitel subsequently expanded into a "billboard" communications system and a general information system, with charges for on-line time. Its most popular and profitable uses involve risqué messages and services. Some have acclaimed Minitel as a model for the computer-interconnected twenty-first century. However, the jury is still out about whether, once it is priced as a commercial good rather than as an experiment, it will be either popular or profitable. I am skeptical about any system that requires people to do something as complicated as type. Although there will always be enthusiasts, I suspect that its popularity will be limited. (See Charon, "Videotext"; Egido, "Home Information Services"; Kramer, "Minitel:

Don't Believe the Hype"; as well as journalistic accounts such as E. Brown, "Minitel," and Tempest, "Minitel.")

13. Telephone predictions: Pool, *Forecasting the Telephone;* "one neighborhood": Carey, "Technology and Ideology"; BBC: Pegg, *Broadcasting and Society,* 159–60; airplane: Corn, *The Winged Gospel,* chap. 5; Rural Free Delivery: Boorstin, *The Americans,* 133; and radio: Wik, "The Radio in Rural America during the 1920s," 346. See also Chapter 1 of this book.

14. For arguments about the historical case in America, see, e.g., various essays in Pool, *The Social Impact of the Telephone.* On arguments for using telephones to spur development, see, e.g., Hudson et al., *The Role of Telecommunications in Socio-Economic Development,* and Saunders et al., *Telecommunications and Economic Development.* The evidence for these claims is largely circumstantial—specifically, that there is a correlation between telephone diffusion and economic development—and anecdotal. Research has not yet, to my knowledge, established a causal connection.

15. Avital Ronnell, quoted in S. Davis, "After the Beep," D–19.

16. Robinson and Converse ("Social Changes Reflected in the Use of Time") report major diversions in time to television from other activities—but not, interestingly, major diversions due to the automobile (pp. 40–44, 52).

17. The books of James J. Flink show how one scholar has wavered on this question (*America Adopts the Automobile; The Car Culture; The Automobile Age*).

APPENDIX B

1. Property and personal income were drawn from R. Easterlin, "State Income Estimates."

2. See Meier, "Innovation Diffusion and Regional Economic Development."

APPENDIX C

1. A fuller analysis of this data will appear in a later publication.

2. Tina Carr helped in this task.

3. Using the SYSTAT logit package (Steinberg, *Logit,* Version 1.0).

4. That is, for households without wives, Dubuque location added .72 to the odds of subscribing. For those with wives, it added $.72 - .52 = .20$ to the odds.

5. Inside Dubuque, the odds were $.20 - .52 = -.32$. However, wifeless households in Dubuque were rare.

APPENDIX E

1. U.S. Bureau of Labor Statistics, "The Cost of Living in the United States"; Inter-University Consortium for Political and Social Research, "Cost of Living in the United States, 1917–1919," (obtained through the Survey Research Center, University of California, Berkeley). In the BLS report, the relatively few nonwhite families were dropped; they are included here. Michele Dillon assisted in the analysis.

2. In 1918, $2000 was roughly the annual pay of a military officer and twice the average earnings of a government worker or a fully employed production worker in manufacturing (U.S. Bureau of the Census, *Historical Statistics,* 176, 170, 167).

3. Inter-University Consortium, "Cost of Living," 11.

4. U.S. Bureau of Labor Statistics, "The Cost of Living," 2, 69.

5. This clustering appears in the distributions of some occupations by city. For example, 35 percent of the "miners" were in either Scranton, Pennsylvania, or Pana, Illinois, and 22 percent of the "laborers" were in Pittsburgh, New Orleans, or Detroit. It also appears in the great variations in telephone subscription rates from city to city, such as 48 percent in Chicago versus 4 percent in New York. (These numbers were calculated from our 20 percent random subsample and based on our own occupational coding.)

6. In the BLS data, residents of New England and the Mid-Atlantic region had very low rates of subscription, whereas the census of telephones for 1917 showed those regions to be about average in per capita telephones. Similarly, whereas the census showed low rates for the South, the BLS data implied that Southerners were about average in subscription. Unless the distribution of business telephones accounts for these discrepancies (which is unlikely), they point to sampling problems in the BLS study. See U.S. Bureau of the Census, *Census of Electrical Industries 1917.*

7. Another concern is that the question about telephone (and automobile) spending came at the end of what must have been an exhausting interview.

8. I used logit because of the skewness on the automobile data and because there are no structural zeros in this data-set. The entries are analogous to regression b-weights. Significance tests are distorted upwards by the cluster sampling but are also underestimated because of the subsampling. In any event, I ran robustness analyses, such as split-sample comparisons. The analysis used the SYSTAT Logit program (Steinberg, *Logit,* Version 1.0).

9. Hershberg and Dockhorn, "Occupational Classification." We initially looked at finer categorizations, but the four-category scheme suffices.

10. Introducing a measure of state-level diffusion into the model ameliorates the regional differences, but residents of the Mid-Atlantic region remained especially low in telephone subscription.

11. The percentages in each urban category reporting *any* telephone spending were, from least to most urban: 31%, 25%, 17%, and 27%. The percentages for spending over $5 were: 30%, 20%, 12%, and 15%. In the logit equations, the effects for living in the largest places are especially high (recall Chicago) and the next most urban category especially low.

12. Removal of the housing variables from the basic equation reduces the explained variance noticeably and increases the effects of income only modestly. Housing value itself was largely a function of the head's earnings, urbanism, and the income contributed by others.

13. Adding those two variables to the equation in Table E–1 increases the Chi-square by 31 and the pseudo-R^2 by .008. The derived logit estimate for automobile is .06 and for electricity is .07, both significant. In ordinary least square models, the correlations among residuals are .06 between telephone and automobile and .10 between telephone and electricity.

APPENDIX F

1. We drew a proportional sample of households from the Mayfield census and added them to the sample from Palo Alto proper in order to maintain geographical continuity from 1900 to 1940. In an initial analysis, segregating the Mayfield sample did not appear to make a substantial difference, so it is lumped together for the most part in this discussion.

2. For each sample year, Barry Goetz read two issues of the town newspapers, that of 15 January or immediately thereafter, and that of 15 May or immediately thereafter. He noted stories reporting on social events or personal notices that occurred in town or obviously included townsfolk. (We excluded stories of business or official activities.) He recorded up to 5 names of participants from each event, until he cumulated 20 names for January and 20 for May. Being named in the town press is our operational definition of being a "social notable." We searched the census lists for the households of the 40 persons and then added up to 25 households to our sample (randomly chosen from the 40, if necessary).

3. Of the households drawn in 1900, 46 percent also appeared in the 1904 listings and of the 1910 sample, 47 percent appeared in 1914. The most distinctive features of those households that "persisted" were: being headed by a male, having adult extended kin at home, and most especially, having owned a home in the prior period. Some households that did not appear in the 1904 or 1914 city directories *did* appear in the telephone directories for those years, indicating again the imperfection of the city directory source.

4. I drew on the following sources for coding suggestions: Hershberg and Dockhorn, "Occupational Classification," and Alba M. Edwards, "A Social-Economic Grouping of the Gainful Workers of the United States."

5. For Palo Alto and San Rafael, about 90 percent of original names, pooling 1920, 1930, and 1936, could be traced forward; for Antioch, about 82 percent could.

6. A statistical analysis for 1900 and 1910 showed that it was largely random which households in the census failed to appear in the secondary sources of the same year, but households that rented their quarters, were headed by women, those that were small, those headed by blue-collar workers, and certain ethnic minorities were most likely to be missed in the directories. Of course, there were probably some households that were listed in neither the census nor the city directory.

7. Kinuthia Macharia photocopied the relevant directories from the Old Directory Library of Pacific Bell. Thanks to Robert Deward of Pacific Bell and the staff of the library for their assistance.

8. Around 1910 in Palo Alto, the number of business telephones rose sharply and then dropped, whereas the number of residential telephones fell and then rose (the latter is illustrated in Figure 9). This was clearly an artifact of many household telephones being listed without the *r* notation. Perhaps, the directory publisher was lax, or more likely, some temporary change in telephone charges made it advantageous to have a business line. I have not been able to explain this artifact. The result was that only seven residential telephones were listed in the entire 1910 Palo Alto sample! I reexamined all the apparent business listings in the

sample. Based on the head of household's occupation, I judged the plausibility that a telephone in the home was truly a business instrument. I converted 34 apparently miscoded business lines to residential ones, which yielded a diffusion rate more comparable to that of the other towns. This new coding correlated with other variables in ways comparable to the other towns. The recoding did, however, yield a few anomalies, which are represented by the control variable, Business Telephone, in the 1910 Palo Alto equations in Table F–2.

9. I used the SYSAT package (Steinberg, *Logit,* Version 1.0).

10. The 1910 census also included a measure of "class" in a more Marxian vein: worker, "own-account" or employer. This variable added nothing to the explained variance.

11. There may be some error in the measurement of the number of employed persons because of people's reluctance to admit that wives were employed. In any event, it was clearly the number of adults at home, rather than of earners, that heightened the probability of subscribing.

12. The total sample equation, which includes many more notables, does fit better (R^2 = .33 versus .29), but this is probably because the dependent variable is more evenly split.

13. Aside from traits noted in the Table F–3, I also coded ethnicity, using a catalog of last names (E. Smith, *New Dictionary of American Family Names*), but little panned out.

14. Adding an interaction term for Town × Number of Family Adults increases the total model R^2 to .24 and shows that the Adults effect was negligible in Antioch in 1930 and 1936, but this interaction term does not alter the basic interpretations.

15. Using maps and suggestions provided by my research assistants, I defined central Antioch as bordered on the north by the river, the west by Kimball (later, F) Street, the south by Rattan (later, 10th) Street, and the east by the city limits; central Palo Alto by San Francisquito Creek, the state highway, Addison Street, and Cowper Street, inclusive; and central San Rafael by a more complex line that roughly described an ellipse around downtown and the Fourth Street corridor.

APPENDIX G

1. Initially, we also coded boxed ads in two additional issues per year, as well. The analysis reported here is based on the twice-a-year sample drawn for all types of ads. Using the extra issues seems to make little difference to this analysis. We also excluded the following from coding: occasional special pages set aside for out-of-town advertisers (unless an in-town retailer had an ad there); similar collective ads, notably the telephone "Information" ads in the *Times,* which grouped together notices for several retailers who had telephone numbers; and ads by mail-order houses from out-of-town.

2. I estimated intrayear reliability by correlating selected counts and percentages, by year, of the January issues with the May issues. The coefficients varied considerably by specific content and town, especially where the categories had small counts. The correlations of the number of ads containing telephone references in January to the number in May ranged from .81 to .86 for the three

towns; the correlations for the percentages of all ads with telephone references ranged from .87 to .90. More specific categories vary. For example, the correlations of the percentage of all professional cards with telephone references in January to that in May were .77, .56, and .96, for Antioch, Palo Alto, and San Rafael respectively. Similarly, I examined the stability over time of various measures. Again, overall references to the telephone were stable, although some more specific measures were more erratic than others in particular towns.

3. On the procedure, see Fredrick Hartwig, *Exploratory Data Analysis.* I used the SERIES module of Wilkenson, *SYSTAT.*

4. Before making the logit transformations, I reset zero percent to .01 and 100 percent to .99.

5. I also explored using regressions weighted for the number of ads in each year, but that number varies greatly by year and is otherwise collinear. The procedure used for Table G–1 seemed to be simpler.

6. In this analysis, the Palo Alto telephone counts comprise all telephones in the Palo Alto area, Mayfield and Stanford included.

APPENDIX H

1. In retrospect, the latter decision was probably an error, since presumably these advertisers were trying to reach out-of-town customers, rather than those in our three towns.

2. Because in-town merchants' ads on out-of-town news pages are included in the denominators of the percentages, that denominator is larger that it would otherwise be, which in turn lowers the percentage of all ads directed to townsfolk by distant merchants. If the number of pages devoted to news of nearby towns increased over time, this would tend to inflate the apparent increase in the percentage of in-town ads. On the other hand, mail-order ads, which we did not count, declined, partly or wholly offsetting the first bias. The results given in this discussion, of declining out-of-town ads, are also confirmed by using the simple number, rather than percentage, of such ads.

3. For a description of how I estimated the number of cars in the towns, see note 7 of this appendix.

4. To check this interpretation, I divided the sample into pre- and post-1915 and dropped "Years" altogether from the second medical and legal cards equation in Table H–1. Before 1915, business telephones had a strong effect ($b = -.91$, $p < .001$), but they had no effect afterwards ($b = .04$, not significant).

5. For several categories of events discussed in Chapter 7, I correlated the number counted in the second January issue with the number counted in the first May issue. These correlations, ranging from zero (the number of "socials" in Antioch in the third week of January by the number in the first week in May) to over .90 (the number of card games in San Rafael in January with the number in early May), understate consistency, since we actually used four issues per year, not just two. Nevertheless, these are quite variable numbers.

6. For example, the number of club meetings reported in Antioch and Palo Alto increased greatly after World War I; so did the number of events reported in the club columns in the newspapers. Did printing such columns lead to

more reports when, in fact, there was no great increase in meetings—an artifactual change? Or did an increase in club activity lead the editors to expand the columns—a substantive change? Only a close historical study of each newspaper, beyond the scope of this project, could answer that question.

7. Unlike telephones, the number of automobiles could not be easily measured in the three towns. There was no effective count or listing. (Antioch did have a list of cars for tax purposes, but examination of those lists at the Antioch Historical Society suggested that they were incomplete. For example, some residents whose cars were noted in the *Ledger* were not listed as owners in the tax rolls.) Thus, I had to construct some rough estimates. These involved taking (a) the occasional reports of town automobiles in the newspapers; (b) statistics on the number of cars in the counties; and (c) other indicators of automobiles, such as town spending on road work and the number of automobile-related businesses in town. I constructed partial composite time series and correlated them with one another and with functions of year. The final mix of indicators varied by town and yielded crude—but the only available—estimates of automobile diffusion for each town.

8. Population would be a fourth collinear variable. I did not use it, treating it as antecedent to telephones and automobiles.

9. For the percentage of all events that were local from 1928 to 1940, $R^2 = .49$, the effect for automobiles is $b = -.15$ (beta $= -3.6$, $p < .05$) and for telephones is $b = 2.3$ (beta $= 4.8$, $p < .06$). For the percentage of meetings that were local, $R^2 = .66$; automobiles, $b = -2.3$ (beta $= -3.4$, $p < .02$); and telephones, $b = 3.9$ (beta $= 4.9$, $p < .05$). For sports, $R^2 = .04$; automobiles, $b = -3.4$ (beta $= -3.2$, $p < .12$); and telephones, $b = 5.5$ (beta $= 4.4$, $p < .20$). Obviously, the collinearity remains severe and the estimates extreme. Such results can only be hints at possible effects.

10. These percentages by decade are based on unweighted means of the percentages for each year in the decades. Although they pool counts from all three towns, no substantial interaction effects with town appeared in the regression analysis.

Bibliography

ABBREVIATIONS

AT&THA   American Telephone & Telegraph, Historical Archives, Warren, New Jersey

IBIC   Illinois Bell Information Center, Chicago, Illinois

MCHS   Marin County Historical Society, San Rafael, California

MIT   Museum of Independent Telephony, Archives, Abilene, Kansas

PAHS   Palo Alto Historical Society, Palo Alto, California

TPCM   Telephone Pioneer Communications Museum of San Francisco, Archives and Historical Research Center, San Francisco, California

Abbott, Philip. 1987. *Seeking Many Inventions: The Idea of Community in America*. Knoxville: University of Tennessee Press.

Abler, R. 1975. "Effects of Space-Adjusting Technologies on the Human Geography of the Future." In *Human Geography in a Shrinking World,* edited by R. Abler, D. Janelle, A. Philbrick, and J. Sommer, 35–36. Belmont, CA: Duxbury Press.

Abler, R., and T. Falk. 1981. "Public Information Services and the Changing Role of Distance in Human Affairs." *Economic Geography* 57 (1): 10–22.

Allen, F. R. 1957. "The Automobile." In *Technology and Social Change,* edited by F. R. Allen et al., 107–32. New York: Appleton-Century Crofts.

Allred, C. E., T. L. Robinson, B. H. Luebke, and S. R. Nesraug. 1937. *Rural Cooperative Telephones in Tennessee*. Rural Research Series Monograph no. 45. Agricultural Experiment Station, University of Tennessee, Knoxville.

American Telephone and Telegraph Company (AT&T). 1908–1921. *Annual Reports of the Directors of American Telephone and Telegraph Company to the Stockholders.* Boston: AT&T.

———. 1932a. "Selected Statistics Relative to Residence and Farm Markets." Chief Statistician's Division (November). Mimeo courtesy of AT&THA.

———. 1932b. Untitled Report. Chief Statistician's Division (27 July). Mimeo courtesy of AT&THA.

———. 1940. "Families and Residence Telephones." Chief Statistician's Division (18 December). Mimeo courtesy of AT&THA.

———. 1941. "Rural Population and Telephones." Chief Statistician's Division (13 June). Mimeo courtesy of AT&THA.

———. 1979a. *A Capsule History of the Bell System.* New York: AT&T.

———. 1979b. *Events in Telecommunications History.* New York: AT&T.

Anderson, G. W. 1906. *Telephone Competition in the Middle West and Its Lesson for New England.* Boston: New England Telephone and Telegraph.

Andrew, Seymour L. c. 1930. "The New America." In "Talks and Papers, Seymour L. Andrew, 1922–1939." Box 2043, AT&THA.

Andrews, Edmund L. 1991. "Regulators Moving to Break Local Monopolies." *New York Times,* 27 December: 1, D3.

Antrim, M. T. 1909. "Outrages of the Telephone." *Lippincott's Magazine* 84 (July): 125–26.

Applebaum, Diana K. 1989. *The Glorious Fourth: An American Holiday, An American History.* New York: Facts on File.

Archer, Margaret A. 1985. "The Myth of Cultural Integration." *British Journal of Sociology* 36 (3): 333–53.

Ariès, Philippe. 1979. "The Family and the City in the Old World and New." In *Changing Images of the Family,* edited by V. Tufte and B. Meyerhoff, 29–42. New Haven, CT: Yale University Press.

Arlen, Michael J. 1980. *Thirty Seconds.* New York: Farrar, Strauss & Giroux.

Armstrong, Christopher, and H. V. Nelles. 1986. *Monopoly's Moment: The Organization and Regulation of Canadian Utilities, 1830–1930.* Philadelphia: Temple University Press.

Arnold, Eleanor, ed. 1985. *Party Lines, Pumps and Privies: Memories of Hoosier Homemakers.* Indianapolis: Indiana Extension Homemakers Association.

Arnold, Joseph L. 1979. "The Neighborhood and City Hall: The Origin of Neighborhood Associations in Baltimore, 1880–1911." *Journal of Urban History* 6 (November): 3–30.

Aronson, Sidney H. 1971. "The Sociology of the Telephone." *International Journal of Comparative Sociology* 12 (September): 153–67.

————. 1977a. "Bell's Electrical Toy: What's the Use? The Sociology of Early Telephone Usage." In *The Social Impact of the Telephone,* edited by Ithiel de Sola Pool, 15–39. Cambridge, MA: The MIT Press.

————. 1977b. "*Lancet* on the Telephone, 1876–1975," *Medical History* 21 (January): 69–87.

Aronson, Sidney H., and R. Greenbaum. 1971. "Prospectus for a Study of the History and Consequences of the Telephone in America." Queen's College, New York. Typescript courtesy of Sidney H. Aronson, 1985.

————. n.d. "Take Two Aspirin and Call Me in the Morning: Doctors, Patients and the Telephone." Queens College, New York. Typescript courtesy of Sidney H. Aronson.

Asmann, Edwin A. 1980. "The Telegraph and the Telephone: Their Development and Role in the Economic History of the United States." Typescript, Lake Forest College.

Atherton, Lewis. [1954]. 1966. *Main Street on the Middle Border.* Bloomington: University of Indiana Press. Reprint. Chicago: Quandrangle Books.

Attali, J., and Y. Stourdze. 1976. "The Birth of the Telephone and Economic Crisis: The Slow Death of Monologue in French Society." In *The Social Impact of the Telephone,* edited by I. de S. Pool, 97–111. Cambridge: MIT Press.

Attman, Artur, Jan Kuuse, and Ulf Olsson. 1977. *L M Ericsson 100 Years,* Volume I: *The Pioneering Years.* Stockholm: L M Ericsson.

Atwood, Roy Alden. 1984. "Telephony and Its Cultural Meanings in Southeastern Iowa, 1900–1917." Ph.D. diss., University of Iowa.

Automobile Manufacturers' Association. 1941. *Automobile Facts and Figures.* 23d ed. New York: Automobile Manufacturers' Association.

Ayer, Harriet H., and Lillie d'A. Berghe, et al. 1906. *Correct Social Usage by Eighteen Distinguished Authors,* 6th Ed. New York: New York Society of Self-Culture.

Balderston, Lydia Ray. 1921. *Housewifery.* Philadelphia: J.P. Lippincott.

Baldwin, F. G. C. 1938. *The History of the Telephone in the United Kingdom.* London: Chapman at Hall.

Ball, D. W. 1968. "Toward a Sociology of Telephones and Telephoners." In *Sociology and Everyday Life,* edited by M. Truzzi, 59–75. Englewood Cliffs, NJ: Prentice Hall.

Bardou, J. P., J. J. Chanaron, P. Fridenson, and J. M. Laux. 1982. *The Automobile Revolution: The Impact of an Industry.* Translated by J. M. Laux. Chapel Hill: University of North Carolina Press.

Bare, W. E. 1930. "Service Outside the Base Rate Area." *Bell System General Commercial Conference, 1930.* Microfiche 368B, IBIC.

Barker, W. H. 1910. "Operating Rural Lines." *Telephony* 19 (5 March): 286.

Barnett, William P. 1988. "The Organizational Ecology of the Early Telephone Industry." Ph.D. diss., School of Business, University of California, Berkeley.

Barrett, R. T. 1931. "Selling Telephones to Farmers by Talking about Tomatoes." *Printers' Ink* (5 November): 49–50.

———. 1940. "The Telephone as a Social Force." *Bell Telephone Quarterly* 19, no. 2 (April): 129–38.

Barsantee, Harry. 1926. "The History and Development of the Telephone in Wisconsin." *Wisconsin Magazine of History* 10 (December): 150–63.

Bartky, Ian R. 1989. "The Adoption of Standard Time." *Technology and Culture* 30 (January): 25–56.

Baxter, Lucia Allen. 1913. *Housekeeper's Handy Book*. New York: Houghton Mifflin.

Bean, Walter. 1952. *Boss Reuf's San Francisco*. Berkeley and Los Angeles: University of California Press.

Belascoe, Warren J. 1979. *Americans on the Road: From Autocamp to Motel, 1910–1945*. Cambridge, MA: The MIT Press.

———. 1982. "Cars versus Trains: 1980 and 1910." In *Energy and Transport*, edited by G. H. Daniels and M. H. Rose, 39–53. Beverly Hills, CA: Sage.

Bell Canada. 1980. *The First Century of Service*. Montreal: Bell Canada.

Bell Telephone Company of Canada. c. 1928. "Selling Service on the Job." Cat. #12223. Montreal: Bell Canada Historical.

Bell Telephone Quarterly. 1928. "The World's Telephones, 1927," 218–30.

Benham, Georgina Corry. 1891. *Polite Life and Etiquette*. Chicago: Louis Benham Co.

Bender, Thomas. 1978. *Community and Social Change in America*. New Brunswick, NJ: Rutgers University Press.

Bennett, Arnold. 1912. "Your United States." *Harper's Monthly* 125 (July): 191–202.

Benyo, Elise Schott. 1972. *Antioch to the Twenties*. Antioch: Antioch Unified School District.

Berger, Michael L. 1979. *The Devil Wagon in God's Country: The Automobile and Social Change in Rural America, 1883–1929*. Hamden, CT: Archon Books.

Berger, Peter. 1979. *The Heretical Imperative*. Garden City, NY: Doubleday.

Bertho, Catherine. 1981. *Télégraphes et Téléphones*. Paris: Livres de Poche.

Berthoff, Rowland. 1971. *An Unsettled People*. New York: Harper & Row.

Bijker, Wiebe E., Thomas P. Hughes, and Trevor Pinch, eds. 1987. *The Social Construction of Technological Systems*. Cambridge, MA: The MIT Press.

Blalock, H. W. 1940. "Streamlining Rural Telephone Service." *Public Utilities Fortnightly* 26 (10 October): 466–73.

Blumin, Stuart M. 1976. *The Urban Threshold*. Chicago: University of Chicago Press.

Bohakel, Charles A. 1984. *Historic Tales of East Contra Costa County*. Vol. 1. Antioch, CA: Printed by author.

Boorstin, D. J. 1973. *The Americans: The Democratic Experience*. New York: Vintage.

Borgmann, Albert. 1984. *Technology and the Character of Contemporary Life: A Philosophical Inquiry*. Chicago: University of Chicago Press.

Borman, R. R. 1936. "Survey Reveals Telephone as a Money Saver on Farm." *Telephony* 111 (11 July): 9–13.

Bornholz, Robert, and David S. Evans. 1983. "The Early History of Competition in the Telephone Industry." In *Breaking Up Bell*, edited by David S. Evans, 7–40. New York: North-Holland.

Bottles, Scott L. 1987. *Los Angeles and the Automobile: The Making of the Modern City*. Berkeley and Los Angeles: University of California Press.

Boudon, Raymond. 1983. "Why Theories of Social Change Fail: Methodological Thoughts." *Public Opinion Quarterly* 47 (Summer): 143–60.

Boyd, R. C., J. D. Howard, Jr., and L. Pedersen. 1957. "A New Carrier System for Rural Service." *Bell System Technical Journal* 3 (March): 349–90.

Brandon, B., ed. 1982. *The Effect of the Demographics of Individual Households on Their Telephone Usage*. Cambridge, MA: Ballinger.

Braverman, H. 1974. *Labor and Monopoly Capital*. New York: Monthly Review Press.

Brilliant, A. E. 1965. "Some Aspects of Mass Motorization in Southern California, 1919–1929." *Southern California Quarterly* 47 (June): 191–208.

Brock, G. W. 1981. *Telecommunications Industry: The Dynamics of Market Structure*. Cambridge, MA: Harvard University Press.

Brock, William R. 1988. *Welfare, Democracy, and the New Deal*. New York: Cambridge University Press.

Brooks, John. 1976. *Telephone: The First Hundred Years*. New York: Harper & Row.

———. 1977. "The First and Only Century of Telephone Literature." In *The social Impact of the Telephone*, edited by Ithiel de Sola Pool, 208–24. Cambridge, MA: The MIT Press.

Brown, Bliss. 1936. *Marin County History*. Marin County, CA: Works Projects Administration. Typescript, Marin County Historical Society.

Brown, D. C. 1980. *Electricity for Rural America: The Fight for the R.E.A.* Contributions in Economics and Economic History no. 29. Westport, CT: Greenwood Press.

———. 1982. "North Carolina Rural Electrification: Precedent of the R.E.A." *North Carolina Historical Review* 59 (April): 109–24.

Brown, Eric. 1986. "Minitel: The French Find the Key to Videotex." *PC World* (November): 335–52.

Brown, Richard D. 1976. *Modernization: The Transformation of American Life, 1600–1865.* New York: Hill & Wang.

————. 1989. *Knowledge is Power: The Diffusion of Information in Early America, 1700–1865.* New York: Oxford University Press.

Brownell, B. A. 1972. "A Symbol of Modernity: Attitudes Toward the Automobile in Southern Cities in the 1920s." *American Quarterly* 24 (March): 20–44.

Brunner, E. de S., and J. K. Kolb. 1933. *Rural Social Trends.* New York: McGraw-Hill.

Buchanan, R. A. 1991. "Theory and Narrative in the History of Technology." *Technology and Culture* 32 (April): 365–76.

Buchner, Bradley Jay. 1988. "Social Control and the Diffusion of Modern Telecommunications Technologies: A Cross-National Study." *American Sociological Review* 53 (June): 446–53.

Builta, F. C. c.1930. "How Can Publicity Assist in Meeting Our Responsibility to the Rural Telephone User?" In "Publicity Conferences of the Bell System, 1921–1934," 73–82. Box 1310, AT&THA.

Bureau of Business Research. 1928. "The Automobile and the Merchant." Bulletin no. 19. College of Commerce and Business Administration, University of Illinois, Champaign, IL.

Busch, Jane. 1983. "Cooking Competition: Technology on the Domestic Market in the 1930s." *Technology and Culture* 24 (April): 222–41.

Calhoun, Craig. 1986. "Computer Technology, Large-Scale Social Integration, and the Local Community." *Urban Affairs Quarterly* 22 (December): 329–49.

California Public Utilities Commission. 1960. "Development of Telephone Service in Rural Areas, 1944–1960." Special Study no. 448. Sacramento: State of California Archives.

Canada, Department of Trade and Commerce. 1925–1940. "Telephone Statistics for 19___." Ottawa: Dominion Bureau of Statistics.

Card, Fred W. 1912. "Cooperative Fire Insurance and Telephones." In L. H. Bailey (Ed.), *Cyclopedia of American Agriculture*, edited by L. H. Bailey, 303–6. Vol. 4. 4th ed. New York: MacMillan.

Carey, J. W. 1983. "Technology and Ideology: The Case of the Telegraph." In *Prospects, An Annual of American Cultural Studies*, edited by J. Salzman, 303–25. New York: Cambridge University Press.

Carr, L. J. 1932. "How the Devil-Wagon Came to Dexter." *Social Forces* 11 (October): 64–70.

Carroll, Raymonde. 1988. *Cultural Misunderstandings: The French-American Experience.* Translated by Carol Volk. Chicago: University of Chicago Press.

Carruthers, K. L. 1945 [1974]. "Rural Telephone Development." Annual Proceedings of the Telephone Association of Canada. Reprinted in *Telephone Oper-*

ation and Development in Canada, 1921–1971, edited by Peter S. Gant, 110–15. Toronto: Telephone Association of Canada.

Carty, J. J. 1922. "Ideals of the Telephone Service." *Bell Telephone Quarterly* 1, no.3 (October): 1–11.

Cashman, Tony. 1972. *Singing Wires: The Telephone in Alberta.* Edmonton: Alberta Government Telephone Commission.

Casson, Herbert N. 1910. *The History of the Telephone.* Chicago: A. C. Mc-Clurg.

Central Union Telephone Company. 1904. *Instructions and Information for Solicitors.* Chicago: Contracts Department, Central Union Telephone Company, IBIC.

Chambers, Clarke A. 1963. *Seedtime of Reform.* Minneapolis: University of Minnesota Press.

Chamber's Journal. 1899. "The Telephone." 76: 310–13.

Chant, A. E. 1923 [1974]. "The Value of Rural Development to Exchange and Toll Earnings." Annual Proceedings of the Telephone Association of Canada. Reprinted in *Telephone Operation and Development in Canada, 1921–1971,* edited by Peter S. Gant, 49–54. Toronto: Telephone Association of Canada.

Chapius, Robert J. 1982. *100 Years of Telephone Switching.* New York: Elsevier.

Charon, Jean-Marie. 1987. "Videotext: From Interaction to Communication." Translated by Jonathan Davis. *Media, Culture and Society* 9 (July): 301–32.

Cheape, C. W. 1980. *Moving the Masses: Urban Public Transit in New York, Boston, and Philadelphia, 1880–1912.* Cambridge, MA: Harvard University Press.

Cherlin, Andrew, and Frank F. Furstenberg, Jr. 1986. *The New American Grandparent.* New York: Basic Books.

Childs, M. 1952. *The Farmer Takes a Hand: The Electric Power Revolution in Rural America.* Garden City, NY: Doubleday.

Claisse, Gérard, and Frantz Rowe. 1988. "The Telephone in Question: Questions on Communication." *Computer Networks and ISDN Systems* 14 (2-5): 207–19.

Clark, D. and K. I. Unwin. 1981. "Telecommunications and Travel: Potential Impact in Rural Areas." *Regional Studies* 15: 47–56.

Clarke, Alfred C. 1952. "An Examination of the Operation of Residential Propinquity as a Factor in Mate Selection." *American Sociological Review* 17 (February): 17–23.

Clymer, Floyd. 1955. *Henry's Wonderful Model T, 1908–1927.* New York: Bonanza Books.

Coates, V. T., and B. Finn. 1979. *A Retrospective Technology Assessment: Submarine Telegraph: The Transatlantic Cable of 1866.* San Francisco: San Francisco Press.

Cohen, Jeffery E. 1991. "The Telephone Problem and the Road to Telephone Regulation in the United States, 1876–1917." *Journal of Policy History* 3(1) 42–69.

Collins, Robert. 1977. *A Voice from Afar: The History of Telecommunications in Canada*. Toronto: McGraw-Hill Ryerson.

Colton, F. Barrows. 1937. "The Miracle of Talking by Telephone." *National Geographic* 72 (October): 395–433.

Congressional Budget Office. 1984. *Local Telephone Rates: Issues and Alternatives*. Staff Working Paper (January). Washington: Congress of the United States.

Consumer Reports. 1984. "Will Your Phone Rates Double?" (March): 154–56.

Conzen, Kathleen Neils. 1980. "Community Studies, Urban History, and American Local History." In *The Past before Us*, edited by Michael Kammen, 270–91. Ithaca: Cornell University Press.

Cook, T. T. 1928. "Advertising of the American Telephone and Telegraph Company." In "Publicity Conference, Bell System, 1921–1934." Box 1310, AT&THA.

Cooke, Maude C. 1896. *Manual of Etiquette*. Cincinnati: Lyons Bros.

Coon, H. 1939. *American Tel & Tel: The Story of a Great Monopoly*. New York: Longman, Green and Co.

Corn, Joseph J. 1983. *The Winged Gospel: America's Romance with Aviation, 1900–1950*. New York: Oxford University Press.

Cowan, Neil M., and Ruth Schwartz Cowan. 1989. *Our Parents' Lives*. New York: Basic Books.

Cowan, Ruth Schwartz. 1983. *More Work for Mother*. New York: Basic Books.

———. 1987. "The Consumption Junction: A Proposal for Research Strategies in the Sociology of Technology." In *The Social Construction of Technology*, edited by W. E. Bijker, T. P. Hughes, and T. Pinch, 261–80. Cambridge, MA: The MIT Press.

Crandall, Robert W. 1987. "Has the AT&T Break-up Raised Telephone Rates?" *The Brookings Review* 5 (Winter): 37–44.

Cross, Ira B. 1927. *Banking in California*. Vol. 4. San Francisco: SJ Clarke Pub Co.

Czitrom, Daniel J. 1983. *Media and the American Mind: From Morse to McLuhan*. Chapel Hill, NC: University of North Carolina Press.

Dale, Daphne. 1891. *Our Manners and Social Customs*. Chicago: Elliott & Beezley.

Danielian, N. R. 1939. *A.T.&T.: The Story of Industrial Conquest*. New York: Vanguard.

Daniels, G. H. 1970. "The Big Questions in the History of American Technology." *Technology and Culture* 11 (January): 1–21.

Davis, Donald Finlay. 1986. "Dependent Motorization: Canada and the Automobile to the 1930s." *Journal of Canadian Studies* 21 (Fall): 106–32.

———. 1988. *Conspicuous Production: Automobiles and Elites in Detroit, 1899–1933*. Philadelphia: Temple University Press.

Davis, Susan. 1990. "After the Beep...," *San Francisco Examiner* (11 February), D–19, D–20.

Dettlebach, Cynthia Golomb. 1976. *In the Driver's Seat: The Automobile in American Literature and Popular Culture.* Westport, CT: Greenwood Press.

Dierkx, K. W. 1983. "San Leandro, California: Technology and Community: The Telephone and Electric Railway." Undergraduate paper. Department of Sociology, University of California, Berkeley.

di Leanardo, Micaela. 1984. *The Varieties of Ethnic Experience.* Ithaca, NY: Cornell University Press.

Dillon, Richard. 1982. *Delta Country.* Novato, CA: Presidio Press.

Dilts, M. M. 1941. *The Telephone in a Changing World.* New York: Longman, Green and Co.

Dobbs, A. E. 1903. "Are Country Telephone Lines Profitable?" *Telephony* 5 (3 April): 248–49; 5 (3 May): 296–98; 5 (3 June): 343–44 .

Donnelly, Florence. 1970. "A History of San Rafael." *Marin Independent-Journal Marin Magazine* 10, 24, and 31 January.

Dordick, Herbert S. 1983. "Social Uses for the Telephone." *Intermedia* 1 (May): 31–35.

———. 1989. "The Social Uses of the Telephone—an U.S. Perspective." In *Telefon und Gesellschaft, Band 1: Beiträge zu einer Soziologie der Telefonkommunikation,* Forschungsgruppe Telefonkommunikation (Hrsg), 221–38. Berlin: Volker Speiss.

———. 1990. "The Origins of Universal Service." *Telecommunications Policy* 14 (June): 223–31.

Douglas, Susan J. 1987. *Inventing American Broadcasting, 1899–1922.* Baltimore: Johns Hopkins University Press.

DuBoff, R. B. 1983. "The Telegraph and the Structure of Markets in the United States, 1845–1850." *Research in Economic History* 8: 253–77.

Duis, Perry R. 1983. *The Saloon: Public Drinking in Chicago and Boston, 1880–1900.* Urbana: University of Illinois Press.

Dykstra, Robert R., and William Silag. 1987. "Doing Local History: Monographic Approaches to the Smaller Community." In *American Urbanism: A Historiographical Review,* edited by Howard Gillette, Jr., and Zane L. Miller, 293–305. New York: Greenwood.

Easterlin, R. 1957. "State Income Estimates." In *Population Distribution and Economic Growth in the United States, 1870–1950,* edited by S. S. Kuznets and D. S. Thomas, 729–59. Philadelphia: American Philosophical Society.

Edwards, Alba M. 1933. "A Social-Economic Grouping of the Gainful Workers of the United States." *Journal of the American Statistical Association* 28 (December): 1–11.

Edwards, W. B. 1923. "Tearing Down Old Copy Gods." *Printers' Ink* 123 (26 April); 65–66.

Egido, Carmen. 1989. "Home Information Services: Why They Have Failed and What Are the Prospects for Success." Paper presented to the American Sociological Association, San Francisco. Morristown, NJ: Bellcore.

Eichler, Lillian. 1921. *Book of Etiquette*. Vol. 2. Oyster Bay, NY: Doubleday.

Ellis, C. T. 1966. *A Giant Step*. New York: Random House.

Emanuels, George. 1986. *Contra Costa: An Illustrated History*. Fresno, CA: Panorama West Books.

Epstein, Ralph C. 1928. *The Automobile Industry: Its Economic and Commercial Development*. Chicago: A. W. Shaw.

Falk, Thomas, and Ronald Abler. 1980. "Intercommunications, Distance, and Geographical Theory." *Geografisk Annaler*, 62B (2): 59–67.

Fancher, A. 1931. "Every Employee is a Salesman for American Telephone and Telegraph." *Sales Management* 28, no.13 (26 February): 450–51, 472.

Faris, H. N. 1926. "Some Lessons Learned from the Past." *Telephony* (22 May): 17.

Farrell, F. D. 1955. *Kansas Rural Institutions: IX, Rural Telephone Company*. Agriculture Experiment Station Circular 326. Kansas State College of Agriculture, Manhattan.

Federal Communications Commission. 1938. *Proposed Report: Telephone Investigation*. Washington: Government Printing Office.

———. 1939. *Investigation of the Telephone Industry: The United States*. Washington: Government Printing Office.

———. 1944. *Preliminary Studies on Some Aspects of the Availability of Landline Wire Communications Service*. Accounting, Statistical, and Tariff Department, Economics Division. Washington, D.C.: Federal Communications Commission, October.

Federal Writers' Project. 1939. *These Are Our Lives*. Chapel Hill, NC: University of North Carolina Press.

Fenwick, Millicent. 1948. *Vogue's Book of Etiquette*. New York: Simon and Schuster.

Field Research Corporation. 1985. *Residence Customer Usage and Demographic Characteristics Study: Summary*. Conducted for Pacific Bell. Courtesy of R. Somer, Pacific Bell.

Fischer, Claude S. 1982a. "The Dispersion of Kinship Ties in Modern Society: Contemporary Data and Historical Speculation." *Journal of Family History* 7 (Winter): 353–750.

———. 1982b. *To Dwell among Friends: Personal Networks in Town and City*. Chicago: University of Chicago Press.

———. 1985. "Studying Technology and Social Life." In *High Technology, Space, and Society*, edited by M. Castells, 284–301. Beverly Hills, CA: Sage.

————. 1987a. "The Revolution in Rural Telephony, 1900–1920." *Journal of Social History* 21 (Fall): 5-26.

————. 1987b. "Technology's Retreat: The Decline of Rural Telephony, 1920–1940," *Social Science History* 11 (Fall): 295–327.

————. 1987c. "'Touch Someone': The Telephone Industry Discovers Sociability." *Technology and Culture* 29 (January): 32–61.

————. 1988. "Gender and the Residential Telephone, 1890–1940: Technologies of Sociability." *Sociological Forum* 3 (2): 211–33.

————. 1990. "Changes in Leisure Activities in Three Towns, 1890–1940." Paper presented to the American Sociological Association, Washington, August.

————. 1991a. "Ambivalent Communities: How Americans Understand Their Localities." In *America at Century's End*, edited by A. Wolfe, 79–92. Berkeley and Los Angeles: University of California Press.

————. 1991b. "Community Cohesion and Conflict, 1890–1940." Paper presented to the American Sociological Association, Cincinnati, August.

Fischer, Claude S., and Glenn Carroll. 1988. "Telephone and Automobile Diffusion in the United States, 1902–1937." *American Journal of Sociology* 93 (March): 1153–78.

Fischer, Claude S., and Stacey Oliker. 1983. "A Research Note on Friendship, Gender, and the Lifecycle." *Social Forces* 62 (September): 124–33.

Fischer, Claude S., Robert Max Jackson, C. Ann Stueve, Kathleen Gerson, Lynne McCallister Jones, with Mark Baldassare. 1977. *Networks and Places: Social Relations in the Urban Setting*. New York: Free Press.

Fischer, D. H. 1970. *Historians' Fallacies*. New York: Harper & Row.

Fishback, Margaret. 1938. *Safe Conduct*. New York: World Publishing.

Flink, James J. 1970. *America Adopts the Automobile, 1895–1910*. Cambridge, MA: The MIT Press.

————. 1976. *The Car Culture*. Cambridge, MA: The MIT Press.

————. 1980. "*The Car Culture* Revised: Some Comments on the Recent Historiography of Automotive History." Special issue of *Michigan Quarterly Review* 19(4)–20(2): 772–78.

————. 1989. *The Automobile Age*. Cambridge, MA: The MIT Press.

Forschungsgruppe Telefonkommunikation (Hrsg). 1989. *Telefon und Gesellschaft, Band 1: Beiträge zu einer Soziologie der Telefonkommunikation*. Berlin: Volker Speiss.

Foster, M. S. 1981. *From Streetcar to Super Highway: American City Planners and Urban Transportation*. Philadelphia: Temple University Press.

Fox, S. 1984. *The Mirror Makers: A History of American Advertising and Its Creators*. New York: William Morrow.

Fredrick, Christine. 1919. *Household Engineering: Scientific Management in the Home*. Chicago: American School of Home Economics.

Frisch, Michael. 1972. *Town into City: Springfield, Massachusetts, and the Meaning of Community, 1840–1880*. Cambridge, MA: Harvard University Press.

Fuhr, Joseph P., Jr. 1990. "Telephone Subsidization of Rural Areas in the USA." *Telecommunications Policy* 14 (June): 183–88.

Fullarton, D. C., and F. D. Goss, eds. 1979. *NTCA: Our First Twenty-five Years*. Washington, D.C.: National Telephone Cooperative Association.

Gabel, David. 1988. "Where Was the White Knight When the Competition Needed One?" Department of Economics, Queens College, New York. Typescript.

Gallaher, Art, Jr. 1961. *Plainville Fifteen Years Later*. New York: Columbia University Press.

Gardner, Dorothy. 1974. "Newspapers of San Rafael." *San Rafael Independent-Journal* (23 March).

Garnet, Robert W. 1979. "The Central Union Telephone Company." Box 1080, AT&THA.

———. 1985. *The Telephone Enterprise: The Evolution of the Bell System's Horizontal Structure, 1876–1909*. Baltimore: Johns Hopkins University Press.

Gary, T. 1928. "The Independents and the Industry." *Telephony* 90 (13 March): 14–18.

Gerstein, Penny. 1990. "Working at Home in the Live-in Office." Ph.D. diss. Department of City and Regional Planning, University of California, Berkeley.

Gherardi, B. and F. B. Jewett. 1930. "Telephone Communication System of the United States." *The Bell System Technical Journal* (January) 1–100.

Giedion, Siegfried. 1948. *Mechanization Takes Command: A Contribution to Anonymous History*. New York: W. W. Norton.

Gifford, Walter S. 1944. "An Address." *Bell Telephone Quarterly* 23 (Autumn): 137–46.

Gilbreth, Lillian. 1927. *The Home-Maker and Her Job*. New York: D. Appleton.

Glauber, R. H. 1978. "The Necessary Toy: The Telephone Comes to Chicago." *Chicago History* 7 (2): 70–86.

Glover, Ellye Howell. 1909. *"Dame Curtsey's" Book of Etiquette*. Chicago: A. C. McClung & Co.

Goist, P. D. 1977. *From Main Street to State Street*. Port Washington, NY: Kennikat.

Gottman, J. 1977. "Metropolis and Antipolis: The Telephone and the Structure of the City." In *The Social Impact of the Telephone*, edited by Ithiel de Sola Pool, 303–17. Cambridge, MA: The MIT Press.

Gray, W. F. 1924. "Typical Schedules of Rates for Exchange Service." In *Bell System General Commercial Engineers' Conference, 1924,* New York City. Microfiche 364B, IBIC.

Greene, Sally. 1987. "Operator—Could You Please Ring? A History of Rural Telephone Service to Kendrick and Juiliaetta, Idaho." *Idaho Yesterdays* 31 (Fall): 2–10.

Griese, Noel L. 1977. "AT&T: 1908 Origins of the Nation's Oldest Continuous Institutional Advertising Campaign." *Journal of Advertising* 6 (Summer): 18–24.

Griswold, George J. 1967. "How AT&T Public Relations Policies Developed." *Public Relations Quarterly* 12 (Fall): 7–16.

Gullard, Pamela, and Nancy Lund. 1989. *History of Palo Alto: The Early Years.* San Francisco: Scottwall Associates.

Gusfield, Joseph. 1975. *Community.* Oxford: Basil Blackwell.

Haire, A. P. 1907. "The Telephone in Retail Business." *Printers' Ink* 61 (27 November): 3–8.

————. 1910. "Bell Encourages Shopping by Telephone." *Printers' Ink* 70 (19 January).

Hall, E. J. 1884. "Notes on History of the Bell Telephone Co. of Buffalo, New York." Box 1141, AT&THA.

Hall, Florence Howe. 1914. *Good Form for All Occasions.* New York: Harper & Bros.

Hall, Trish. 1989. "With Phones Everywhere, Everyone is Talking More." *New York Times* (11 October): 1.

Haltman, Kenneth. 1990. "Reaching out to Touch Someone? Reflections on a 1923 Candlestick Telephone." *Technology in Society* 12 (3): 333–54.

Hanselman, J. J., and H. S. Osborne. 1944–1945. "More and Better Telephone Service for Farmers." *Bell Telephone Magazine* (Winter): 213–26.

Harger, C. M. 1907. "The Automobile in Western Country Towns." *World Today* 13 (December): 1277–79.

Harrell, J. E. 1931. "Residential Exchange Sales in the New England Southern Area." In *Bell System General Commercial Conference on Sales Matters, June.* Microfiche 368B, IBIC.

Harriman, Mrs. Oliver. 1942. *Book of Etiquette.* New York: Greenberg.

Hartwig, Fredrick, with Brian E. Dearing. 1979. *Exploratory Data Analysis.* Sage University Paper Series on Quantitative Applications in the Social Sciences, no. 07–016. Beverly Hills, CA: Sage.

Hawkins, Richard, and J. Greg Getz. 1986. "Women and Technology: The User's Context of the Automobile." Paper presented to the American Sociological Association, New York City, (August).

Heer, Clarence. 1933. "Trends in Taxation and Public Finance." In *President's Commission on Recent Social Trends,* 1331–90. New York: McGraw-Hill.

Helms, W. C. 1924. "The Relation between Telephone Development and Growth and General Economic Conditions." In *Bell System General Commercial Engineers' Conference, 1924.* Microfiche 364B, IBIC. (Also published in *Bell Telephone Quarterly* 4 (January 1925): 8–21.)

Hershberg, Theodore, and Robert Dockhorn. 1976. "Occupational Classification." *Historical Methods Newsletter* 9 (March/June): 59–98.

Hibbard, A. 1941. *Hello-Goodbye: My Story of Telephone Pioneering.* Chicago: A. C. McClurg Co.

Hine, R. V. 1980. *Community on the American Frontier.* Norman, OK: University of Oklahoma.

History of Contra Costa County. 1926. Los Angeles: Historic Records Co.

Hohlmayer, Earl. 1991. *Looking Backward: Tales of Old Antioch and Other Places.* Visalia, CA: Jostens Publishing; distributed by E & N Hohlmayer, Antioch, CA.

Holcombe, A. N. 1911. *Public Ownership of Telephones on the Continent of Europe.* Cambridge, MA: Harvard University Press.

Holland, J. D. 1938. "Telephone Service Essential to Progressive Farm Home." *Telephony* 114 (19 February): 17–20.

Holt, Emily. 1901; 1923. *Encyclopedia of Etiquette.* Garden City, NY: Doubleday and Page.

Horowitz, Daniel. 1985. *The Morality of Spending: Attitudes Toward the Consumer Society in America, 1875–1940.* Baltimore: Johns Hopkins University Press.

Horton, Tom. 1987. "Against All Odds." *San Francisco Focus* 34, no. 5 (May): 50–54; 89–97.

Houghtelling, Leila. 1927. *The Income and Standard of Living of Unskilled Laborers in Chicago.* Chicago: University of Chicago Press.

Hounshell, David. A. 1975. "Elisha Gray and the Telephone: On the Disadvantages of Being an Expert." *Technology and Culture* 16 (April): 133–61.

———. 1984. *From the American System to Mass Production, 1900–1932.* Baltimore: Johns Hopkins University Press.

Howe, F. L. 1979. *This Great Contrivance: The First Hundred Years of the Telephone in Rochester.* Rochester, NY: Rochester Telephone Corp.

Hower, R. M. 1949. *The History of an Advertising Agency: N. W. Ayer & Son at Work, 1869–1949.* Cambridge, MA: Harvard University Press.

Hoyt, D. R., and N. Babchuck. 1983. "Adult Kinship Networks." *Social Forces* 62 (September): 84–101.

Hudson, H. D., D. Goldschmidt, E. B. Parker, and A. Hardy. 1979. *The Role of Telecommunications in Socio-Economic Development: A Review of the Literature with Guidelines for Further Investigation.* Keewatin Communications Group.

Hughes, Thomas P. 1987. "The Evolution of Large Technological Systems." In *The Social Construction of Technological Systems,* edited by Wiebe E. Bijker, Thomas P. Hughes, and Trevor Pinch, 51–82. Cambridge, MA: The MIT Press.

―――. 1989. *American Genesis: A Century of Invention and Technological Enthusiasm.* New York: Penguin.

Hugill, P. J. 1982. "Good Roads and the Automobile in the United States." *Geographical Review* 72 (July): 327–49.

Hunter, Albert. 1975. "The Loss of Community: An Empirical Test through Replication." *American Sociological Review* 40 (October): 537–52.

―――. 1979. "The Persistence of Local Sentiments in Mass Society." In *The Handbook of Contemporary Urban Life,* edited by D. Street et al., 133–62. San Francisco: Jossey-Bass.

Hurt, R. Douglas. 1986. "REA: New Deal for Farmers." *Timeline* (December): 320–47.

Hyde, Anne F. 1990. "From Stagecoach to Packard Twin-Six: Yosemite and the Changing Face of Tourism." *California History* 69 (Summer): 154–69.

Illinois Telephone and Telegraph Company. 1931. *Sales Manual.* Chicago: Illinois Bell Commercial Department. Microfiche, IBIC.

Independent Telephone Institute, Inc. 1945. "The Farm Telephone Story." Chicago: Independent Telephone Institute, Inc. Photocopy from MIT.

Interrante, J. 1979. "You Can't Go to Town in a Bathtub: Automobile Movement and the Reorganization of Rural American Space, 1900–1930." *Radical History Review* 21 (Fall): 151–168.

―――. 1980. "The Road to Autopia: The Automobile and the Spatial Transformation of American Culture." Special issue of *Michigan Quarterly Review* 19(4)–20(2): 502–17.

Inter-University Consortium for Political and Social Research. 1986. "Cost of Living in the United States, 1917–1919," Study no. 8299. Ann Arbor, MI.

Iowa State Planning Board. 1935. "Telephone Communication in Iowa." *Telephony* 109 (2 November): 7–11.

Jackson, Kenneth C. 1985. *Crabgrass Frontier: The Suburbanization of the United States.* New York: Oxford University Press.

Janowitz, Morris. 1967. *The Community Press in an Urban Setting.* 2d. ed. Chicago: University of Chicago Press.

Jeannin, André, Michel Bassand, Christophe Jaccoud, Dominique Joye, Fabio Lorenzi, and Roger Perrinjaquet. 1986. "Pratiques et Representations télécommunicationelles des Ménages Suisses." MANTO-Spezialstudie 3.21. Ecole Politechnique Fédérale de Lausanne, Switzerland.

Jensen, Richard. 1979. "The Lynds Revisited." *Indiana Magazine of History* 75 (December): 303–19.

Johnson, Dirk. 1991. "Where Phone Lines Stop, Progress May Pass." *New York Times* (18 March): 7, 12.

Judson, C. H. 1909. "Unprofitable Traffic—What Shall be Done with It?" *Telephony* 18 (11 December): 644–47.

Katz, James E. 1988. "US Telecommunications Privacy Policy." *Telecommunications Policy* (December): 353–67.

Katz, Michael. 1986. *In the Shadow of the Poorhouse*. New York: Basic Books.

Keegan, Frank L. 1987. *San Rafael: Marin's Mission City*. Northridge, CA: Windsor Publications.

Keller, Suzanne. 1968. *The Urban Neighborhood*. New York: Random House.

Kemp, R. F. 1905. "Telephones in Country Homes." *Telephony* 9 (5 May): 432–43.

Kern, Richard. 1976. "In Advertising, the Telephone Is the Message." *Telephony* (5 July): 212–15.

Kern, S. 1983. *The Culture of Time and Space, 1880–1918*. Cambridge, MA: Harvard University Press.

Kildegaard, I. C. 1966. "Telephone Trends." *Journal of Advertising Research* 6 (June) 56–60.

Kingsbury, J. E. 1915. *The Telephone and Telephone Exchanges*. London: Longman, Green, and Co.

Kingsbury, N. C. 1915. "Advertising Viewed as an Investment." *Printers' Ink* 93 (19 December): 37–40.

———. 1916. "Results from the American Telephone's National Campaign." *Printers' Ink* 95 (29 June): 182–84.

Kirkpatrick, E. L. 1926. "The Farmers' Standard of Living." Bulletin no. 1466. U. S. Department of Agriculture. Washington: Government Printing Office.

Kneeland, Hildegarde. 1929. "Is the Modern Housewife a Lady of Leisure?" *Survey* 62 (June): 301–2, 331, 336.

Knights, Peter. 1969. *The Plain People of Boston*. New York: Oxford University Press.

Kolb, J. H., and R. A. Polson. 1933. *Trends in Town-Country Relations*. Research Bulletin no. 117. Agricultural Experimentation Station, University of Wisconsin.

Kominski, Robert. 1991. *Computer Use in the United States, 1989*. Current Population Reports, P-23, Special Studies no. 171. Washington: United States Bureau of the Census.

Kramer, Richard. 1990. "Minitel: Don't Believe the Hype." *Antenna* [publication of Annenberg School, University of Pennsylvania] 2 (February): 2.

Kranzberg, Melvin, and Carroll W. Purcell. 1967. *Technology and Western Civilization*. New York: Oxford University Press.

Kyrk, H., D. Monroe, D. S. Brady, C. Rosenstiel, and E. D. Rainboth. 1941. *Family Expenditures for Housing and Household Operation. Five Regions.* Consumer Purchases Study, Farm Series. Bureau of Home Economics, Miscellaneous Publication no. 457. Washington: U.S. Department of Agriculture.

Kyrk, H., D. Monroe, K. Cronister, and M. Perry. 1941. *Family Expenditures for Housing and Household Operation. Five Regions.* Consumer Purchases Study, Urban and Village Series. Bureau of Home Economics, Miscellaneous Publication no. 432. Washington: U.S. Department of Agriculture.

Langdale, J. V. 1978. "The Growth of Long-Distance Telephony in the Bell System, 1875–1907." *Journal of Historical Geography* 4 (2): 145–59.

LaPorte, T. 1984. "Technology as Social Organization." Studies in Public Organization Working Paper no. 84–1. Institute of Governmental Studies, University of California, Berkeley.

Larson, Carl F. W. 1987. "A History of the Automobile in North Dakota to 1911." *North Dakota History* 54 (4): 3–24.

Larson, O. F., and T. F. Jones. 1976. "The Unpublished Data from Roosevelt's Commission on Country Life." *Agricultural History* 50 (October): 583–99.

Lauderback, H. C. 1930a. "Introduction." In *Bell System General Commercial Conference, 1930,* 1. Microfiche 368B, IBIC.

———. 1930b. "Residence Exchange Service Sales." In *Bell System General Commercial Conference, 1930,* 33–35. Microfiche 368B, IBIC.

———. 1930c. "Bell System Sales Activities." *Bell Telephone Quarterly* 9 (October): 262–69.

Layton, Larry. 1987. *The Community in Urban Society.* Chicago: Dorsey.

Lazare, Daniel. 1991. "Collapse of a City." *Dissent* (Spring): 267–75.

Lebergott, Stanley. 1976. *The American Economy: Income, Wealth, and Want.* Princeton, NJ: Princeton University Press.

Lee, T. R. 1980. "The Resilience of Social Networks to Changes in Mobility and Propinquity." *Social Networks* 2 (December): 423–37.

Lewis, D. L. 1980. "Sex and the Automobile: From Rumble Seats to Rockin' Vans." Special issue of *Michigan Quarterly Review* 19(4)–20(2): 518–28.

———, ed. 1980. "The Automobile and American Culture." Special issue of *Michigan Quarterly Review* 19(4)–20(2).

Liberatore, P. 1985. "Home Computers: Why the Craze Is Fizzling Out." *San Francisco Chronicle* (5 July): 1.

Lipartito, Kenneth. 1989. *The Bell System and Regional Business: The Telephone in the South, 1877–1920.* Baltimore: Johns Hopkins University Press.

———. 1989. "System Building at the Margin: The Problem of Public Choice in the Telephone Industry." *Journal of Economic History* 49 (June): 323–36.

Literary Digest. 1912. "Fallacies as to the High Cost of Living" (24 February): 400–42.

———. 1925. "More Automobiles Now than Telephones" (25 April): 68-9.

Litwak, Eugene. 1985. *Helping the Elderly*. New York: Guilford.

Loeb, G. H. 1977. "The Communications Act Policy toward Competition: A Failure to Communicate." Harvard University Program on Information Resources Policy Publication, no. 77–3. Cambridge, MA.

Long, Larry H. 1976. "The Geographical Mobility of Americans." *Current Population Reports*. Special Studies, ser. P–23, no. 64. Washington: U.S. Bureau of the Census.

———. 1988. *Migration and Residential Mobility in the United States*. New York: Russell Sage Foundation.

Long, Norton. 1937. "Public Relations of the Bell System." *Public Opinion Quarterly* (October): 5–22.

Lundberg, George A., Mirra Komarovsky, and Mary Alice McInerny. 1934. *Leisure: A Suburban Study*. New York: Columbia University Press.

Lynd, Robert S. 1933. "The People as Consumers." In *Recent Social Trends*, President's Research Committee on Social Trends. Vol. 2, 857–911. New York: McGraw-Hill.

Lynd, Robert S., and Helen M. Lynd. 1929. *Middletown*. New York: Harcourt Brace Jovanovich.

MacMeal, H. B. 1934. *The Story of Independent Telephony*. Chicago: Independent Pioneer Telephone Association.

Mahan, Gary P. 1979. "The Demand for Residential Telephone Service." Public Utilities Paper. Michigan State University.

Mahon, Ralph L. c.1955. *The Telephone in Chicago, 1877–1940*. Typescript, IBIC.

Marchand, Roland. 1980. "Creating the Corporate Soul: The Origins of Corporate Image Advertising in America." Paper presented to the Organization of American Historians, University of California, Davis.

———. 1985. *Advertising the American Dream: Making Way for Modernity, 1920–1940*. Berkeley and Los Angeles: University of California Press.

Martin, Michèle. 1987. "Communication and Social Forms: A Study of the Development of the Telephone System, 1876–1920." Ph.D. diss., Department of Sociology, University of Toronto.

———. 1988. "'Rulers of the Wires'? Women's Contribution to the Structure of Means of Communication." *Journal of Communication Inquiry* 12 (Summer): 89–103.

———. 1991. *"Hello Central?" Gender, Technology, and Culture in the Formation of Telephone Systems*. Montreal: McGill-Queen's University Press.

Marvin, Carolyn. 1989. *When Old Technologies Were New: Thinking about Electric Communication in the Late Nineteenth Century.* New York: Oxford University Press.

Marx, Leo. 1964. *The Machine in the Garden.* New York: Oxford University Press.

Mason, C. F. 1945. "America's Farm Telephone Program Is Well on Its Way." *Public Utilities Fortnightly* 36, no. 2 (19 July): 85–98.

Mason, Jack. 1971. *Early Marin.* Petaluma, CA: House of Printing.

Masters, R. S., R. C. Smith, and W. E. Winter. 1927. *An Historical Review of the San Francisco Exchange.* San Francisco: Pacific Telephone and Telegraph Co.

————. 1927. *An Historical Review of the East Bay Exchange.* San Francisco: Pacific Telephone and Telegraph Co.

Mathias, P. 1934. "The Rural Telephone Situation." *Telephony* 107 (22 December): 8–11.

Mayer, M. 1977. "The Telephone and the Uses of Time." In *The Social Impact of the Telephone,* edited by Ithiel de Sola Pool, 225–45. Cambridge, MA: The MIT Press.

Mayntz, Renata, and Thomas Hughes, eds. 1988. *The Development of Large Technical Systems.* Boulder, CO: Westview Press.

McDaniel, Charles M. c.1929. "Reminiscences." Typescript, MIT.

McGaw, Judith. 1982. "Women and the History of American Technology." *Signs* 7 (4): 798–828.

McGinn, Robert E. 1991. *Science, Technology, and Society.* Englewood Cliffs, NJ: Prentice-Hall.

McHugh, Keith S. 1931. "Introductory Review of Sales." *Bell System General Commercial Conference on Sales Matters.* Shawnee-on-Delaware, PA. Microfiche 368B, IBIC.

McKay, J. P. 1976. *Tramways and Trolleys: The Rise of Urban Mass Transport in Europe.* Princeton, NJ: Princeton University Press.

McKenzie, R. 1921 [1968]. "The Neighborhood." Reprinted in *Rodrick D. Mc Kenzie on Human Ecology,* edited by A. Hawley, 51–93. Chicago: University of Chicago Press.

McLuhan, Marshall. 1964. *Understanding Media: The Extensions of Man.* New York: McGraw-Hill.

McShane, C. 1979. "Transforming the Use of Urban Space: A Look at the Revolution in Street Pavements, 1880–1924." *Journal of Urban History* 5 (May): 279–307.

Mead, Margaret. 1976. "Looking at the Telephone a Little Differently." *Bell Telephone Magazine* 54 no. 1 (January-February): 12–14.

Meier, Avionoam. 1981. "Innovation Diffusion and Regional Economic Development: The Spatial Diffusion of Automobiles in Ohio." *Regional Studies* 15 (2): 111-22.

Melvin, Patricia Mooney. 1985. "Changing Contexts: Neighborhood Definitions and Urban Organization." *American Quarterly* 37 (3): 357-68.

Meret, Charles. c.1936. *A Chronological History of Marin County, 1542–1936.* 3 vols. Marin County Library, Marin County Civic Center, San Rafael, CA.

Merrill, Francis E. 1950. "The Highway and Social Problems." In *Highways in Our National Life,* edited by Jean Labatut and Wheaton J. Lane, 135–43. Princeton, NJ: Princeton University Press.

Merrill, Robert S. 1968. "The Study of Technology." *International Encyclopedia of the Social Sciences.* Vol. 15, 577–89. New York: MacMillan.

Meyerowitz, Joshua. 1985. *No Sense of Place: The Impact of Electronic Media on Social Behavior.* New York: Oxford University Press.

Michelson, W. 1971. "Some Like It Hot: Social Participation and Environmental Use as Functions of the Season." *American Journal of Sociology* 76 (May): 1072–83.

Miller, C. E. 1980. "Telecommunications/Transportation Substitution." *Socio-Economic Planning Sciences* 14 (4): 163–66.

Miller, Guy C. 1952. *Palo Alto Community Book.* Palo Alto, CA: Crawston.

Miller, Spencer, Jr. 1950. "History of the Modern Highway in the United States." In *Highways in Our National Life,* edited by Jean Labatut and Wheaton J. Lane, 88-119. Princeton, NJ: Princeton University Press.

Miller, Zane L. 1981. "The Role and Concept of Neighborhood in American Cities." In *Community Organization for Urban Social Change: A Historical Perspective,* edited by R. Fisher and P. Romanofsky, 3–32. Westport, CT: Greenwood.

Miss Manners [Judith Martin]. 1989. "Calling for a Few Simple Phone Rules." *San Francisco Chronicle* (3 April).

———. 1990. "You Don't Have to Talk to a Machine." *San Francisco Chronicle* (22 January).

———. 1991. "Proper Use of the Telephone." *San Francisco Chronicle* (24 June).

Mitchell, B. R. 1980. *European Historical Statistics, 1750–1975.* New York: Facts on File.

Mitchell, Gary. 1984. "Some Aspects of Telephone Socialization." In *Studies in Communication,* edited by Sari Thomas, 249–53. Vol. 1. Norwood, NJ: Ablex.

Mock, Roy D. 1916. "Fundamental Principles of the Telephone Business: Part V: Advertising and Publicity." *Telephony* 71 (22 July): 21–23.

Modell, John. 1983. "Review Essay: *The Inner American.*" *Journal of Social History* 17 (Fall): 139–46.

Moline, Norman T. 1971. *Mobility and the Small Town, 1900–1930: Transportation Change in Oregon, Illinois.* Research Paper no. 132. Department of Geography, University of Chicago.

Monaco, Cynthia. 1988. "The Difficult Birth of the Typewriter." *American Heritage of Invention and Technology* 4 (Spring/Summer): 11–21.

Monroe, Day, Dorothy S. Brady, June F. Constantine, and Karl F. Benson. 1941. *Family Expenditures for Automobile and Other Transportation. Five Regions.* Consumer Purchases Study, Urban, Village, and Farm Series. Bureau of Home Economics, Miscellaneous Publication no. 415. Washington: U.S. Department of Agriculture.

Morgan, Winona L. 1939. *The Family Meets the Depression.* Minneapolis, MN: University of Minnesota Press.

Morlan, Robert L. 1984. "Municipal versus National Election Voter Turnout: Europe and the United States." *Political Science Quarterly* 99 (Fall): 457–70.

Moyal, Ann M. 1984. *Clear Across Australia.* Melbourne: Nelson.

———. 1989a. "Woman and the Telephone in Australia: Outline of a National Study." In *Telefon und Gesellschaft, Band 1: Beiträge zu einer Soziologie der Telefonkommunikation,* Forschungsgruppe Telefonkommunikation (Hrsg), 283–93. Berlin: Volker Speiss.

———. 1989b. "The Feminine Culture of the Telephone. People, Patterns and Policy." *Prometheus* (Australia) 7 (June): 5–31.

Moyer, J. A. 1977. "Urban Growth and the Development of the Telephone: Some Relationships at the Turn of the Century." In *The Social Impact of the Telephone,* edited by Ithiel de Sola Pool, 342–70. Cambridge, MA: The MIT Press.

Mueller, J. N. 1928. "The Automobile: A Sociological Study." Ph.D. diss., Department of Sociology and Anthropology, University of Chicago.

Mueller, Milton. 1989. "The Switchboard Problem." *Technology and Culture* 30 (July): 534–60.

———. 1989. "From Competition to 'Universal Service': The Emergence of Telephone Monopoly in the U.S., 1907–1921." Typescript, University of Pennsylvania.

Mulrooney, J. B. 1937. "Some Aspects of Rural Telephone Service." *Telephony* 112 (9 January): 14–16.

Mumford, L. 1972. "Two Views of Technology and Man," In *Technology, Power, and Social Change,* edited by C. A. Thrall and J. M. Starr, 1–16. Carbondale, IL: Southern Illinois University Press, 1972.

Murphy, John Allen. 1917. "How the Automobile Has Changed the Buying Habits of Farmers." *Printers' Ink* 101 (29 November): 3–6, 98–101.

N. W. Ayer & Son. n.d. *Fifty Years of Telephone Advertising, 1908–1958.* New York: N. W. Ayer & Son.

Nagel, J., and M. Nagel. n.d. *Talking Wires: The Story of North Dakota's Telephone Cooperative.* Aberdeen, ND: North Dakota Association of Telephone Cooperatives.

Nash, Anedith Jo Bond. 1983. "Death on the Highway: The Automobile Wreck in American Culture, 1920–1940." Ph.D. diss., University of Minnesota.

Newmark, J. H. 1914. "Have Automobiles Been Wrongly Advertised?" *Printers' Ink* 86 (5 February): 70–72.

————. 1918. "The Line of Progress in Automobile Advertising." *Printers' Ink* 105 (26 December): 97–102.

New York State. 1910. *Report of the Committee of the Senate and Assembly Appointed to Investigate Telephone and Telegraph Companies.* Albany: New York State.

Nisbitt, Florence. 1918. *Household Management.* New York: Russell Sage Foundation.

Noble, Grant. 1989. "Towards a 'Uses and Gratifications' of the Telephone." In *Telefon und Gesellschaft, Band 1: Beiträge zu einerSoziologie der Telefonkommunikation, Forschungsgruppe Telefonkommunikation (Hrsg),* 298–307. Berlin: Volker Speiss.

Nye, David E. 1990. *Electrifying America: Social Meanings of a New Technology, 1880–1940.* Cambridge, MA: The MIT Press.

Ogburn, William F. 1922. *Social Change.* New York: Heubsch.

————. 1937. "National Policy and Technology." In *Technological Trends and National Policy,* National Resources Committee, 3–14. Washington: Government Printing Office.

————. 1946. *The Social Effects of Aviation.* New York: Houghton Mifflin.

————. 1957. "How Technology Causes Social Change." In *Technology and Social Change,* edited by F. R. Allen et al., 12–26. New York: Appleton Century Crofts.

Ogburn, William F., with S. C. Gilfillan. 1933. "The Influence of Invention and Discovery." In *Recent Social Trends,* President's Research Committee on Social Trends, 122–66. New York: McGraw-Hill.

Ogle, E. B. 1979. *Long Distance Please: The Story of the TransCanada Telephone System.* Toronto: Collins.

Olsen, Michael L. 1986. "But It Won't Milk the Cows: Farmers in Colfax County Debate the Merits of the Telephone." *New Mexico Historical Review* 61 (June): 1–13.

Ornati, Oscar. 1966. *Poverty Amid Affluence.* New York: Twentieth-Century Fund.

Our Society Blue Book. 1899. San Francisco.

Page, A. W. 1928. "Public Relations and Sales." In *General Commercial Conference (Bell System),* Shawnee-on-Delaware, PA. Microfiche 368B, IBIC.

———. 1941. *The Bell Telephone System*. New York: Harper and Bros.

Paine, Albert Bigelow. 1929. *Theodore N. Vail*. New York: Harper and Bros.

Paine, Robert. 1967. "What Is Gossip About?" *Man* 2 (June): 278–85.

Parker, F. E. 1938. "Cooperative Telephone Association, 1936." *Monthly Labor Review* 46 (February): 392–413.

Parkerson, Donald H. 1982. "How Mobile Were Nineteenth-Century Americans?" *Historical Methods* 15 (Summer): 99–109.

Patten, William. 1926. *Pioneering the Telephone in Canada*. Montreal: Telephone Pioneers.

Pegg, M. 1983. *Broadcasting and Society, 1918–1939*. London: Croom-Helm.

Peixotto, Jessica B. 1927. *Getting and Spending at the Professional Standard of Living*. New York: MacMillan.

———. 1929. "Cost of Living Studies II: How Workers Spend a Living Wage." *University of California Publications in Economics* 5 (3): 161–245.

People's Mutual Telephone Company. 1976. *It's for You: A History of People's Mutual Company Co.* Gretna, VA: People's Mutual, at MIT.

Perry, C. R. 1976. "The British Experience, 1876–1912: The Impact of the Telephone." In *The Social Impact of the Telephone*, edited by Ithiel de Sola Pool, 69–96. Cambridge, MA: The MIT Press.

Pike, Robert M. 1988. "Adopting the Telephone: The Social Diffusion and Use of the Telephone in Urban Central Canada, 1876 to 1914." Typescript, Queen's University, Kingston, Ontario.

———. 1989. "Kingston Adopts the Telephone: The Social Diffusion and Use of the Telephone in Urban Central Canada, 1876 to 1914." *Urban History Review* 18 (June): 32–47.

Pike, Robert M., and Vincent Mosco. 1986. "Canadian Consumers and Telephone Pricing." *Telecommunications Policy* 10 (March): 17–32.

Pool, Ithiel de Sola. 1977a. "The Communication/Transportation Tradeoff." *Policy Studies Journal* 6 (Autumn): 74–83.

———. 1977b. "Foresight and Hindsight: The Case of the Telephone." In *The Social Impact of the Telephone*, edited by Ithiel de Sola Pool, 127–58. Cambridge, MA: The MIT Press.

———, ed. 1977. *The Social Impact of the Telephone*. Cambridge, MA: The MIT Press.

———. 1983. *Forecasting The Telephone*. Norwood, NJ: Albex.

Pope, D. 1983. *The Making of Modern Advertising*. New York: Basic Books.

Pope, Jesse L. 1912. "Rural Communication." In *Cyclopedia of American Agriculture*, edited by L. H. Bailey, 312–40. Vol. 4. 4th ed. New York: MacMillan.

Post, Emily. 1923; 1942; 1955. *Etiquette*. New York: Funk & Wagnalls.

Pound, Arthur. 1926. *The Telephone Idea: Fifty Years After.* New York: Greenberg.

Prairie Farmers' Home and County Directory of Dubuque County. 1924. Chicago: The Prairie Farmer.

Pred, A. R. 1973. *Urban Growth and the Circulation of Information: The United States System of Cities, 1790–1840.* Cambridge, MA: Harvard University Press.

Preston, Howard Lawrence. 1979. *Automobile Age Atlanta.* Athens, GA: University of Georgia Press.

———. 1991. *Dirt Roads to Dixie: Accessibility and Modernization, 1885–1935.* Knoxville, TN: University of Tennessee Press.

Printers' Ink. 1906a. "Results by Telephone." 57 (24 October): 35.

———. 1906b. "Advertising the Automobile." 57 (7 November): 47–50.

———. 1910. "Bell Encourages Shopping by Telephone." 70 (19 January).

———. 1911. "Broadening the Possible Market." 74 (9 March): 20.

———. 1933. "Telephone Company Works with Retailers on Campaign." 163 (4 May): 41.

Purcell, Mae Fisher. 1948. *History of Contra Costa County.* Berkeley, CA: Gillick Press.

R. O. Eastman, Inc. 1927. *Zanesville and 36 Other American Communities.* New York: The Literary Digest.

Rae, John B. 1965. *The American Automobile: A Brief History.* Chicago: University of Chicago Press.

———. 1971. *The Road and the Car in American Life.* Cambridge, MA: The MIT Press.

———. 1984. *The American Automobile Industry.* Boston: Twayne.

Rainwater, Lee. 1974. *What Money Buys.* New York: Basic Books.

Rakow, Lana F. 1987. "Gender, Communication, and Technology: A Case Study of Women and the Telephone." Ph.D. diss., Institute of Communications Research, University of Illinois at Urbana-Champaign.

———. 1991. *Gender on the Line: Women, the Telephone, and Community Life.* Champaign, IL: University of Illinois Press.

Rankin, J. O. 1928. *The Use of Time in Farm Homes.* Nebraska Agricultural Experiment Station Bulletin no. 230. University of Nebraska, Lincoln.

Reich, Leonard. 1985. *The Making of American Industrial Research: Science and Business at GE and Bell, 1876–1926.* New York: Cambridge University Press.

Reinhold, Francis L. 1936. "Federal-Municipal Relations: The Road So Far." *National Municipal Review* 25 (August): 452, 458–64.

Renshaw, E. F. 1962. "Experience of the United States in Bringing Modern Telephone Service to Unserved Areas." In *Science, Technology and Devel-*

opment, edited by Agency for International Development, 67–81. Vol. 12, *Communications*. PB–207 505. Washington, D.C.: National Technical Information Service.

Reynolds, A. P. 1906. "Selling a Telephone." *Telephony* 12 (6 November): 280–81.

Richardson, Anna Steese. 1930. "Telephone Manners: Why Not?" *Successful Farming* (March): 46–47.

Rippey, James Crockett. 1975. *Goodbye, Central; Hello, World: A Centennial History of Northwestern Bell*. Omaha: Northwestern Bell.

Robertson, J. H. 1947. *The Story of the Telephone: A History of the Telecommunications Industry of Britain*. London: Pitman and Sons.

Robertson, L., and K. Amstutz. 1949. *Telephone Problems in Rural Indiana*. Bulletin 548 (September). Purdue University Agricultural Experiment Station, Lafayette, IN.

Robinson, John P., and Philip E. Converse. 1972. "Social Change Reflected in the Use of Time." In *The Human Meaning of Social Change*, edited by Angus Campbell and Philip E. Converse, 17–86. New York: Russell Sage Foundation.

Rogers, E. 1983. *Diffusion of Innovations*. 3d ed. New York: Free Press.

Ronnell, Avital. 1989. *The Telephone Book: Technology, Schizophrenia, Electric Speech*. Lincoln, NE: University of Nebraska Press.

Roos, C. F., and Victor von Szeliski. 1939. "Factors Governing Changes in Domestic Automobile Demand." In *The Dynamics of Automobile Demand*, edited by General Motors, 21–95. New York: General Motors.

Rose, M. 1984. "Urban Environments and Technological Innovation: Energy Choices in Denver and Kansas City, 1900-1940." *Technology and Culture* 25 (July): 503–39.

Rosenzweig, Roy. 1985. *Eight Hours for What We Will: Workers and Leisure in an Industrial City, 1870–1920*. New York: Cambridge University Press.

Rosher, A. 1968. "Residential Telephone Usage among the Chicago Civic-Minded." Master's thesis, Department of History, University of Chicago.

Ross, Ellen. 1983. "Survival Networks: Women's Neighborhoods Sharing in London before World War I." *History Workshop* 15 (Spring): 4–27.

Rossi, Peter. 1972. "Community Social Indicators." In *The Human Meaning of Social Change*, edited by Angus Campbell and Philip E. Converse, 87–126, New York: Russell Sage.

Rothman, Ellen. 1984. *Hands and Hearts: A History of Courtship in America*. New York: Basic Books.

Rothschild, Joan. 1983. "Technology, Housework, and Women's Liberation: A Theoretical Analysis." In *Machina Ex Dea: Feminist Perspectives on Technology*, edited by Joan Rothschild, 79–93. New York: Pergamon.

Rubin, P. 1976. "Urbanization," In "Working Papers for Retrospective Technology Assessment of the Telephone," Ithiel de Sola Pool et al. Report to the National Science Foundation, Vol. 5. Typescript, Cambridge MA: Massachusetts Institute of Technology.

Ruges. J. F. 1970. *Le Téléphone pour Tous*. Paris: Editions du Seuil.

Rural Electrification Association. 1974. *Twenty-five Years of Progress: Rural Telephone Service USA*. Washington: Government Printing Office.

Rutman, Darrett B. 1980. "Community Study." *Historical Methods* 13 (Winter): 29–41.

Rutter, Derek R. 1987. *Communicating by Telephone*. Oxford: Pergamon.

Sabin, John I. 1895–1896. "Special Reports." Letters to John E. Hudson, President, American Bell Telephone Co. Box 1278, AT&THA.

St. Clair, D. J. "The Motorization and Decline of Urban Public Transit." *Journal of Economic History* 41 (September): 579–600.

Sales Management. 1937. "Average Person Finds Letter Writing More Difficult than Telephoning or Wiring." 41 (20 October): 60.

Salomon, Ilan. 1985. "Telecommunications and Travel: Substitution or Modified Mobility?" *Journal of Transport Economics and Policy* 19 (September): 219–35.

———. 1986. "Telecommunications and Travel Relationships: A Review." *Transportation Research* 20A (3): 223–38.

Sangster, Margaret. 1904. *Good Manners for All Occasions*. New York: Cupples & Leon.

Saunders, Robert J. and Jeremy J. Warford. 1978. "Evaluation of Telephone Projects in Less Developed Countries." Public Utility no. 37 (July). Energy, Water, and Telecommunications Department, The World Bank.

Saunders, Robert J., Jeremy J. Warford, and Bjorn Wellenius. 1983. *Telecommunications and Economic Development*. Baltimore: Johns Hopkins University Press.

Scharff, Virginia A. 1986. "Reinventing the Wheel: American Women and the Automobile, 1910–1930." Paper presented to the Organization of American Historians, New York City.

———. 1991. *Taking the Wheel: Women and the Coming of the Motor Age*. New York: Free Press.

Schlesinger, Arthur M. 1933. *The Rise of the City*. New York: MacMillan.

———. 1947. *Learning How to Behave*. New York: MacMillan.

Schmidt, S. 1948. "The Telephone Comes to Pittsburgh." Master's thesis, Department of History, University of Pittsburgh.

Schmookler, J. 1962. "Economic Sources of Inventive Activity." *Journal of Economic History* 22 (March): 1–20.

Schoonmaker, C. B. 1929. "When Sales Lagged, Direct Mail Saved the Day." *Printed Salesmanship* 53 (August): 527–29.

Schudson, Michael. 1985. *Advertising: The Uneasy Persuasion.* New York: Basic Books.

Schwartz, H. 1985. "Its a Bird, a Plane—No, a Phone Bill." *New York Times* (30 January): 23.

Segal, H. P. 1982. "Assessing Retrospective Technology Assessment: A Review of the Literature." *Technology in Society* 4: 231–46.

Seymour, Jim. 1990. "The 'Home-PC Market' Myth Rises Again." *PC Magazine* 9 (30 October): 85–6.

Shannon, Gary W., and John D. Nystuen. 1972. "Marriage, Migration, and the Measurement of Social Interaction." In *International Geography 1972,* edited by W. Peter Adams and Frederick M. Helleiner, 491–95. Toronto: University of Toronto Press.

Sharpless, John B., and Ray M. Shortridge. 1975. "Biased Underenumeration in Census Manuscripts." *Journal of Urban History* 1 (August): 409–39.

Shaw, J. M. 1934. "Buying by Telephone at Department Stores." *Bell Telephone Quarterly* 13 (July): 267–88.

Shenk, N. O. 1944. "Small Exchanges Struggle for Survival." *Telephony* 127 (7 October, 11 November, 9 December): 20–23, 13–14, 34–36.

Sherwood, Mrs. John. [1897] 1975. *Manners and Social Usage.* New York: Harper & Bros. Reprint. New York: Arno.

Sichter, J. W. 1977. "Separations Procedures in the Telephone Industry." Harvard University Program on Information Resources Policy Paper, no. 77–82. Cambridge, MA.

Sidwell, E. H. 1906. "Some Phases of Rural Telephony." *Telephony* 11 (6 April): 238–39.

Sifianou, Maria. 1989. "On the Telephone Again! Differences in Telephone Behavior: England versus Greece." *Language in Society* 18 (December): 527–44.

Silverman, Carol J. 1981. "Negotiated Claim." Ph. D. diss., Department of Sociology, University of California, Berkeley.

Simonds, W. A. 1958. *The Hawaiian Telephone Story.* Honolulu: Hawaiian Telephone Company.

Singer, Benjamin D. 1981. *Social Functions of the Telephone.* Palo Alto, CA: R & E Research Associates.

Smith, Eldson C. 1973. *New Dictionary of American Family Names.* New York: Harper & Row.

Smith, George David. 1985. *The Anatomy of a Business Strategy: Bell, Western Electric, and the Origins of the American Telephone Industry.* Baltimore: Johns Hopkins University Press.

Smith, Helena H. 1937. "Intrusion by Telephone." *Readers Digest* 31, no. 187 (November): 34-37.

Smith, Julian. 1981. "A Runaway Match: The Automobile in the American Film, 1900–1920." Special issue of *Michigan Quarterly Review* 19(4)–20(2): 574–87.

Smith, R. A. 1972. *A Social History of the Bicycle.* New York: McGraw-Hill, American Heritage Press.

Solnick, Steven L. 1991. "Revolution, Reform and the Soviet Telephone System, 1917–1927." *Soviet Studies* 43 (1): 157–76.

Sorokin, Pitrim, and Clarence Q. Berger. 1939. *Time-Budgets of Human Behavior.* Cambridge, MA: Harvard University Press.

Spacks, Patricia Meyer. 1986. *Gossip.* Chicago: University of Chicago Press.

Spasoff, I. M., and H. S. Beardsley. 1922; 1930. *Farmers' Telephone Companies: Organization Financing and Management.* Farmers' Bulletin no. 1245. Washington, DC: U. S. Department of Agriculture.

Stack, Jennifer Daryl. 1984. *Communication Technologies and Society: Conceptions of Causality and the Politics of Technological Intervention.* Norwood, MA: Ablex.

Starr, Kevin. 1973. *Americans and the California Dream, 1850–1915.* New York: Oxford University Press.

Steel, G. O. 1905. "Advertising the Telephone." *Printers' Ink* 51, no. 2, (12 April): 14–17.

Steffens, Marec Bela. 1990. "Bewildering Range of Topics: International Symposium on the Sociology of the Telephone." *Telecommunications Policy* 14 (April): 176–77.

Stehman, J. W. 1925. *The Financial History of the American Telephone and Telegraph Company.* Boston: Houghton Mifflin.

Stein, Maurice. 1962. *The Eclipse of Community.* New York: Harper & Row.

Steinberg, Dan. 1985. *Logit: A Supplementary Module for SYSTAT.* San Diego, CA: Salford Systems.

Steiner, J. F. 1933. "Recreation and Leisure Time Activities." In *Recent Social Trends,* President's Research Committee on Social Trends, 912–957. Vol. 2. New York: McGraw-Hill.

Stone, Alan. 1991. *Public Service Liberalism: Telecommunication and Transitions in Public Policy.* Princeton, NJ: Princeton University Press.

Strasser, Susan. 1982. *Never Done: A History of American Housework.* New York: Pantheon.

———. 1989. *Satisfaction Guaranteed: The Making of the American Mass Market.* New York: Pantheon.

Sullivan, G. L. 1915. "Forces that Are Reshaping a Big Market." *Printers' Ink* 92 (29 July): 26–28.

Summerfield, C. H. 1930. "Kansas Company's [United Telephone] Campaign Record." *Telephony* 98 (14 June): 14–18.

Sundstrom, Geoff. 1985. "The Paving of America: Mud and Macadam to Superhighways." *Automotive News* (30 October): 107–14.

Synge, J., C. J. Rosenthal, and V. W. Marshall. 1982. "Phoning and Writing as a Means of Keeping in Touch in the Family of Later Life." Paper presented to the Canadian Association on Gerontology, Toronto.

Tarkington, Booth. 1918. *The Magnificent Ambersons*. Garden City, NY: Doubleday, Page, and Co.

Tarr, Joel A. 1984. "The Evolution of the Urban Infrastructure in the Nineteenth and Twentieth Centuries." In *Perspectives on Urban Infrastructure,* National Academy of Sciences, 4–66. Washington, D.C.: National Academy Press.

Tarr, Joel A., and Gabriel Dupuy, eds. 1988. *Technology and the Rise of the Networked City in Europe and America*. Philadelphia: Temple University Press.

Tarr, Joel A., with Thomas Finholt and David Goodman. 1987. "The City and the Telegraph: Urban Telecommunications in the Pre-Telephone Era." *Journal of Urban History* 14 (November): 38–80.

Taylor, I. D. 1980. *Telecommunications Demand*. Cambridge, MA: Ballinger.

Telephony. 1904. "The Farmer and the Telephone." 8 (4 December): 520.

———. 1905a. "Facts Regarding the Rural Telephone." 9 (5 April): 303.

———. 1905b. "Swearing over the Telephone." 9 (5 May): 418.

———. 1905c. "Doctors and Rural Telephones." 9 (June): 492.

———. 1905d. "John I. Sabin." 10 (5 November): 362.

———. 1909. "Report of Secretary Ware." 18 (11 December): 651–54.

———. 1913. "Educating the Public to the Proper Use of the Telephone." 64 (21 June): 32–33.

———. 1914. "Limiting Party Line Conversations." 66 (2 May): 21.

———. 1922. "Limiting 'Talkers' in Jamestown, N.Y." 82 (7 January): 16.

Temin, Peter. 1990. "Cross Subsidies in the Telephone Network after Divestiture." *Journal of Regulatory Economics* 2 (December): 349-62.

Temin, Peter, with Louis Galambos. 1987. *The Fall of the Bell System*. New York: Cambridge University Press.

Tempest, Rone. 1989. "Minitel: Miracle or Monster." *Los Angeles Times* (24 October): 1, 10.

Thernstrom, Stephen. 1968. *The Other Bostonians*. Cambridge, MA: Harvard University Press.

Thompson, John. 1958. "The Settlement Geography of the Sacramento-San Joaquin Delta, California." Ph.D. diss., Stanford University.

Tichi, Cecelia. 1987. *Shifting Gears*. Chapel Hill, NC: University of North Carolina Press.

Tilly, Charles. 1984. *Big Structures, Large Processes, Huge Comparisons*. New York: Russell Sage Foundation.

Tobin, G. A. 1976. "Suburbanization and the Development of Motor Transportation: Transportation Technology and the Suburbanization Process." In *The Changing Face of the Suburbs,* edited by B. Schwartz, 95–112. Chicago: University of Chicago Press.

Tomblein, R. L. 1931. "Population Changes in Small Communities and in Rural Areas." *Bell Telephone Quarterly* 10 (April): 115–23.

Toogood, Anna Coxe. 1980. *Historic Resource Study: A Civil History of Golden Gate National Recreation Area and Point Reyes National Seashores, California.* Historic Preservation Branch, National Park Service. Washington: U.S. Department of the Interior.

Tropea, Joseph L. 1989. "Rational Capitalism and Municipal Government: The Progressive Era." *Social Science History* 13 (Summer): 137–58.

Umble, Diane Zimmerman. 1989. "The Telephone Comes to Pennsylvania Amish Country: A Study of Resistance to Technology at the Turn of the Century." Paper presented to International Communication Association, San Francisco.

United Media Enterprises. 1983. *Where Does the Time Go?* New York: Newspaper Enterprises Association.

United States Bureau of Labor Statistics. 1924. "The Cost of Living in the United States." Bulletin no. 357. U.S. Department of Labor. Washington: Government Printing Office.

———. 1945. "Family Spending and Saving in Wartime." Bulletin no. 822. House Document no. 147A, 79th Cong., 1st sess. Washington: Government Printing Office.

United States Bureau of the Census. 1892. "Statistics of Manufactures, 1890: Operating Telephone Companies." *Census Bulletin* no. 196 (25 June).

———. 1906. *Special Reports: Telephones and Telegraphs 1902.* Washington: Government Printing Office.

———. 1910. *Special Reports: Telephones: 1907.* Washington: Government Printing Office.

———. 1915. *Telephones and Telegraphs and Municipal Electric Fire-Alarm and Police-Patrol Signalling Systems 1912.* Washington: Government Printing Office.

———. 1920. *Census of Electrical Industries: 1917: Telephones.* Washington: Government Printing Office.

———. 1930. *U. S. Statistical Abstract 1929.* Washington: Government Printing Office.

———. 1932. *U. S. Census of Agriculture, 1930.* Vol. 4, *General Reports.* Washington: Government Printing Office.

———. 1934. *Census of Electrical Industries, 1932: Telephones and Telegraphs.* Washington: Government Printing Office.

———. 1939. *Census of Electrical Industries, 1937: Telephones and Telegraphs.* Washington: Government Printing Office.

————. 1942. *U. S. Census of Agriculture, 1940.* Vol. 3, *General Reports.* Washington: Government Printing Office.

————. 1975. *Historical Statistics of the United States, 1790–1970.* Washington: Government Printing Office.

————. 1988. *Statistical Abstract of the United States 1988.* Washington: Government Printing Office.

————. 1990. *Statistical Abstract of the United States 1990.* Washington: Government Printing Office.

————. 1991. *Statistical Abstract of the United States 1991.* Washington: Government Printing Office.

United States Congress. House. 1934. "Federal Communications Commission: Hearings before the Committee on Interstate and Foreign Commerce." 73d Cong., 2nd sess. Washington: Government Printing Office.

United States Congress. Senate. 1909. *Report of the Country Life Commission.* 60th Cong., 2d sess., Senate Document no. 705. Washington: Government Printing Office.

United States Department of Agriculture. 1915. *Social and Labor Needs of Farm Women.* Report no. 103. Washington: Government Printing Office.

————. 1938–1939. "Income Parity for Agriculture. Part 3, "Prices Paid by Farmers for Commodities and Services." Section 2, "Rates for Electricity for the Farm Home" (Preliminary, September 1938). Section 3, "Telephone Rates for Farmers" (Preliminary, December 1938). Section 5, "Index Numbers of Prices Paid by Farmers for Commodities" (Preliminary, May 1939). Washington: U. S. Department of Agriculture.

————. 1944. "The Time Costs of Homemaking: A Study of 1500 Rural and Urban Households." Mimeograph, Agricultural Research Administration, Bureau of Human Nutrition and Home Economics.

Urquhart, M. C., and K. A. H. Buckley. 1965. *Historical Statistics of Canada.* Toronto: MacMillan.

Valentine, F. P. 1926. "Some Phases of the Commercial Job." *Bell Telephone Quarterly* 5 (January): 34–43.

Verba, Sidney, and Norman H. Nie. 1972. *Participation in America.* New York: Harper & Row.

Vietor, Richard H. K. 1989. "AT&T and the Public Good: Regulation and Competition in Telecommunications, 1910–1987." In *Future Competition in Telecommunications,* edited by Stephen P. Bradley and Jerry A. Hausman, 27–104. Boston: Harvard Business School Press.

Voice Telephone Magazine (United Communications). 1926. "Telephone Etiquette." (December): 1. MIT.

von Auw, Alvin. 1983. *Heritage and Destiny: Reflections on the Bell System in Transition.* New York: Praeger.

von Hippel, Eric. 1988. *The Sources of Innovation*. New York: Oxford University Press.

Wade, Margaret. 1924. *Social Usage in America*. New York: Thomas Crowell.

Wallace, C. 1930. "Service Outside the Base Rate Area and Rural Service, Introduction." In *Bell System General Commercial Conference, 1930*, 30–31. Microfiche 386B, IBIC.

Walsh, J. L. 1950. *Connecticut Pioneers in Telephony*. New Haven, CT: Morris F. Tyler Chapter of Telephone Pioneers of America.

Ward, F. E. 1920. "The Farm Women's Problems." U.S. Department of Agriculture Circular no. 148. Washington: Government Printing Office.

Wasserman, Neil H. 1985. *From Invention to Innovation: Long-Distance Telephone Transmission at the Turn of the Century*. Baltimore: Johns Hopkins University Press.

Weinstein, B. 1976. "The Telephone in Popular Journals, 1908-1913." In "Working Papers for Retrospective Technology Assessment of the Telephone," Ithiel de Sola Pool et al. Report to the National Science Foundation, Vol. 5. Typescript, Cambridge, MA: Massachusetts Institute of Technology.

Weiss, Ellen. 1986. *Telephone Time: A First Book of Telephone Do's and Don'ts*. New York: Random House.

Wellenius, B. 1977. "Telecommunications in Developing Countries." *Telecommunications Policy* 1 (September): 289–97.

———. 1978. "The Role of Telecommunications Services in Developing Countries." Workshop on Special Aspects of Telecommunications Development in Isolated and Underprivileged Areas of Countries, Ottawa, Canada, June.

Wellman, Barry. 1979. "The Community Question: The Intimate Networks of East Yorkers." *American Journal of Sociology* 84 (5): 1201–31.

———. 1989. "Reach Out and Touch Some Bodies: How Telephone Networks Connect Social Networks." Paper presented to the American Sociological Association, San Francisco, August.

West, J. [C. Withers]. 1945. *Plainville, U.S.A.* New York: Columbia University Press.

Westrum, Ron. 1983. "What Happened to the Old Sociology of Technology?" Paper presented to Society for Sociological Study of Science, Blacksburg, VA, November. Typescript, Eastern Michigan State University.

———. 1991. *Technologies and Society: The Shaping of People and Things*. Belmont, CA: Wadsworth.

White, Annie Randall. 1900. *Twentieth-Century Etiquette*. No publisher indicated.

White, Lyn, Jr. 1962. *Medieval Technology and Social Change*. New York: Oxford University Press.

———. 1974. "Technology Assessment from the Stance of a Medieval Historian." *American Historical Review* 79 (February): 1–13.

Wiebe, Robert. 1967. *The Search for Order, 1877–1920.* New York: Hill & Wang.

Wilkenson, Leland. 1990. *SYSTAT.* Evanston, IL: SYSTAT, Inc.

Wik, R. M. 1972. *Henry Ford and Grass-Roots America.* Ann Arbor: University of Michigan Press.

———. 1977. "The Radio in Rural America during the 1920s." *Agricultural History* 55 (October): 339–50.

Wilkins, James. 1915. *Early History of Marin.* San Rafael, CA: Marin County Library, Marin County Civic Center.

Willey, Malcolm M., and Stuart A. Rice. 1933. *Communication Agencies and Social Life.* New York: McGraw-Hill.

Williams, Faith M., and Alice C. Hanson. 1941. "Money Disbursements of Wage Earners and Clerical Workers, 1934–36. Summary Volume." Bulletin no. 638. Bureau of Labor Statistics. Washington: Government Printing Office.

Williams, Raymond. 1975. *Television, Technology, and Cultural Form.* New York: Schocken.

Willmott, Peter. 1987. *Friendship Networks and Social Support.* London: Policy Studies Institute.

Wilson, L. B. (Chair). 1928a. "Sales Activities." In *General Commercial Conference, Bell System, 1928,* 1–18. Microfiche 368B, IBIC.

———. 1928b. "Promoting Greater Use of Toll Service." In *General Commercial Conference, Bell System, 1928.* 38–45. Microfiche 368B, IBIC.

Wilson, Margarey. 1947. *The New Etiquette.* New York: Lippincott.

Wiltsee, Ernest A. 1933. "The City of New York of the Pacific," *California Historical Society Quarterly* 12 (March): 25–34.

Winner, L. 1977. *Autonomous Technology: Technics-out-of-Control as a Theme in Political Thought.* Cambridge, MA: The MIT Press.

Wisconsin State Telephone Association. 1985. *On the Line.* Madison, WI: Wisconsin State Telephone Association.

Wisener, P. 1984. "'Put Me on to Edenville': One Hundred and Six Years on the Telephone." *Mind and Nature* 3 (1): 23–31.

Wolfe, L. M. 1979. "Characteristics of Persons with and without Home Telephones." *Journal of Marketing Research* 16 (August): 421–25.

"A Woman of Fashion." 1909. *Etiquette for Americans.* Rev. ed. New York: Duffield.

Wonbacher, G. F. 1908. "Proper Development of the Rural Telephone." *Western Telephone Journal* 12 (July): 242.

Wood, D. E. 1939. *History of Palo Alto.* Palo Alto, CA: Crawston.

Woolgar, Steve. 1991. "The Turn to Technology in Social Studies of Science." *Science, Technology, and Social Values* 16 (Winter): 20–50.

Wurtzel, A. H., and C. Turner. 1977. "Latent Functions of the Telephone." In *The Social Impact of the Telephone,* edited by Ithiel de Sola Pool, 246–61. Cambridge, MA: The MIT Press.

Wyman, A. F. 1936. "A Study of Contemporary Unemployment and of Basic Data for Planning a Self-Help Cooperative in Palo Alto, California, 1935–6," WPA Project 65–3–3454. Palo Alto Historical Society.

Yago, G. 1983. *The Decline of Transit.* New York: Cambridge University Press.

Young, H. B. 1929. "The Elements of a Balanced Advertising Program." In "Publicity Conference, Bell System, 1921–1934." Box 1310, AT&THA.

Zunz, Olivier. 1985. "The Synthesis of Social Change: Reflections on American Social History." In *Reliving the Past: Worlds of Social History,* edited by Olivier Zunz, 53–114. Chapel Hill, NC: University of North Carolina Press.

Index

413

H

G

I